HUMANKIND

HUMANKIND

HOW BIOLOGY AND GEOGRAPHY SHAPE HUMAN DIVERSITY

ALEXANDER H. HARCOURT

PEGASUS BOOKS

NEW YORK LONDON

HUMANKIND

Pegasus Books LLC
80 Broad Street, 5th Floor
New York, NY 10004

Copyright © 2015 by Alexander H. Harcourt

First Pegasus Books cloth edition June 2015

Interior design by Maria Fernandez

Library of Congress Cataloging-in-Publication Data is available.

ISBN: 978-1-60598-784-2

10 9 8 7 6 5 4 3 2 1

Printed in the United States of America
Distributed by W. W. Norton & Company

Dedicated to my sisters,
Sylvia, Caroline, and Elizabeth,
in memory of our many travels together.

CONTENTS

1   **PROLOGUE**
    Where we are going—and why   1

2   **WE ARE ALL AFRICAN**
    The birthplace of humankind   9

3   **FROM HERE TO THERE AND BACK AGAIN**
    A mostly coastal route out of Africa—across the world?   43

4   **HOW DO WE KNOW WHAT WE THINK WE KNOW?**
    The science behind the "facts"   72

5   **VARIETY IS THE SPICE OF LIFE**
    Where we are affects *what* we are   90

6   **GENE MAPS AND ROADS LESS TRAVELED**
    Barriers to movement maintain diversity   126

7   **IS MAN MERELY A MONKEY?**
    Human cultural diversity varies across the globe in the
    same way and for the same reasons as biological diversity   148

8   **ISLANDS ARE SPECIAL**
    Size and metabolism in a small environment   173

9   **WE ARE WHAT WE EAT**
    Our diet affects our genes, and different regions eat different foods   195

10   **WHAT DOESN'T KILL US HALTS US OR MOVES US**
Other species influence where we can live   213

11   **MAD, BAD, AND DANGEROUS TO KNOW**
We are bad for many species, even if we help a few   230

12   **CONQUEST AND COOPERATION**
Humans are bad for each other, even if we occasionally help one another   261

13   **EPILOGUE**
Are we going to last the distance?   278

Citations   281

Sources   293

Some Suggested Reading   317

Index   321

HUMANKIND

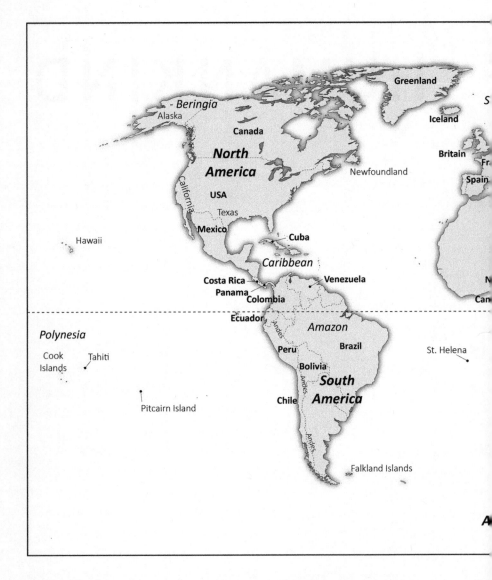

Map of the world showing regions and sites mentioned in the text. *Credit: John Darwent.*

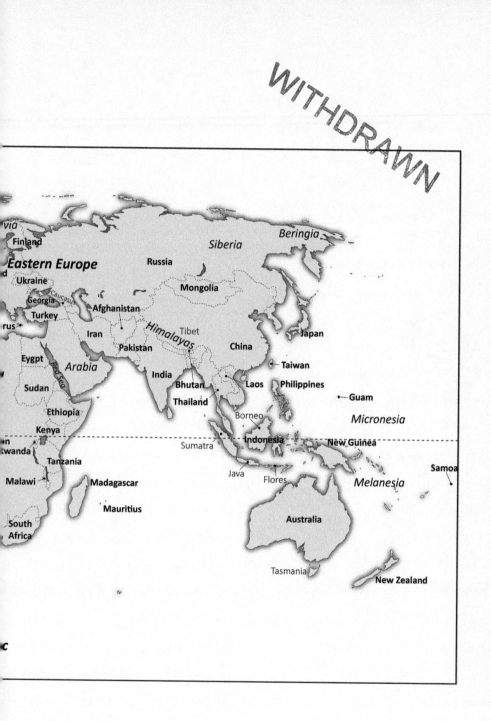

CHAPTER 1

PROLOGUE

Where we are going—and why

"In the beginning . . ." Genesis 1:1.

I was born in East Africa. Kenya, to be exact. Nevertheless, nobody would mistake me for a native-born Kenyan. My skin is white, because I am at the end of a long line of British-born ancestors. And if we go back far enough, I could even have a Norman ancestry. A few hundred years ago, the Normans invaded Britain—1066 and all that—and in Normandy lies a town named Harcourt. In my birth country of Kenya, my father ran a cattle-and-sheep ranch. Although the peoples of over fifty cultures live in Kenya, the workers on the ranch were mostly Maasai. They were taller, longer-legged, and darker-skinned than my parents. They of course spoke a different language from us. They dressed differently. And their customs differed from ours—except that at one time they too had entered Kenya as an invading power. My family eventually moved back to Britain, and three decades later I moved to California with my Californian wife, whom I met while working in Rwanda.

This brief description of my life history touches on several of the themes of this book: the movement of humankind across the globe, the physical and cultural differences between peoples from different regions, Western imperialism, and the great variety of cultures in the tropics. And if, for some, the mention of Africa conjures visions of the many lethal diseases of the tropics, the effect of diseases on humankind is no less relevant to the question of why we are what we are where we are.

Why are we what we are where we are? That is not just one question, it is a broad set of questions. Where did the human species originate? How did humans spread across the globe? Why do people from different regions differ? Does this variation exist because people had different places of origin, or is it because over time we have adapted to the environment in which we live? Some people have adapted to the tropics, others to cold, some to mountains, others to the plains, some to a diet of fish and meat, others to a diet of starch. What effects have other species had on the global distribution of humans? What effect have humans had on theirs? How have human populations affected each other's distribution? These questions encompass much of the discipline termed biogeography—in short, the biology behind the geography of species distributions. And that is the topic of this book.

In humanity's distribution around the world, and in the biological explanations for that distribution, humans show many of the same patterns as other animals. Hence the apt title of Robert Foley's book about human evolution and ecology, *Another Unique Species*. All species are unique—they would not be separate species if they were not. But as regards the biology behind our species' global distribution, our biogeography, we humans are often not obviously unique. If I had adopted Jared Diamond's description of us as "the third chimpanzee," and substituted "modern ape" for "human," it might have taken some time before the reader realized that I was writing about humans.

Through the book, I write about biological and cultural differences between people who live in or come from different parts of the world. The issue of race necessarily raises its divisive head. Humans are disturbingly good at noticing differences between themselves and others, and stigmatizing the others accordingly. Consequently, some anthropologists and others do not want to recognize any biological differences between peoples from different regions of the world.

In one sense, they might be correct. Obvious as are the superficial differences between me and a Maasai man, those differences are only partial. As Darwin wrote in *The Descent of Man*, "It may be doubted whether any character can be named which is distinctive of a race and is constant." Humans change gradually from region to region, and usually in only minor ways. Such gradual, minor change is not surprising, given Anne Stone and her co-authors' finding that all humans are so genetically similar to one another that each *subspecies* of chimpanzee is more genetically varied than the *whole* human species.

On the other hand, nobody can possibly question the fact that people from different parts of the world differ biologically and culturally in various ways. Nobody would mistake me for a Maasai, just as nobody would mistake a Maasai for someone of European ancestry. The differences are real. They can be seen. They can be measured. Refusing to try to understand why people from different parts of the world are different is not going to make discrimination against others go away. It is the discrimination that we have to halt, not the understanding.

Because no "character can be named which is distinctive of a race and is constant," the concept of "race" applied to humans is largely a sociopolitical term. I'll say here what I will describe in more detail in chapter 2, namely that we find more genetic variation within the so-called African race than we do in all the rest of the world put together. In other words, in no biological sense is there such an entity as "the African race." "Race" as normally used is a meaningless term in the context of this book. It hides diversity. In a book that celebrates diversity by trying to understand it, the concept of "race" has no place.

Knowledge and understanding of our diversity can be vital in practice. Famine programs, for example, used to ship milk powder across the world as food aid. Sounds fine? No, it was not. Adults in much of the world outside western Europe and sub-Saharan Africa cannot digest milk. Worse than that, it can make them ill.

The prevalence of many diseases differs from one part of the world to another. The English doctor in the London clinic examining a west African suffering from flu-like symptoms needs to know that the patient might be as likely to have malaria as to have the flu.

The doctor also needs to know that people from different parts of the world react differently to different drugs. Given all the other differences between peoples from different parts of the world, it would be surprising if they did not. Take ACE inhibitors (ACE stands for *angiotensin-converting enzyme*). While people of European origin with congestive heart failure respond well to ACE inhibitors, patients of African origin do not. Luckily, another drug, BiDil (a combination of two drugs with long Latin/Greek names), does seem to work in people of African origin. Indeed, the combination apparently worked so well that the study that compared its efficacy with a previous therapy was ended early, in order that the patients receiving the previous therapy could be quickly put onto BiDil. For the moment, BiDil has not been tested on patients of non-African origin, so we do not know why it works better for Africans than do ACE inhibitors, nor indeed why ACE inhibitors do not work well for Africans.

Despite the fact that a drug had been found for people of African origin who could not as easily be otherwise helped, complaints appeared of "race-based" science. The logic of the complaints sometimes escaped me; but, as far as I could tell, the objection was that scientists had considered the possibility that people from different regions might be biologically different.

In fact, a medical problem for people with origins other than European is that most drug testing is confined to subjects of European origin. So indeed a racial bias exists, but it is not the one that most people think. For instance, geneticists Anna Need and David Goldstein recorded the country of origin of people included in investigations of the genetic bases of disease, and found that over a million and a half of them were of European ancestry compared to just seventy-five hundred with African origins. Given the fact of regional differences in susceptibility to disease, Need and Goldstein, along with several others, argue that we need more regionally based studies of disease genetics in order that more people from more parts of the world can be helped by modern medicine.

Happily, many doctors are now aware of the fact that people from different parts of the world can have different diseases, and that they can respond differently to the same drugs, even if we still know far too little about the regional variation. At the same time, categorization still tends to be done by race, rather than region of origin. "Black" or "African-American" covers people from both the Caribbean and the whole of Africa—and yet we know that the diseases to which the peoples of those regions are susceptible or resistant differ between the two regions, and differ within just Africa alone.

To influence medical practice is far from my main hope for the book, though. My hope and aim is simply to increase knowledge and interest in the biogeographical reasons for the diversity of the human species, both with regard to our biological diversity and our cultural diversity. Yes, indeed, I follow many others in arguing that biology can explain some of the geographic diversity of cultures.

I will say here that throughout the book, I occasionally spend time on disagreement between scientists, or on our lack of full understanding of the biology behind certain patterns in the geography of humankind. I do so for two reasons.

Firstly, textbooks tend to present science as if everything is known. Here is how it works, they say. Readers, perhaps especially young readers, must then inevitably get the impression that nothing is left for them to discover. However, an immense amount is still unknown. Much remains to be found out. That is what makes science so exciting.

The second reason has to do with political inertia. Because textbooks so often present the facts as if all is known, the public is left thinking that we either know or do not know a fact. Consequently, when scientists honestly say that they cannot be one hundred percent certain, because true certainty is almost impossible in so many areas of science, politicians can say that the scientists do not know anything, and therefore the politicians can justifiably do nothing. I am talking here, of course, of climate-change deniers.

I start the book with our origins in Africa, and our subsequent spread throughout the world. The whole human race was once African. So recently did a small number of us leave Africa, so recently have humans in different parts of the world differentiated from each other, that we are *all* still fundamentally "African." At the same time, human populations from different parts of the world have become dissimilar because they have adapted biologically and culturally in a variety of ways to the diverse environments across the world since people first departed from Africa perhaps sixty thousand years ago.

And where do we see the greatest variety of cultures, but in the same place as we see the greatest variety of plant and animal species—the tropics. Some of the biology behind that geographic correlate of biodiversity explains the great human cultural diversity of the tropics, from the Aweti of Brazil to the Mbama of Gabon and the Zorop of Indonesia. Similarly, we see fewer cultures on the smaller islands for some of the same reasons that we see fewer species on small than on large islands. Mention of small islands raises the topic of the miniature Flores Island "hobbit." The report in 2004 of this new species of our genus *Homo* rocked the scientific world. But here too, biology can explain why on a small island the hobbit was so small. In other words, despite our big brain, our religions, our philosophies, our consciousness, our self-awareness—all the abilities that should separate us from the animals—we humans nevertheless often turn out biogeographically to be just another species.

I am going to add here that this aspect of human biogeography, this fact that in the distribution of our cultures, humans reflect the distribution of other species, and do so for the same biogeographical reasons, is a topic that I have not seen in anthropology textbooks. I do not know why it is not there. Are the findings too recent? Are the findings too contentious? The title of one of the chapters in which I write about these findings indeed is "Is Man Merely a Monkey?" This question alone still ignites controversy.

A crucial aspect of the environment affecting what we are where we are is the presence of other species, especially the microbes that cause diseases. Our bodies have adapted by natural selection to fight the diseases in our environment, which means that as disease organisms differ from region to region, so also does our immune system. Additionally, humans in turn have affected and affect the distribution of other species. As we have driven and are driving hundreds of species to extinction, including perhaps other hominins such as the Neanderthals, so too do more powerful and populous human cultures overrun the weaker and less populous. Yet many species benefit from humans, and have extended their geographic range because of us. Many cultures also have extended their range not through conquest, but via peaceful interaction.

How much longer will we continue to influence the geography of the earth and its biota, and vice versa? I suspect that biogeography has no answer to that question.

Before I get to the acknowledgements for the various ways in which a variety of people have helped me in the writing of this book, a word about the measuring system I use. The word is addressed especially to readers in the USA. I use the rest of the world's metric system of measures, not the USA's system. Everything divisible by multiples of ten in the metric system is so very much easier than the Imperial system's outdated divisions by sixteen, twelve, fourteen, three, 5280, and so on.

The metric system's one kilogram is 2.2 pounds; its tonne is 2,200 pounds, in other words close to the US system's ton; one liter is 2.1 pints; one meter is 3.3 feet, in other words a yard as near as makes no odds; and a kilometer is five eighths of a mile, which for the distances described in this book might as well be a mile. The freezing point of water is 0° centigrade, or Celsius; and the boiling point of water is 100° centigrade or Celsius. To get from Fahrenheit to centigrade over the normal range of experienced daily temperatures, subtract 30 and halve the remainder, and you will be close enough. Fahrenheit and Celsius are capitalized, because they are named after the scientists who developed the respective scales, Daniel Fahrenheit, a German, and Anders Celsius, a Swede.

I want to emphasize that few of the research findings that I describe and discuss are my own. The bulk of the book is based on the work of hundreds of other people. My role has largely been to bring all their work together in what I hope is a cohesive story.

The collation involves a lot of searching in an extensive literature. I will therefore start the acknowledgements by thanking for their assistance the knowledgeable, friendly, and helpful staff of my university's Shields Library.

Tima Farmy might not remember it, but it was she who told me that I should convert my scientific book on the topic into a popular book. I've enjoyed the process and thank her for the stimulus.

Thanks also to the generosity of all those who put their photographs into Wikimedia Commons, and so enabled me to easily illustrate this book.

Their photographs have been credited as in "WikiCommons." Alfredo Carrasco Valdivieso, my brother-in-law, allowed me free use of his photographs of native South American peoples. Many thanks to all.

John Darwent generously produced all the maps, asking in payment just a sample of one of Scotland's grandest products.

The book benefited greatly from commentary on all chapters by Elizabeth Harcourt, Mona Houghton, and François von Hurter, on some chapters by Sylvia Harcourt, and from incisive, knowledgeable criticism of both style and content of the whole book by Kelly Stewart. I could not have asked for a better editor.

Donald Lamm provided valuable general editorial advice. Mona Houghton put me onto Peter Riva and Sandra Riva, who in turn introduced me to Jessica Case and Pegasus Books. The book benefited immensely from Jessica Case's expert editing. Alex Camlin designed the cover. Maria Fernandez did the typesetting and interior layout. Phil Gaskill did the copyediting and proofreading. I thank them all for their help.

On the scientific content of the book, Brian Codding helped my understanding of Californian cultural diversity. Geoffrey Clark helped me understand how limited are the measures of a culture that many of us use, such as language, or type of tool. David G. Smith was extremely helpful with my many genetics questions. For commentary on a previous technical version of this book, without which this book would have been less accurate, I thank Robert Bettinger, Chris Darwent, Victor Golla, Mark Lomolino, Frank Marlowe, David G. Smith, Teresa Steele, John Terrell, and Tim Weaver. I and the book benefited enormously from their advice and help.

Finally, I thank Kelly Stewart, my wife, for her love and support. I could not have written the book without her.

My share of royalties from the sale of this book will go to Survival International, an organization dedicated to helping *"tribal peoples defend their lives, protect their lands and determine their own futures,"* http://www.survivalinternational.org/. Registered charity no. 267444 | 501(c)(3).

Alexander H. Harcourt, Professor Emeritus
University of California, Davis, CA 95616, USA.
ahharcourt@ucdavis.edu

CHAPTER 2

WE ARE ALL AFRICAN

The birthplace of humankind

Genesis has God putting man, Adam, "eastward" in the Garden of Eden. Eastward of the Dead Sea caves where some of the earliest biblical texts were found lies Iraq and the fertile valley of the Tigris and the Euphrates. It is not at all impossible that the people whose history the Dead Sea Scroll writers recorded had their origins there. After all, that's the rich land of Sumer, created by the gods Enki and Ninhursag. The earliest recorded writing, Sumerian, comes from this region. But the biblical writers got their compass direction wrong for where humankind evolved. That is, unless they mean agricultural humans. One of the regions of origin of agriculture was indeed in the vicinity of the Tigris and Euphrates.

The earliest signs so far of our species, humans, come from three thousand kilometers to the south of the Middle East, from southern Ethiopia.

These signs are bones dated to nearly two hundred thousand years ago. Later genetic work on human origins brackets this date. Other genetic studies, which I will mention later, indicate that fully modern humans too, humans essentially completely indistinguishable from us, came from somewhere near or in Ethiopia. Africa is the birthplace of humankind. It is from African "dust of the ground" (Genesis 2:7) that we are made. We are all African. All pictures of Adam and Eve should show Africans, not the usual white Caucasians.

If one of the 200,000-year-old early humans were dressed in modern clothes, we would not look twice at them as we passed them in the street. Nevertheless, the skeletons of these early humans are slightly different from ours. Detailed scientific descriptions remark (in the usual Latin of science-speak) differences in the so-called occipital torus, the interparietal keel, the canine fossa, and so on. To the uninitiated, the thicker brow ridges of the oldest humans might be one of the more obvious features of the skull that distinguish them from us. Also, the bones of their arms and legs are slightly more robust than ours. As a result of these slight differences, anthropologists distinguish between early modern humans and fully or late modern humans. Yet all are considered to be "anatomically modern humans," as the terminology goes.

Different writers mean different things by the word "human." Scientists describe all animals (and plants) with two names. These two names tend to be in Latin or Greek, or both. All readers of this book, fully modern humans, are *Homo sapiens*. We are members of the genus *Homo* and the species *sapiens*. *Homo* is Latin for "man"; *sapiens* is Latin for "wise." We are "Wise Man." The *sapiens* part of the two-word name is unique to us. There exists no other species that we call *Homo sapiens*. So I use the word "human" for only *Homo sapiens*, only us.

However, scientists recognize and name several other species in the genus *Homo*. And here is where confusion can arise, because some scientists use the word "human" for all of these too. For them, all *Homo* species are humans. So Neanderthal (also spelled Neandertal), *Homo neanderthalensis*, is human. So are, going back in time, *Homo heidelbergensis*, *Homo erectus*, *Homo ergaster*, *Homo rudolfensis*, and *Homo habilis*. In English, these would be Heidelberg Man, Upright

Man, Working Man, Rudolf Man, and Handy Man. If all of these are "human," then to refer to ourselves, *Homo sapiens*, we have to use the mouthful "fully modern human."

Before *Homo*, several *Australopithecus* species lived, one of which is likely to have been the ancestor of subsequent *Homo* species. *Australopithecus* is Latin and Greek for "Southern Ape," so-called because South Africa is where the first fossils of it were found. Good reason exists to call the species of this genus an ape. *Australopithecus* brains were barely larger than a chimpanzee's.

Nevertheless, *Australopithecus* is definitely one of our ancestors, definitely closer to us than to what we might call a real ape, namely the ancestors of gorillas and chimpanzees. For a start, *Australopithecus* walked on two legs. So all *Australopithecus* species, all other *Homo* species, and humans are lumped under the term "hominin." We are all hominins.

The hominin genus *Australopithecus* lived only in Africa, as far as we know. It appeared roughly four million years ago, and lasted for two and a half million years. *Australopithecus* might have used stone tools, but scientists do not yet know whether any *Australopithecus* actually manufactured stone tools. Only with *Homo habilis*, Handy Man, do we know that our ancestors started to make stone tools, perhaps two and a half million years ago. Indeed, part of the reason why anthropologists classify Handy Man as *Homo*, not *Australopithecus* is the evidence that Handy Man *made* stone tools. Also, it is with Handy Man that our ancestors' brains began to exceed the size of chimpanzees' brains in an obvious way.

From then, brains rapidly increased in size. Neanderthals had the largest brains. But then they also weighed more than do modern humans. Elephants have larger brains than humans too. The way to get a sense of the size of a species' brain compared to other species is to relate it to the size of the body. Larger bodies have larger brains, other things being equal. Compare our ratio of average brain size to average body size, and we modern humans have the largest brains of all the hominin species. The phrase that scientists use to term this measure of brain size compared to size of body is "encephalization quotient."

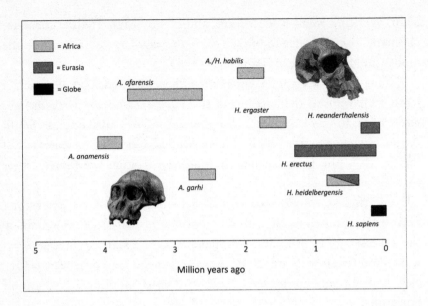

Human ancestry. The length of the bars indicates how long the species lasted. Their shading shows where they lived. Note, all the earliest species originated in Africa (if *Erectus* is considered to be a later *Ergaster*, then it too originated in Africa). A. = *Australopithecus*; H. = *Homo*. Credit: *John Darwent. Picture credits: Early* Australopithecus - *B.Eloff; L.Berger; Wits Univ, Wikicommons;* A./H. habilis - *Daderot, WikiCommons.*

This description reads all very cut and dried, as if everyone agrees. In fact, the scientists involved argue all the time. Quite often we do not agree on terminology, as I indicated at the start. You will find that some people include Neanderthal in *Homo sapiens*, and many have Handy Man as *Australopithecus habilis*, not *Homo habilis*. Even with modern humans, scientists do not completely agree on terminology. For instance, some keep "modern" for humans from around fifty thousand years ago, reserving "early modern" for humans before that.

Also, the number of our ancestors and their relatives keeps changing with new finds and new arguments about existing finds. That is the nature of science. For instance, the team of David Lordkipanidze, Director of the Georgian National Museum in Tbilisi, published in 2013 a description of a new Erectus skull from the important Dmanisi site in southern Georgia, just north of the border with Armenia. It dates at close to one million eight hundred thousand years ago, and is the most complete skull yet found there. Indeed, nobody has yet found *anywhere* a more complete skull of the same age.

It turns out that this new skull and the four other skulls from the site differ so much from each other that if they had been found in different sites in Africa, they might have been classed as different species. Nevertheless, Lordkipanidze and his team show that the variation within this one Dmanisi population of just five skulls is about the same as we see among modern humans, or among either of the two species of modern chimpanzee. And that is with just five skulls. Imagine how much more variation we would see with a larger sample. In other words, we might have to rethink the number of *Homo* species in Africa. Rather than the three or more species that some anthropologists have argued for, maybe just one species, *Homo erectus*, Upright Man, lived there around two million years ago.

The point of these last few paragraphs is to emphasize that science is not about how textbooks often present our knowledge of the world: a tidy list of facts. Science is a groping toward understanding. Science is a process of collecting evidence to test whether the current explanations are correct—or, rather, *might* be correct. Science is all about coming up with reasonable explanations about the world that can be shown to be wrong (this is called being falsifiable). It is a means, more than an end.

Scientists are still gathering information, working out what it tells us, and disagreeing quite a lot of the time. But if the new data and ideas prove to be better than the old, then the old will eventually disappear. Eventually most of us come to the same view, as we gather enough information and agree that some interpretations have fewer problems than do others. Sometimes we have knowledge. We know the earth is round. We know we evolved. But we are usually working toward knowledge. Additions, tweaks, alterations, and rejections of current ideas will continue throughout this process.

Let us get back to the distribution of ourselves and our ancestors across the world, back to biogeography. Of the list of hominins that I gave a few paragraphs earlier, Neanderthal, Heidelberg Man, and Erectus were the only ones to leave—or to live outside—Africa. All the *Australopithecus* species lived in only Africa. At least, that is the only continent on which paleoanthropologists have found their fossil remains. Handy Man, *Homo habilis*, the first *Homo* species, also lived only in Africa, as far as we know. The current earliest human *Homo sapiens* fossils are from Africa. Humans' closest ape relatives, the gorilla and chimpanzee, live only in Africa. It was an African ape species whose sperm or egg carried a mutation that began the long line leading to us. The Garden of Eden was not eastward of the Middle East. It was in Africa—and in sub-Saharan Africa, to be more precise.

Just where in sub-Saharan Africa is another question. Some argue that the human species originated in the region of the earliest human skeletal remains, Ethiopia. Others, using genetic evidence, suggest the region of Cameroon. Yet others, also using genes, suggest southern Africa.

But we are talking of a process that began two hundred thousand years ago. In even a thousand years, people and their genes can move thousands of kilometers. So, ten thousand years after the origin of the first modern human (and of course that process took time and many humans, and therefore many genes), people will have moved and bred, and they and their descendants moved and bred again. Is trying to discern a precise region of human origins within Africa therefore an impossible task? Maybe not.

Analysis of the genetic makeup of the Khoi-San hunter-gatherers (also Khoe-San, Khwe, Khoi, Kxoe) of southern Africa indicates that they might be descendants of some of the earliest Africans. That finding is in part why some suggest southern Africa as humankind's cradle. But maybe the Khoi-San did not always live in southern Africa. We know that hunter-gatherers have been marginalized into some of the poorest land by later pastoralists and agriculturalists, including Europeans. Could the Khoi-San have been pushed from eastern to southern Africa by these later arrivals? Indeed they could, but the genetic evidence is that they did not. If they had, we should see, as Brenna Henn and colleagues point out, greater genetic variety among eastern African hunter-gatherer populations than among southern African ones. Yet we do not. That is not to say, though, that some did not migrate from eastern Africa, as I will describe later when I raise again the topic of "race."

However, the genetic evidence above so far comes from populations still living in eastern and southern Africa. What if the eastern Khoi-San died out, or nearly died out, perhaps outcompeted by the dispersing Bantu? Then might it be the case that we would not find great genetic variety among the indigenous hunter-gatherer peoples of eastern Africa? In other words, I am not so sure that from the genes of populations living today, we can definitively conclude a region of origin within southern Africa.

Humans were not the first of our evolutionary line to leave Africa. At least two other species did so before us. The newly named *Homo antecessor*, which might be a form of *Homo heidelbergensis*, left Africa just about eight hundred thousand years ago, judging by the date of the earliest remains in Europe. Indeed, Antecessor/Heidelberg reached southern Britain during what was a warm period then in the Northern Hemisphere. But warm is only relative, for winters were dark and cold, with temperatures certainly reaching the freezing point. The well-known Neanderthal of Europe and western Asia could well have evolved from Antecessor/Heidelberg. And then, of course, we have the recent findings of probably another hominin species in Eurasia (Europe plus continental Asia), the Denisova hominin, and maybe even yet another.

And a million years before Antecessor/Heidelberg left Africa, *Homo erectus* exited. We know that from the wonderful archeological site of Dmanisi in beautiful south-central Georgia, which I have already mentioned. In fact, Erectus got as far as eastern China.

The bible and, therefore, creationists have the original diaspora as Adam and Eve's banishment from Eden. Where they went, the bible does not immediately say. In fact, the earliest indications so far of the human species outside Africa are stone tools and bones at archeological sites in the Middle East, including southeastern Arabia. They are approximately a hundred and twenty-five thousand years old. Fifty thousand years later, though, humans seem to have disappeared from the Middle East. If so, they must have either died there, or retreated back into Africa, or moved on into western south Asia. We have next to no further evidence yet of which of these three scenarios happened.

Some have suggested that the advancing ice age and its associated aridity is what drove us back out of the Middle East. A problem with the idea of cold inducing death or retreat then is that Neanderthals moved into the Middle East as modern humans left. For instance, Kristin Hallin and her co-authors report Neanderthal remains in Israel sixty thousand years ago. But if Neanderthals could live there then, why not humans with a culture and tool kit more advanced than the Neanderthals'? The question is especially pertinent given that the environment was apparently wetter, and therefore more congenial, than when humans were there around a hundred thousand years ago.

Of course, absence of evidence is not evidence of absence, as the famous saying goes, and as presumably the most junior detective is told. In the first place, the chances of bones becoming fossilized are small enough. They need to be undisturbed by, for example, hyenas or floods. They need to stay intact, which is less likely in warm and wet environments than cold or dry ones. And they preferably need to be buried in order to remain undisturbed and intact. If they do remain undisturbed and intact, then the chemical conditions have to be right for the calcium phosphate compound of which they are formed (strictly, calcium hydroxylapatite) to convert to a more durable chemical, such as calcium carbonate, also known as calcite. Then the fossils have to be found. And the Middle East is not the easiest place to be a fossil hunter. In sum, I will not be surprised if we later discover evidence that humans did not in fact leave the Middle East.

The next signs of humans in the Middle East—in other words, signs of humans living outside of Africa—are from sixty thousand years ago, give or take five thousand years or so. This diaspora was the one that resulted in the peopling of the rest of the world. Human technology was becoming more sophisticated all the time, and we have to assume that our better tools and skills enabled us to cope with an environment even worse than the one that had previously hindered our spread. I return in chapter 6 to the issue of barriers to the global movement of humans across the world.

What stimulated these exoduses from Africa? Many argue that the answer is a drying of the continent. However, when humans first left Africa, maybe a hundred and twenty-five thousand years ago, the world was between ice ages. Africa was warm and wet then, as it is now. The massive tropical forest of the Congo Basin shows how warm and wet the continent is at present—we do not get thick green tropical forest without lots of rain.

Certainly, the world subsequently and rapidly headed into a major ice age—the Last Glacial Maximum, as it is known. By the end of it, African forests had nearly disappeared and desert had expanded. But if aridity drove us out of Africa halfway to the depth of the last glaciation sixty thousand years ago, why is there no evidence of a surge of emigration twenty-five thousand years ago, when the ice age was at its maximum, and Africa would presumably have been at its driest?

Countering the aridity-exodus hypothesis is evidence from the bottom sediments of Lake Malawi dated to around seventy thousand years ago. That is just a few thousand years before humans finally left Africa. The sediments indicate that the climate then was becoming wetter—in other words, more favorable to humans. And that evidence leads to the suggestion that the consequent increase in population of humans was in fact the stimulus for our exodus.

That seems more sensible to me. Yes, animals and humans will flee a bad environment if they can. But on the scale of human movement, maybe twenty kilometers a day at most, if it is bad here, it will be bad there. In other words, the emigrants probably will not survive. On the other hand, if the fields are green here, then just over the horizon, they are probably green there too. And so in the face of population pressure here, why not move there?

Indeed, a study by Anders Eriksson and eight others seems to confirm this idea. They tied high rates of plant growth ("high productivity," in scientific jargon) to the size of populations and their expansion into new regions once they reached a threshold density where they currently were.

To model the time and place of the human global diaspora, they used genetic similarities and differences among fifty-one populations from across the globe. These comparisons allowed them to draw a branching tree of human movement, with not only the shape of the tree, but also the timing of the growth of its branches. In chapter 4, I explain in a bit more detail the production of the gene tree.

The model produced dates of expansion and arrival of humans throughout the world that quite nicely fit the archeological record—with two exceptions. In the model, humans arrive in western Europe too early, sixty thousand years ago instead of forty-five thousand. And they get to southern South America too late, only five thousand years ago instead of fifteen thousand.

Models, like all scientific explanations, should always be as simple as possible. This one by the Eriksson group is admirably simple. At the same time, simple models omit, out of necessity, several potentially important factors. The modelers are trying to produce the least complicated hypothesis to explain the highest proportion of the facts. In this case, maybe the late arrival of humans in southern South America in their model could be explained by humans' exploitation of coastal seafood. That source of food would have been missed in the Eriksson model, which used only a measure of plant growth to indicate suitability of the environment. The Sahara would have been at times supremely unsuitable for humans, but its eastern coast along the Red Sea could have been supremely suitable to shellfish eaters.

The Eriksson study did not model movements in Africa before a hundred and twenty thousand years ago, because the information on climate then was not sufficiently detailed. Subsequent to that time, we see slight movement as, for example, the Sahara expands and contracts. By contrast, Margaret Blome and co-workers argued from one of the most detailed studies to date that the climate across Africa was so uneven in space and time from a hundred and fifty thousand years ago that arguments relating climate to human evolution or movement in the continent are difficult, even impossible, to validate.

Moreover, the Blome team suggested that with a little topography, local altitudinal movement seems a more likely means to cope with changing climate than large-scale migration. If it is hot, move to the nearest mountainside. If it is cold, descend into the plains. The trouble with this latter idea is that moving up a mountain means moving into a smaller area, which means less land, which means smaller populations, which means higher likelihood of extinction. This process is one of the effects of global warming on animal and plant species today.

If the exodus from Africa that led to the peopling of the world was a one-time event, even if it was a trickle that lasted just a few thousand years, we have a major difficulty in tying climate to the departure of some of us from Africa (whether we are talking cold and dry, or warm and wet). The problem is that single events are effectively anecdotes, potentially mere coincidence. How with one event do we test the idea that climate change resulted in the exodus? One answer is to suggest that for humans, successfully tying climate to their exits and their entrances in other parts of the world can validate the suggestion of an influence of climate on the diaspora from Africa.

Another potential problem of tying climate to our exodus (or indeed evolution) is the fact that an effect of climate on the environment and ourselves would not be immediate. The climate changes, but the terrestrial environment—and humans—react hundreds or thousands of years later. How then do we tie climate to our history?

Finally, we might not even expect an obvious influence of any but the most severe climate change on the evolution or movements of modern humans, because our intelligence and tools can buffer the effects of climate. Witness the advance of humans into Europe starting from perhaps forty-five thousand years ago as the peak of the last ice age approached. This is in the Aurignacian period, characterized by bone tools, elegant stone flake tools, and figurines carved of bone.

Nevertheless, even with our big brains and advanced tools, severe climate still affects decisions of where to live and when to move. For example, extreme crashes in temperature in the Arctic correlated with humans abandoning the region over the last few thousand years, including the latest abandonment of Greenland by Norse settlers. They left roughly seven hundred years ago, coinciding with the severe cold of the Little Ice Age.

At the same time, however, European economies were changing. We have to ask, therefore, whether the Norse Greenlanders emigrated because they could no longer cope with the climate, because sea ice prevented contact with Europe, or because they wanted to seek better opportunities elsewhere. We do not know, but a combination of all these influences seems likely.

One interesting candidate for a climatic influence on our exodus from Africa is the massive explosion of Mt. Toba in northern Sumatra seventy-four thousand years ago. The beautiful Lake Toba, a hundred kilometers long and thirty kilometers wide, over twice the surface area of Lake Geneva, is all that is left of the former volcano. The explosion might have expelled over fifty times the amount of volcanic ash and pumice that Indonesia's still-active Krakatau volcano did in 1883: at that time, Britain, halfway around the world, experienced beautiful sunsets and, for five years or so, world temperatures dropped by over a degree.

Mt. Toba would have had a larger effect, of course. The anthropologist Stanley Ambrose argues for a cooling lasting hundreds of years, with some evidence that for a few decades, temperatures in the Northern Hemisphere dropped by a whopping ten degrees. However, the argument for a link between Mt. Toba and our exodus from Africa has problems. The earth was already cooling then toward the last glacial maximum, so how do we separate that global trend from any effect of Mt. Toba? The cooling was not steady. Sometimes the temperature increased. But even if the Toba explosion coincided exactly with the most severe of the global cooling periods (and it does seem to have done that), it did not coincide with the others, and neither does any other past large volcanic eruption that we know of. Even locally on Sumatra, lasting effects of the explosion on mammals is difficult to detect. For instance, orangutans did not apparently die out.

The other problem with tying Mt. Toba to any cooling and aridity that might have driven humans out of Africa is that temperatures did not change evenly across the globe. Even if Europe experienced a volcanic winter, Africa might not have done so. Indeed, Caroline Lane and her co-workers argue that at least in Malawi in eastern Africa, no volcanic winter resulted, even though the Toba ash clearly reached the region. The ash is there in the same Lake Malawi sediments that I mentioned previously in reference to aridity driving humans from Africa. However, these sediments show no signs of unusual cooling: for example, no signs of the expected changes of microorganisms or nature of sedimentation.

Moreover, to paraphrase Malthus in the approximate words of Parkinson's Law, populations contract or expand to fit the environment available. People of a small population in a bad (dry) environment could feel as much pressure to move out as people of a large population in a good (wet) environment.

This is not to say that no Southeast Asian volcanoes affect western continents. The greatest volcanic explosion in recorded history, that of Mt. Tambora in Indonesia, probably did cause the summer's winter of 1816 in both North America and Europe. Described as the year without a summer, the east coast of the United States experienced frost and snow in June, crops failed along the Atlantic coast of North America, and thousands died in

Europe from the resultant combination of starvation and disease. Tens of thousands died in Indonesia.

One of the signs of the extent of the effect of Mt. Tambora is peaks of sulphate concentration in ice cores in Antarctica and Greenland. The same ice cores that show evidence of the Mt. Tambora explosion show evidence of an equally large volcanic explosion six years previously. As Stephen Oppenheimer points out, that volcano is yet to be identified. Twenty-five years previously, the Icelandic volcano Laki caused famine in Iceland that killed maybe twenty-five percent of the island's population. It too could have affected weather and agriculture across the Northern Hemisphere, as Alexandra Witze and Jeff Kanipe describe in their *Island on Fire*.

Geneticists have weighed in on whether and when African populations were expanding or contracting. Their results are both vague and sometimes contradictory, in part because the range of dates associated with genetic signatures of population size are so wide. However, supporting the idea of an expanding population pushing humans out of Africa, Richard Klein and Teresa Steele show that the size of shellfish in middens (waste heaps of shells) decreases fairly abruptly fifty thousand years ago, give or take a few, which is near the time humans expanded from Africa. An interpretation of the finding is that an expanding population of humans was exploiting the shellfish so intensively that the shellfish had no time to grow to full size before they were taken. We see exactly the same effect now with world fisheries. Witness the western Atlantic cod, which experienced in the twenty years from 1975 a drop in average length from sixty to forty-five centimeters.

In sum, then, arguments and their support go both ways, or no way. Poor climate, good climate cause the African exodus; decreasing, increasing populations cause it; or we cannot detect a change in population size, or any obvious stimulus to leave Africa. The jury is undecided on any single cause.

More precisely, the jury is undecided on any external, environmental cause. But maybe we do not need an external cause. By the time humans left the continent, especially for the second time, they/we were making sophisticated stone tools. Why would intelligent, tool-making humans not simply go where no human has gone before? Population pressure and lack of food did not take us to the moon. Curiosity was why scientists were happy to have brave astronauts go there. Animals explore. Why not humans?

Indeed, one study has found that the farther populations are from Africa, the greater the proportion in the populations of certain genes (strictly "alleles," as they are termed) associated with exploration and risk-taking. I view the finding with a pinch of salt. After all, Africa's environment is hardly stable, and African neighboring groups are surely hardly less predictable than Eurasia's or the Americas'. Moreover, once a group has moved outside its ancestral range, is further movement of descendants into other unknown regions really helped by yet more exploratory impulse? Would not the same level of exploratory urge as initiated the out-of-Africa movement keep a population moving on?

Be that as it may, what would the explorers find but populations of mammals and birds unused to their weapons and traps, and therefore easily hunted? Greener grass, metaphorically, and more meat seem sufficient causes to move on out.

In sum, we do not have a firm answer regarding the relation between climate and the human exodus from Africa. Frankly, if we cannot even agree on whether aridity drove humans out of Africa, or clement wet weather allowed people to leave by making the Sahara crossable, we hardly have an answer at all. We are still searching. The topic of human biogeography is loud and alive and going on now.

The arguments do not mean, though, that we fundamentally disagree about the overall picture. Do we know to the year when the first *Homo sapiens* left Africa? No, we do not. Do we know the exact spot where they crossed to the Arabian Peninsula? No, we do not. Do we know whether it was raining or burning sun for most of the year at that spot? No, we do not. But the ignorance of the details does not mean that we know nothing. Knowledge and understanding will certainly change, but we know quite a bit already about the human exodus from Africa.

Before I delve into the details, let me stress that of course not all modern humans left Africa, and of course humans did not spread outside of only Africa. At the same time as the human species began to people the rest of the world, so too did it disperse through Africa.

Genetic evidence indicates old dispersals within Africa, perhaps a hundred thousand years ago, from Ethiopia south and north, and from Cameroon eastward and south. These were migrations of hunter-gatherers.

Genes indicate also that the pygmy peoples of Africa might have separated from the populations that became agriculturalists around sixty thousand years ago, long before the development of any agriculture. Then twenty-two thousand years ago, or thereabouts, the eastern and western pygmy populations separated. The last ice age was at its peak then, and the great Congo Basin forest had shrunk to remnants on the mountains on the east and west edges of the Basin. Near the same time, people from the Middle East might have moved into the region of Ethiopia.

More recently still, maybe three thousand years ago, we see the expansion of the Bantu peoples from the Nigeria-Cameroon region. They were by now modern agriculturalists, using iron. Large-scale movements of people through Africa continued into historical times. The first European explorers in East Africa in the nineteenth century encountered Maasai who had moved in from the north as part of a general spread of Nilotic peoples from southern Sudan dispersing south and west.

For the rest of the story regarding our expansion across the Old World, I will assume that humans first got past Arabia about sixty thousand years ago. In other words, I am ignoring the claim by Hugo Reyes-Centeno, Silvia Ghirotto, and others that people got beyond the Arabian Peninsula all the way to Australia over a hundred thousand years ago. Let me quickly say that I am not rejecting that claim: it is just that the paper making the claim is so recent—published in mid-2014—that nobody else has yet had the time to test the results of the study with further research. The order of the story roughly matches the timing of our arrival in the various regions of the world, Arabia first, and the Pacific islands last.

Whether or not humankind's first foray out of Africa at around a hundred and twenty-five thousand years ago was a bust, our second exodus was a smashing success. But it became a success only after a period of maybe ten thousand years when it looked to be another failure. It seems that humans got stuck in the Middle East, because it is not until forty-eight thousand years ago that we get archeological signs of humans east of Arabia.

By approximately forty-five thousand years ago, humans had reached New Guinea and northern Australia, and even down into southwest Australia. By contrast, in southeast Australia, Tasmania, earliest signs of our presence date to only approaching thirty-five thousand years ago.

The earliest New Guineans and Australians might have had the continent to themselves for the next twenty thousand years, even forty thousand years. I say "the continent" because, at the time, the sea level had dropped sufficiently that dry land joined New Guinea and Australia into one landmass, Sahul. Even so, genetic data indicate that people did not mingle between even New Guinea and Australia over that period.

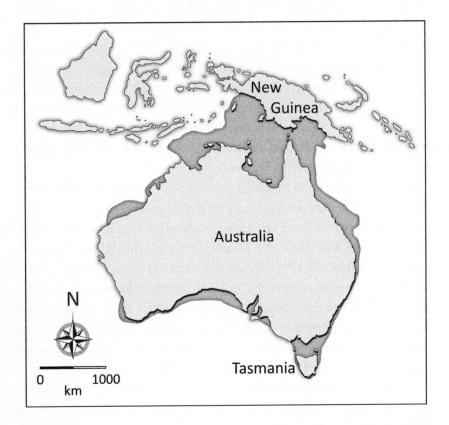

Sahul, the landmass of New Guinea, Australia, and Tasmania, from about 50,000 to 10,000 years ago when sea levels were up to 100 meters lower than today during the peak of the last ice age. *Credit: John Darwent.*

Irina Pugach and her co-workers opt for the 40,000-year separation of Australia. They suggest that it might not have been until a little over four thousand years ago that the next immigrants arrived there. Some geneticists say the genetic data indicate an influx from New Guinea. Others, including Irina Pugach, see Indian ancestry in this next immigration. Whichever, approaching four thousand years ago is also when we first see dingo remains in archeological sites in Australia. Dingoes originated in Asia, and could not have swum to Australia, and thus are incontrovertible proof of human arrival.

The most clement route to Australia will have taken people through India. Nobody has yet confirmed humans there much earlier than forty-five thousand years ago. If humans got to Australia by then, they were presumably in India earlier than that. Certainly, the claim exists that humans made the 70,000-year-old stone tools discovered in southern India. However, the claim is disputed, because nobody has yet found bones associated with the tools. Michael Petraglia and Paul Mellars are currently the main names involved in the argument about the earliest presence of humans in India. Petraglia argues for our early presence, Mellars for late. Both of course reckon they have enough evidence to substantiate their contentions. At the same time, both acknowledge that it would be great to have more evidence. As with so much of our understanding of this aspect of human biogeography, these dates and routes of our diaspora, we simply have to wait for more finds. If it turns out that modern humans did not make the tools, we might have to accept that Neanderthals did. If so, the tools would be the first record of Neanderthals in India.

The date of forty-five thousand years ago for the arrival of modern humans in India is also the date for our spread over much of the rest of Eurasia. Humans reached southeastern Asia by then too. We know this from a site in Laos that has produced a cranium (head minus lower jaw) dated to at least forty-six thousand years ago by three quite different methods.

It seems that it was not until forty thousand years ago that humans penetrated eastern Asia. They could have reached there by continuing along the southern coast of Asia and then heading north. Alternatively, a route through Mongolia north of the Tibetan Plateau is not impossible. Perhaps because of Tibet's high altitude and the protective wall of the Karakorams and Himalayas to the east and south, the Taklamakan Desert to the north, and the Qilian mountains to the north and east, humans apparently did not get into Tibet until maybe thirty thousand years ago, according to an extremely thorough genetic investigation by Xuebin Qi and co-investigators. The earliest archeological signs are hand- and footprints in hot-spring mud at an altitude of forty-two hundred meters dated to around twenty-three thousand years ago.

Japan is about as far east as one can get in eastern Asia. Its northern island, Hokkaido, is closest to southeast Siberia, while its southern one, Kyushu, is close to South Korea. So humans had two spots of possible entry into Japan. Hokkaido was connected to southeast Siberia near the height of the last ice age, when sea levels were more than one hundred meters lower than now, but separated then—and now—from the rest of Japan by the deeper Tsugaru Strait. Fitting this geography, the oldest-yet identified DNA from ancient bones and teeth indicates both a strong connection between the Homon peoples of Japan and the Amur region of southeast Siberia, but no connection to southeast Asia, and a possible date of arrival of the Siberians into Hokkaido at about the time of the last glacial maximum, twenty to twenty-five thousand years ago. Other genetic data indicate that around the same time, southern Japan, in fact the Okinawa islands, received an influx from Korea, the descendants of which barely reached Hokkaido in northern Japan, as judged by a near-absence of the Korean genes there.

Evidence for a northern route into eastern Asia lies in finds of human presence approximately forty-five thousand years ago in the Altai Mountains of western Mongolia and southern Russia, and farther north too in west-central Russia. The migrations moved west as well as east, for that period is also when we find signs of humans in western Europe. Artifacts of this time in Europe are far more common than skeletal or other datable remains, but human teeth have been dated to then in Italy, as have shell beads. The time marks the start of the Aurignacian period in Europe.

Despite the approach of the peak of the last ice age, humans even got to Britain then. We know this from carbon-14 dating in 2011 of collagen from a tooth in part of a jaw first dug up in 1927 near the holiday town of Torquay in Devon. The Torquay jaw and its dating are common to quite a few other finds discovered around the same time. All of the specimens have sat in museums for decades, and we are now benefiting from the greater precision and accuracy of the latest dating methods. I cannot resist adding that Torquay might be more familiar to many of us as the setting for John

Cleese's hysterically funny television series, *Fawlty Towers*, than as the site of the oldest modern human in Britain.

In fact, humans continued to move north. By thirty-two thousand years ago, our species had arrived well into the Eurasian Arctic. We know this thanks to the Yana Rhinoceros Horn archeological site near the Arctic Sea coast in central northern Siberia. The site was first identified in 1993 with the discovery of a spear foreshaft made of rhinoceros horn. (The spear blade is attached to the foreshaft, which itself is attached to the main shaft.) Eight years later, a research group directed by Vladimir Pitulko of the Russian Academy of Sciences at St. Petersburg began serious archeological excavations there.

The several other bone and stone tools found during the excavations, along with accurate dating of the finds, showed that humans were in the Arctic twice as long ago as previously thought. At five hundred kilometers north of the Arctic Circle and close to the peak of the last ice age, the living must have been extraordinarily difficult. The average air temperature there in July these days, the height of the summer, is ten to fifteen degrees Celsius. The annual January average is minus thirty-five degrees. And that is nowadays, when the Northern Hemisphere is between glacial periods, warmer than it has been for many tens of thousands of years.

It appears that humans disappeared from Britain nearly thirty thousand years ago. Perhaps our retreat is not surprising, given that the European ice cap might have reached the Midlands of Britain. We reappear again fifteen thousand years later, as the ice age comes to an end.

Some archeologists suggest that as humans abandoned Britain, so we abandoned northeastern Eurasia—in other words, soon after we had arrived there. However, others argue that we could have hung on in northeastern Siberia even through the peak of the last ice age.

The reason humans might have been able to remain as far north as the Yana site is that it, along with much of the rest of northeastern Siberia, was relatively dry. Therefore, it was not only ice-free, but in fact supported a rich tundra vegetation. Mammals do well in both environments. Red deer, for instance, survived in Siberia from maybe more than fifty thousand years ago throughout the peak of the last ice age.

In addition to food, fire would have been needed for survival in the Arctic winter. The tundra's small shrubs would have provided enough wood to start fires. Once a hot fire is going, bone then burns well. A population of large mammals unused to being killed at a distance would have provided an abundance of bones. Humans, then, might have lived in northeastern Siberia throughout the last glaciation. If so, and if others west and south of them retreated, we would have an explanation for how the genetic makeup of northeastern Siberians became different enough from other populations' types to be identifiable as the source of the earliest Americans.

However, we know from both 24,000-year-old and 17,000-year-old human bones found in Mal'ta in south-central Siberia, which I write about in the next chapter, that populations probably persisted in at least southern Siberia through the last ice age. Given the large geographic range over which peoples of northern temperate regions move, maybe xenophobia needs to be invoked to explain the apparent isolation of the American founders in northeast Siberia for centuries before they entered the Americas. Xenophobia can be a more common barrier to movement of peoples than it is often given credit for. I write about it in a later chapter.

Northeast Asia (Siberia) is across the Bering Sea from northwest America (Alaska). The two regions and the Bering Strait are, collectively, Beringia. Crucially, the Bering Strait is very shallow, less than fifty meters deep. It was dry land during the last ice age, when sea levels were more than a hundred meters lower than they are now. However, at the time that people were at the Yana Rhinoceros Horn site, an ice sheet even larger than the one in northwest Eurasia probably blocked overland movement south out of Beringia into the Americas.

It might not have been until maybe sixteen and a half thousand years ago, with the melting of the American ice sheets, that humans moved out of Beringia south to the rest of the Americas. So far, the earliest well-confirmed date for human presence in the Americas is fifteen and a half thousand years ago, way down in Texas, at Buttermilk Creek, roughly halfway between Fort Worth and Austin. There, archeologists have found over fifteen thousand stone tools and associated debris from their making, which they could directly date, using the so-called optically stimulated luminescence method.

By fourteen and a half thousand years ago, humans had reached southern Chile. That date comes from the Monte Verde site on the west coast of the country one hundred and eighty kilometers or so south of Valdivia. Archeologists have found there the remains of wood-framed tents, along with weapons, animal bones, medicinal plants, and even a human footprint. Monte Verde is one of the most important archeological sites in the Americas because of its early date so far south.

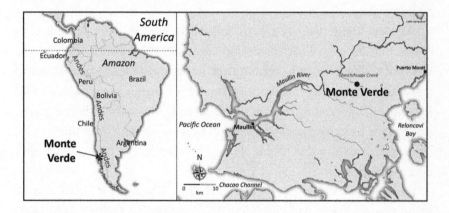

Monte Verde site in southern Chile. Dated to 14,500 years ago, it is one of the earliest in the Americas, and well before arrival of the Clovis culture in North America. *Credit: John Darwent.*

Discovered in the 1970s, the site is notorious as well as famous. For many archeologists in the Americas, its date was way earlier than they could accept, and they expressed their disbelief loudly. The date contravened a long-accepted earliest arrival in the Americas of a little over thirteen thousand years ago. Now, though, with more and better evidence from Monte Verde, the disbelief has evaporated.

The people thought to be the earliest Americans, the people of the Clovis culture, arrived thirteen thousand years ago. This is the culture that used to be considered the Americas' first peoples. "Clovis first" was how the idea came to be known in the arguments about which peoples were in fact the first immigrants. "Clovis first" was the entrenched position against which the main excavator of the Monte Verde site, Thomas Dillehay, had to fight.

Note that I use the phrase "Clovis culture," not "Clovis people." Archeologists identify the people not from their skeletons, not from their DNA, but from their stone tools. It is as if we identified a Western European people of the late 1800s from their paintings, and termed them the Impressionist culture. Archeologists identify the Clovis culture by its elegantly made stone arrow and spear heads. These have a characteristic broad groove at the base for fitting into the weapon's shaft. The culture's name comes from the town in New Mexico near which the first of the stone tools were found, in the early 1900s.

We now know that the Clovis culture developed among people who came from Siberia. That conclusion comes in part from a study of the DNA of a Clovis infant's skeleton from about twelve and a half thousand years ago in Montana. Genetically, the Clovis people seem to be similar to the pre-Clovis immigrants to the Americas. The main difference is the stone-blade technology, which seems to be a North American innovation. The characteristic stone blades are not in Siberia, and only just get into South America.

The Clovis culture was highly successful. Their stone blades can be found all over North America. Here might be a twist on the "Clovis first" story. It could in fact be true in some places. The original immigrants to the Americas traveled fast down the west coast. If they went south, rather than east, while the later Clovis culture spread with people east rather than south, then maybe people using the Clovis stone-tool culture did in fact reach some parts of eastern North America before the pre-Clovis people did. The otherwise moribund "Clovis first" hypothesis might still be alive.

Buttermilk Creek and Monte Verde are at less than two hundred meters altitude. The town of Clovis is at thirteen hundred meters. In the Americas, as in Asia, it took humans a bit of time to move into the higher mountains of the region. The earliest well-dated higher site is at two and a half thousand meters in Peru. People settled there a little over twelve thousand years ago. Two and a half thousand meters is not exactly high. If one hiked at that altitude, most of us would not be able to distinguish any effects of altitude from simple tiredness. However, just about six centuries later, people were at thirty-eight hundred meters, again in Peru. Now we are talking headaches for the first few nights there.

The influx of people into the Americas from Siberia did not end with the Clovis culture. Two further immigrations from Siberia peopled North America something like five thousand years ago—after the disappearance of the major ice caps there. These immigrants gave rise to the present-day Eskimo-Aleut and the Na-Dene speakers of Canada and Greenland.

If people can move one direction, they can also move back again. And so we find that the peoples in eastern Beringia—i.e., Alaska—moved back into Siberia. Far to the south, some then-South American peoples moved back into Central America, and yet others moved from one side of Central America to the other.

When humans entered the Americas by moving out of Beringia, they moved rapidly south into warmer climates, bypassing the northern ice sheet. By contrast, in Europe people had to wait for the Scandinavian ice sheet to melt before they could move back into the northwest part of the continent. Consequently, despite their late start in the Americas, humans covered much of those two continents about five thousand years before they finally reached northern Scandinavia, perhaps ten thousand years ago.

People were able to move back into western Europe a few thousand years before they could inhabit Scandinavia. A variety of studies indicate that hunter-gatherers re-entered Britain from the south as early as fourteen and a half thousand years ago, near the end of the so-called Upper Paleolithic period in Europe. There, despite the cold, they flourished on animals naïve to human hunters, and on an abundant marine life, including seabirds.

Following the Paleolithic is the Mesolithic. It is characterized by extremely small manufactured stone tools, maybe only a centimeter long, hafted onto shafts or handles of various sorts. Finally came the Neolithic, characterized by beautiful smoothly polished stone tools, as opposed to the flaked tools of previous ages. The Neolithic is also when livestock began to be domesticated and agriculture developed. The favored sites for the origins of Middle-Eastern domestication have changed over time, but by early this century Anatolia in Turkey was the winning site.

Because these periods are characterized by their tools—the "lithic" part of the name means "stone"—and because people moved with their tools from the south to the north, arriving sometimes several thousand years later in the north, the periods begin and end earlier in the south than they do in the north. Think of tides rising up a beach. Thus the Eurasian Paleolithic began in the Middle East fifty thousand years ago and ended in northern Europe a little over ten thousand years ago. By then, the Middle Eastern Mesolithic had just about ended, having begun twenty thousand years ago. As the Mesolithic ended in the Middle East, so it began in Europe. The Neolithic that began ten thousand years ago in the Middle East did not appear in Europe until approximately seven thousand years ago, and not in western Europe until a little after six thousand years ago.

Genes and tools indicate that the early Mesolithic people in Britain came in large part from northern Spain. That certainly makes sense. Southerly Spain stayed warmer during the last ice age than did more northerly parts of western Europe. Spain will have warmed earlier than elsewhere in western Europe, and so, perhaps, an expanding population spilled out of the Iberian Peninsula earlier than it did elsewhere in western Europe.

European immigration to Britain continued, with Celts preceding Anglo-Saxons. The Celts went largely into western Britain and the Anglo-Saxons into eastern Britain, from where they probably pushed some of the Celts farther west.

We do not necessarily have to postulate a mass migration of Anglo-Saxons to explain the spread of their culture through Britain. In 2014, Susan Hughes and her co-authors reported the results of chemical analysis of the tooth enamel of nineteen skeletons in an Anglo-Saxon cemetery in Oxfordshire. The question was: Were these people Anglo-Saxon immigrants from the continent, or were they locals who had adopted Anglo-Saxon culture?

To answer that question, they compared the ratio of oxygen to strontium in the teeth to the ratio in local soils and local animal bones. Eighteen of the people showed such similar ratios to those in the soil and animal bones that Hughes and her crew inferred that the people had been born in the area and adopted the culture. Only one came from the continent, they concluded.

This study brings my account of humankind's worldwide diaspora to 500 A.D., or as academics tend to non-denominationally term it nowadays, not

A.D. but C.E. The latter stands for Common Era, but it can equally stand for Christian Era, which then is not much different from Anno Domini.

Toward the end of the first millennium C.E., Vikings from Scandinavia began to father children over much of Britain, as we can tell from genetic studies of buried bones, because Viking genetic makeup is different from Celtic or Anglo-Saxon genetic patterns. Then came the Normans of France in 1066. The year 1066 is one of the best-known dates in British history, at least to my generation. I do not know the year that the current Queen Elizabeth succeeded to the throne—either 1952 or 1953—but I do know when William the Conqueror arrived.

This summary of Britain's peopling does a disservice to all that we know about the complexity of the process, and especially to all the scientists who have elucidated what we do know. Anyone who wants to find out more should go to Stephen Oppenheimer's masterly *The Origins of the British*. There, he marshals mostly genetics, but also archeology, history, and literature in a 500-page exposition, which at one point he describes as abbreviated.

Oppenheimer effectively stops at 1066 and the arrival of the Normans. But Britain has experienced subsequent influxes of people. In my lifetime, many thousands have arrived from Britain's former colonies, for example Kenya, Uganda, and Tanzania in East Africa. They began coming in large numbers in the mid-1900s, three hundred years or so after the emigration *from* Britain that established a former colony, the USA.

The Neolithic revolution's development of agriculture led to a population increasing in size at approaching five times the previous rate. Agriculture and agriculturalists expanded throughout Eurasia over the next few millennia. Some of these agriculturalists were seafarers too. We know this because by ten and a half thousand years ago, we see signs of farmers and their domestic animals, including cats and dogs, in Cyprus, which lies more than fifty kilometers out to sea from Turkey and Syria.

The advent of agriculture and the ensuing increase in the human population happened later in Europe than in the Middle East. Matching previous archeological indications, a genetic analysis by Christopher Gignoux, Brenna Henn, and Joanna Mountain indicates agriculture and the population increase starting about eight thousand years ago in eastern and central Europe. Northwestern Europe had to wait another two to three thousand years before it was warm enough for agriculture there.

A warming climate, warming later in the north than the south, does not explain all differences between regions in the timing of the agrarian revolution. The Gignoux, Henn, and Mountain study showed that tropical southeast Asian populations did not increase until approaching five thousand years ago. That is about the same time that the increase happened in northwestern Europe, despite southeast Asia's more clement climate. Why the delay? What happened five thousand years ago in southeast Asia? That is when people started to farm rice. The human population increased in west Africa at more or less the same time, which matches the fact that there, too, archeologists find a near-simultaneous appearance of signs of the adoption of agriculture.

The interesting question here is how agriculture takes over a region, how it supplants the former hunter-gatherer lifestyle. Does the culture spread as local hunter-gatherers adopt agriculture? Or does it spread because agriculturalists are on the move, taking the practice with them? If the latter, what happened to the previous residents? Did they become agriculturalists too, did the agriculturalists absorb them by mating with them, or did the agriculturalists push them out? We know from recent history that African pastoralists and the later European agriculturalists have marginalized resident hunter-gatherer peoples on the continent.

Luigi Luca Cavalli-Sforza is a famous geneticist born in Genoa, Italy. He moved, via bacterial genetics in Cambridge, England, to human genetics at Stanford, California. He has consistently argued that his studies and others' showed that the people and their culture move together. Agriculture arrives in western Europe because agriculturalists move into western Europe. Some of the strongest evidence for Cavalli-Sforza's idea is that maps of the presence and movements of languages in Europe closely match maps of the presence and movement of genes. The same holds for the spread of humans and their cultures across the Pacific. In these cases, ideas (agriculture) did not spread independently of the people who had the idea initially.

Nevertheless, subsequent work both substantiates Cavalli-Sforza's contention, and indicates more complexity. For instance, one of the few quantitative tests of the idea showed that on balance—indeed, more often than not—increasing geographic distance between populations correlated with both increasing genetic distance and increasing linguistic distance. However, disconnects existed. The correlation was nowhere near as obvious for languages as for genes, implying that genes (people) and language could move somewhat independently. A specific example that I described a little while ago is the inferred adoption by fifth-century c.e. Thames Valley residents in Oxfordshire of the continental Anglo-Saxon culture. They in effect went from being Italians—i.e., Romans—to being Germans. The change will have been most visible among the rich, where masonry houses returned to timber and thatch.

The analyses of the timing of the adoption of agriculture and its relation to the increase in population size in Europe, Asia, and Africa that I described a few paragraphs ago indicate a rapid increase in population size among the dispersers, the new people in the region, the agriculturalists, but not among the resident hunter-gatherers. The implication is that the residents did not adopt agriculture. Here, then, culture—i.e., agriculture— moved with genes, with people.

But what happened to the resident hunter-gatherers? That depends. One study of central European populations found that hunter-gatherer genes disappeared as agricultural genes arrived. The three-continent study of the previous paragraph found that the residents' genes persisted, even if they did not increase. Similarly, another study of three 5,000-year-old skeletons of hunter-gatherers and one farmer in Scandinavia found that the farmer's genes were most similar to those of the people now around the Mediterranean, whereas the hunter-gatherers' genes were most similar to those of extant northern Europeans. Indeed, it looked as though northern hunter-gatherers might have remained largely genetically separate from the immigrant agriculturalists for a thousand years, before the two genotypes finally blended. So there the hunter-gatherers hung on.

Marie-France Deguilloux and her colleagues argue that the sudden appearance of agriculture through much of western Europe, and yet the persistence in the populations of large percentages of the genes of pre-agricultural residents, implies that for the most part, the probably more numerous residents married with the arriving agriculturalists at the same time the residents very quickly adopted agriculture. Deguilloux did not define "large percentage." However, maybe as much as three quarters of at least the mitochondrial genes could have come from the resident western European population. (Mitochondrial genes are passed down the generations through only the female line, because a sperm is too small to contain any.) Such a high proportion of mitochondrial genes from the resident hunter-gatherers has to mean that they adopted agriculture from a small number of arriving immigrants, as opposed to being swamped by a large population of arriving agriculturalists.

We have different studies by different scientists with different methods looking at different parts of the world. So, are we coming up with the contrasting findings because some studies are wrong? Or do different populations do different things? Both happen all the time. At present, most of the studies that I have been describing are so recent that we are going to have to wait for confirmation. My strong bet, though, is that all the answers—adoption, absorption, replacement—will turn out to be true, depending on the region of the world.

The Mayflower bringing Pilgrims to the future USA in the fall of 1620 c.e. carried one third as many women as men: seventy-eight men and just twenty-four women. That piece of information is a sudden break from the previous topic of the arrival of agriculture in Europe, I know. However, I use it to highlight the fact that I have not so far talked about differences between the sexes in terms of which of them migrated to the new parts of the world.

As might be expected, some studies show that resident females accepted invading males and their culture. We see this from the fact that both language and Y-chromosome genes of the immigrants, carried by only males, are in the resident population, whereas the mitochondrial genes, passed through females, are largely those of the residents, not the immigrants. An example in the Americas is Uto-Aztecan, a language family of the western USA and Mexico. It maps onto the distribution of Y-chromosome genes, but not mitochondrial genes. The males moved with their language, but females stayed where they were, differing genetically from region to region, and adopted the language of the immigrants.

A classic case of a region's language being that of male invaders is Iceland. Icelandic is a variant of Scandinavian languages, Norse. But while

Icelandic Y chromosomes are predominantly Scandinavian, Icelandic mitochondrial (female) DNA is predominantly British. Why the contrast? Because Iceland, especially the west, was largely peopled by Vikings bringing with them women captured from the British Isles. In the east, the Vikings came direct from Scandinavia, and we see more equal proportions of the male and female genes. Further evidence of the peopling of the west via Britain is its higher incidence of place names of British origin.

The pattern of male immigrants and female residents is not ubiquitous, however. Greenland's language and mitochondrial DNA are Inuit, whereas half the Y chromosome is European. Here the language and the mitochondrial DNA reflect the residents, while the Y-chromosome genes reflect passing visitors, probably mostly whalers.

I mentioned the Viking influence on the British and Icelandic genetic landscapes. The Vikings moved through much of the rest of Europe too. We can pick up the Viking expansion into eastern Europe from the fact that a quarter to three quarters of the genes from the peoples there—Bulgaria, Romania, Poland, and so on—come from northwest Europe. Most of the other three quarters to a quarter come from the southeast, from the region of Greece and Turkey. And then a small percentage reflects northeast Asian ancestry.

These findings about the origins of present-day eastern Europeans come from a worldwide study by Garrett Hellenthal and his colleagues of events within the last four millennia. They speculated that the genetic signature of eastern European immigration could reflect the movement of peoples pushed by invaders from the east, while the small percentage of northeast Asian ancestry could arise from the invaders: the Huns and Mongols, among others.

Central Asia, roughly Afghanistan, seems to be even more of a mixing pot. There, we see in the genetic makeup of the people that in approximately the middle of the last millennium, around the time of the Tudors in Britain and in the centuries bracketing the time of European arrival in the Americas, contributions come from all surrounding regions, including eastern and northeastern Asia. A little farther to the east, we find that the Kalash of Pakistan are about one quarter western European, including even a percentage of Scottish genetic heritage.

European Mediterranean populations have a large genetic contribution from Africa, as do Arabian Sea populations. Part of that contribution is a result, almost certainly, of the slave trade. That conclusion comes from the fact that historical records show that each region got their African genetic contribution from the same regions as they got their slaves. In the case of the European Mediterranean, both came from west Africa. Arabia got both from east and south Africa.

The Hellenthal study did not analyze the Americas. However, of course they too show a strong influence of the slave trade, along with, again of course, massive European influence. One cultural exemplar of that European influence is the United States' celebration of Columbus Day. The odd thing about Columbus Day in the USA is that Columbus never set foot there. At least Hawaii and South Dakota recognize their first immigrants with Discoverers' Day and Native American Day. It is surely time that the rest of the United States followed suit, scrapping Columbus Day, like Oregon, or commemorating the real first settlers, the Native Americans.

I wrote in the first chapter about how irrelevant the concept of "race" is in the context of human biogeography. How does one define a "race" when the people are, as I have just been describing, often a genetic mix from many regions?

Let me give an example from the Hellenthal worldwide study. One of their diagrams shows the genetic constitution of the San Khomani and the San Namibia of southern Africa. These two peoples are almost neighbors. To an outsider such as I am, they look alike. Many people would term them both "bushmen," and they have historically been treated as primitive. And yet while the San Namibia have three detectable genetic sources, one major and two very minor, the San Khomani have a dozen detectable sources, at least a third of them from Eurasia.

From another study, we know why we find these differences. Something like three thousand years ago, a migration from southwestern Eurasia moved to eastern Africa, and then, around fifteen hundred years later, another migration took them and their language into southern Africa, alongside resident Khoi-San peoples. And so some of these Khoi-San peoples have over ten percent of European genes in them from east African populations that might have had a quarter of their genes from Europe. We humans are all so obviously and so recently of African origin regardless of our skin color, we almost all so obviously originated somewhere other than where we are now, and we are almost all of us such a mixture of peoples from different regions that to define any one of us as being of a particular race is, scientifically, an utterly empty statement—unless we are all defined as Africans.

In the spread of our species across the world from Africa, we arrived latest in those places reachable only across wide stretches of ocean. The Pacific is the widest of the world's oceans, and so the Pacific islands were some of the last places where humans landed. We spread across the Pacific from Asia, not from America, and so the western Pacific islands were our first conquests in the region. We arrived there something like thirty thousand years ago. We might even have arrived in the Solomon Islands by then. The Solomons are eight hundred kilometers east of southern New Guinea, but we could have reached them from New Guinea via New Britain and Bougainville with a maximum crossing of a little over two hundred kilometers.

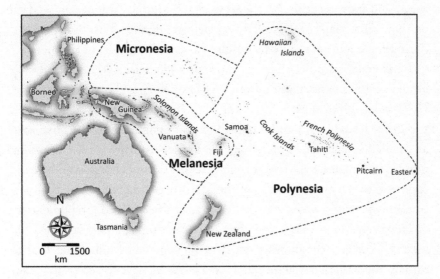

Map of the Pacific, showing the regions of Melanesia, Micronesia, and Polynesia. *Credit: John Darwent.*

Humans remained in the western Pacific for the next twenty-six thousand years, so great are the distances to the central and eastern Pacific islands, and so small are those islands. Then a little after four thousand years ago, a rapid expansion took humans across Melanesia (east and southeast of New Guinea, out to Fiji) and Micronesia (north of Melanesia). One way we know about this expansion is from the beautifully patterned pottery left by the peoples of the Lapita culture, named for the site in New Caledonia where the archeologists first discovered the pottery. The remains are oldest in the western islands, and youngest in the eastern.

Finally, a little over two thousand years ago, humans suddenly began to move across the rest of the Pacific. We reached Polynesia and the farthest eastern Pacific islands, such as Tahiti and the Tuamotus, maybe two thousand years ago, Hawaii perhaps fifteen hundred years ago, and, finally, New Zealand only a scant seven hundred years ago.

An obvious explanation for the 26,000-year hiatus in our spread across the Pacific is that humans could not go any farther until they designed large ocean-going canoes. The canoes had to be sufficiently big to carry enough people and their supplies, including livestock, to begin a new community on a distant island. What were these pioneers thinking when they set out? All they could see was ocean. All they saw for day after day was ocean. And yet on they went. Many of those early exploring emigrants must have perished during these journeys of many hundreds, even thousands, of kilometers.

Early genetic studies showing little variation within Māori DNA suggested a founding population from very few people and very few landings. No wonder, given how perilous the long voyage to New Zealand (from the Cook Islands, two and a half thousand kilometers to the northeast, as detailed in chapter 3) must have been. One estimate for the founding number of females, based on current diversity in mitochondrial DNA and estimates of the rate of change of the DNA over time, indicated that perhaps only seventy women contributed to the subsequent Māori population.

However, Māori oral history has always spoken of many voyages, and indeed, new genetic analyses confirm previous ones indicating that to explain the level of variation in Māori DNA, we have to accept that they accomplished several successful voyages. The descriptions of the number of voyages as "very few," "many," and "several" are vague, I know—but so far the genetic results do not allow anything more precise. Nevertheless, they do allow an estimate of the eventual total number of successful founders: those crucial seventy women of the previous paragraph.

The only islands in the Pacific that I have visited are two of the Hawaiian islands, and New Zealand. When I visited Hawaii, I was so ignorant of its geography that I asked at the Honolulu airport where I could hire a car to drive to Hilo, and how long it would take me to get there—not knowing, to my shame, that Honolulu and Hilo are on different islands.

About the same time as humans finally expanded into Polynesia, doing so with a far better sense of geography than I displayed in Hawaii, they reached Madagascar. Or at least that is what everyone thought until earlier this century. But then, in a cave in northern Madagascar, researchers found bones with cut marks on them in sediments dated to at least thirty-four hundred years ago.

But was this early settlement successful? Did it lead to the peopling of Madagascar? Perhaps not, given that all the other genetical and archeological evidence so far indicates no sustained population of humans in Madagascar until at least as late as two thousand years ago. The date for a permanent human presence there might even be later, maybe even not until fifteen hundred years ago. Indeed, some argue that not until a fairly extensive maritime trade out of Indonesia twelve hundred years ago was there a permanent human presence on Madagascar.

Indonesia? Yes. Close as Madagascar is to Africa and humanity's place of birth, genetic and linguistic evidence indicates that the first successful settlers in Madagascar came from Indonesia. We do not know the origins of the earlier ones. The same sort of calculations that produced an estimate of only seventy women in the founding population of New Zealand indicated perhaps as few as thirty women in Madagascar's founding population. The estimates were made with mitochondrial DNA, which, remember, is passed down via only eggs, not sperm, and so gives information about only female ancestry.

Bantu peoples from Africa crossed the Mozambique Channel to Madagascar either at the same time as the Indonesians who successfully colonized Madagascar arrived, or maybe up to five hundred years later. The final wave of settlers, the French, arrived five hundred years ago.

My wife and I had the privilege of visiting Madagascar in 2001. The separate influences of the Indonesians, and the Bantu and French peoples and culture, was gloriously evident. The houses reminded us of Bornean houses, and the rice paddies and rice meals of course took us straight back to a previous visit to southeast Asia. The zebu cattle and the dry landscape raised in our minds memories of our many trips to Africa. And the breakfast baguettes in the bicycle panniers could not have been anything but French.

∽

An African presence in Madagascar is understandable. After all, the island lies just a few hundred kilometers off the coast of southwestern Africa. French in Madagascar is understandable with a little bit of knowledge of French exploration of the region. But Indonesians as the first inhabitants of Madagascar must be a surprise to most people. That surprise introduces us to the next topic of discussion and the next chapter, namely the exact routes that humans took to get to where humans went.

CHAPTER 3

FROM HERE TO THERE
AND BACK AGAIN

A mostly coastal route out of Africa—across the world?

W e humans left Africa sixty thousand or so years ago in the exodus
that peopled the globe. We reached eastern Asia, Australia, and
western Europe fifteen thousand years later, but did not get
to the Americas until a little over fifteen thousand years ago, and not to
most eastern Pacific islands until two thousand years ago. We landed on
one of the last major land masses, New Zealand, in historical times, just
seven hundred years ago. With that quick summary of the chronology of
our worldwide spread, I turn now to what we know of the exact routes that
those venturous humans took as they dispersed.

What barriers did these early travelers face? Where and how did they
overcome them? Did they leave Africa via the Sinai across the northern end
of the Red Sea on the same route as followed by Moses? Or did they leave
from Eritrea via Arabia at the junction at the south end of the Red Sea and

the Gulf of Aden? Whichever of those two routes, where then did they go on their way to all the other continents and islands of the world? Did they move into eastern Asia along a route north of the Himalayas through Mongolia, cold and high as it was, or did they disperse there via southern Asia? Did anyone in prehistoric times manage to reach the Americas across the vast inhospitable expanses of the Atlantic or Pacific oceans? Or was the trip exclusively across Beringia?

I give in the pages to come what I see as the current best answers to these questions. But nobody should think that anywhere near complete consensus exists among the experts. Archeologists, geneticists, geographers—professional and amateur—have argued and are still arguing the answers, and undoubtedly will continue to argue them for a long time. As I said in the last chapter, the lack of agreement does not mean *no* agreement. We do know some answers. But the fact that so many questions and other possible answers remain makes the search all the more interesting.

I will start with the exodus from Africa. Although I use the word "exodus," remember that apart from the fact that we are talking about a small number of people over a relatively short time, we do not know whether we are describing a sudden wave, several wavelets, or a steady trickle. And as I said in the previous chapter, of course it was not a complete exodus. Many people remained in Africa.

I am going to leave those phrases from the last paragraph—"a small number" and "a relatively short time"—largely undefined. We possibly need to think in terms of a few thousand people over the course of, at most, a few millennia. However, so many factors affect estimates of just the size of the founding populations, let alone anything else about them, that even "a few" might be too exact.

A word of warning here to anyone who goes to the papers that produce these estimated sizes of the founding populations. Geneticists distinguish between the *effective* population size and the *real* population size. The former is, put simply, the number of individuals in the population that leave descendants. It is often at least half what the non-geneticists among us would consider to be the number of adults in the population.

Assuming that humans went on foot when they left Africa, they could have taken one or both of two routes out. Starting from roughly Ethiopia, one route goes up the west side of the Red Sea, almost reaches the Mediterranean, and then turns east across the top of the Sinai Peninsula, and on out across the Middle East and into Asia, with several possible routes to choose from. The other route from Africa goes across the southern end of the Red Sea, what is now the Bab el Mandeb Strait between southern Eritrea and southwest Yemen, and on out—with several possible subsequent routes to choose from. How do we decide which of the several routes was the more likely one?

In 2005, Julie Field and Marta Lahr used a mathematical model to work out the possible routes out of Africa all the way to Australia. The model is essentially an algorithm—when people reach spot X, they keep going straight if the environment is suitable. They turn only if it is unsuitable. Grassland or savannah was suitable; desert was not. If a choice between inland or coast presented itself, the model incorporated a preference for moving along the coast, because humans would have found lots of easily catchable shellfish there. Knut Fladmark had argued the same point over twenty years before Field and Lahr when suggesting a coastal route from Beringia into North America. For both the nature of the environment and for the coastal route, Field and Lahr took reasonable account of the estimated environment approximately fifty thousand years ago, including the fact that sea levels were on the order of eighty meters lower than now.

"Easily catchable" shellfish is the key to the coastal route. They are on all coasts of the world, even on the edge of deserts. Witness the abundant half-shells of oysters on the Eritrean coast from a hundred and twenty-five thousand years ago, along with stone tools presumably used to open them, as well as the huge mounds of shells—"middens," the mounds are termed—along coasts worldwide. Around the same time in South Africa, shellfish remains were several times more abundant than ungulate remains in archeological sites, according to Teresa Steele and Richard Klein.

These shellfish mounds are almost unimaginably large. In North America, Amanda Taylor, Julie Stein, and Stephanie Jolivette describe "small" shell middens as under three thousand square meters. That is one hundred meters by thirty meters, which is not much smaller than an American football field or a non-American football (soccer) pitch. Some of the North American shellfish middens are over a meter tall. In Australia, some shell middens are up to five meters high, with trees growing on them.

Even now, on the heavily populated California coast north of San Francisco, you can easily pick a plateful of shellfish off the stony rocks. Make sure you get a permit to do so, though. A good friend of my wife's and mine told us how, back in the late 1960s, he and his friends would go camping on the north California coast. They would easily pick their limit of abalone by feeling for them with their bare hands under the exposed kelp and sea grass-covered rocks. On their return to camp, they would dig for their limit of horseneck clams on a large exposed sandbar, and then head off to collect steamer clams. Add a sackful of black mussels, and in half a relaxed day, they had, with the most primitive of tools, collected enough to feed their families at home for a couple of days.

A major advantage of shellfish over other animal food is that children can rapidly harvest shellfish, even if not as efficiently as can adults. Betty Meehan recorded adults in Arnhem Land in northern Australia harvesting an average of ten kilograms or so per trip, and children nearly three kilograms. Doug and Rebecca Bird saw that children in the Torres Strait between New Guinea and Australia could collect a kilogram of shellfish meat per hour. That is a meal for four people, according to the recipe books in my house. How proud and thankful would modern Western parents be if their child could collect, on the way back from school, the family's evening meal!

Field and Lahr concluded that the likeliest route out of Africa and onward was up the west coast of the Red Sea to the Sinai, across above the Dead Sea to the upper reaches of the Euphrates, and then on down that river to the eastern side of the Persian Gulf. Rivers, like coasts, have easily catchable food in them, so much so that two million years ago in the Turkana Basin in Kenya, hominins were eating large amounts of turtle and catfish.

Although Field and Lahr's hypothesized coastal routes are appealing, some of the other routes are not impossible. Two years ago, Anders Eriksson and eight others also modeled the migration of humans out of Africa and across the world. They, like Field and Lahr, incorporated the nature of the environment into their model. However, the Eriksson group's model used climate and environmental productivity to predict growth of the human population. From that, they plotted expansion of the population across the land. Unlike Field and Lahr, they did not stipulate a preference for moving along the coast. This is the same team whose genetic work on timing of the human diaspora I wrote about in the previous chapter. The Eriksson team's model suggested a southern route across the Bab el Mandeb Strait between southern Eritrea and Yemen, and on across the bottom of the Arabian Peninsula, as the more likely of the two routes out of Africa.

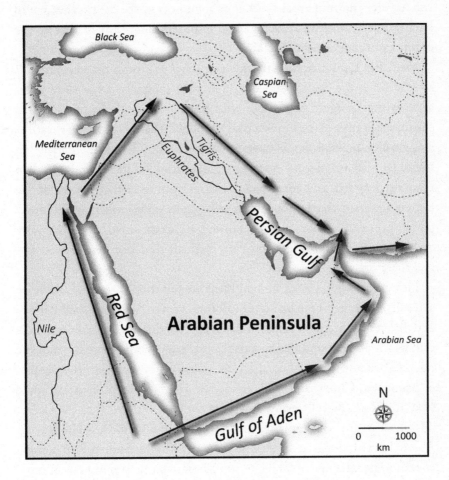

Map of the Middle East, and potential routes out of Africa. *Credit: John Darwent.*

To cross the Bab el Mandeb Strait, humans will probably have needed boats. The deepest part of the Strait is currently at three hundred meters. If the dispersing humans continued east along the Arabian coast, their first sight of Iran might have been across the Strait of Hormuz. It is not impossible that the Strait then was less than twenty kilometers wide. Someone from the dispersing party who climbed to the top of the cliffs above the beach would easily have seen across the Strait. If male teenagers were among the party, they might even have swum across, notoriously dismissive of danger as male teenagers are. Even if the travelers did not cross by boat of some sort at the Strait, they could have crossed dry-shod at nearly eight hundred kilometers farther up the coast, from modern-day Qatar.

Field and Lahr's detailed routing continues east from the Persian Gulf mostly along the coast, occasionally moving inland to cross large rivers such as the Ganges. A problem with testing a hypothesis holding that the diaspora from Africa took a coastal route is that much of the coastline along which humans would have trekked is now under water. Starting on the order of ten thousand years ago, global sea level has risen more than a hundred meters as a consequence of the melting of the ice caps of the last ice age. Any coastal route is therefore now under water, and in many places several kilometers out to sea from the current coastline. Underwater archeology exists, but it is a lot more difficult than dry-land archeology, and therefore rarely practiced.

However, stones, bones, and middens are not the only evidence that we can call on. Genetics can be highly informative too. Analyses of similarities and differences in DNA between the peoples of different regions can indicate which populations are the most closely related, and therefore indicate potential routes. For instance, a close relationship between the peoples of Mongolia, China, and Malaysia would indicate a central Asian route (the Silk Road route) toward Australia. In fact, we find that the closest relationships are from India through Malaysia and on to New Guinea and Australia via Indonesia, as we would expect if humans dispersed along a coastal route. (In the next chapter, I elaborate on how genes reveal more history of human movement through the ages.)

Field and Lahr's model gives us two potential routes from Asia to Australia. One is a southern route along the Indonesian peninsula, from Sumatra, through Java, Bali, and East Timor, and then a jump to Australia. The other is a northern route via Borneo, Sulawesi, and New Guinea.

By the time humans reached Australia, perhaps fifty thousand years ago, the shallow Torres Strait between New Guinea and Australia and the equally shallow Bass Strait between Australia and Tasmania were then above water, because of the drop of sea level during the last ice age. So, if humans went by the northern route to New Guinea, they could have carried on to Australia and Tasmania without getting their feet wet.

Nevertheless, it looks as though separate populations peopled New Guinea and Australia. The peoples of these two areas are very different in their genetics and their languages. According to some genetic DNA analyses, they might have separated around fifty thousand years ago. Whatever the time and place of separation, after moving along the coast as far as Indonesia, one lot branched into New Guinea and mostly stayed there, while another went to Australia.

Beware. This last paragraph hides a lot of complex findings and interpretations that are continually being updated. As I write, one of the more complete reviews of the peopling of New Guinea and Australia is Sheila Pellekaan's. It takes up twelve dense pages of text. Not only is the story very involved, but we still know too little. Indeed, as far as Australia and Tasmania in particular are concerned, we might never know enough, at least from genetics, given the European settlers' appalling treatment of the native aboriginal peoples, even up to total genocide. If a people is wiped out, as we wiped out the Tasmanians, we have next to no genes from which to deduce their origins. In effect, we have killed off part of our own heritage.

To reach New Guinea, or to reach Australia without going via New Guinea, humans at roughly fifty thousand years ago would have had to cross a hundred kilometers of open sea from one of the islands of eastern Indonesia. Perhaps it was because of this water barrier that the first New Guineans and Australians had, as far as anyone can tell, no contact with Asia for the next twenty thousand years, maybe even forty-five thousand years in the case of Australia, as I described in chapter 2.

No water barrier existed between New Guinea and Australia, though, so we cannot explain the apparent lack of contact for tens of thousands of years between those two lands. That is, unless we accept the suggestion that in fact the first exodus from Africa into the Arabian Peninsula in fact continued east and led to the peopling of the Old World at that time. Then,

New Guinea and Australia would have been separated by what is now the Torres Strait. The difficulty with that idea is that all the other sea crossings that would have gotten them to either of those landmasses would have been even longer, because the world was then at the height of an inter-glacial, when sea levels are high. The authors of the suggestion of the early arrival in Australia did not address this potential problem with their conclusions.

The earliest archeological signs of humans in Australia to date are in the north, as might be expected, but they also exist in the far southwest. Between the two regions is a large expanse of desert that, then as now, spread to the western coast. A coastal route from first arrival down to the bottom left of the continent therefore seems likely.

The lack of early remains in the east of Australia implies the southern Indonesian route to the continent. The more northerly route via New Guinea would get them to the northern end of the Cape York Peninsula, from which eastern Australia is an easy coastal route away. However, eastern coastal Australia has an essentially warm, wet tropical climate, in which bones do not preserve well. The earliest humans on the continent could have passed that way without leaving any record of their passage.

India is on the way from Africa to Australia. Genetics and linguistics indicate three main regional populations in the subcontinent. The oldest, and now the fewest, are the rather scattered Austro-Asiatic speakers of eastern and central India who might be descendants of the first people to arrive. Then we have the Dravidian speaking peoples of southern India, Indo-European speakers in northern India, and in northeastern India the Tibeto-Burman speakers. Some genetic analyses reveal that a large proportion of the Dravidian speakers have genes indicating that, like the Austro-Asiatic speakers, their ancestors were some of the earliest arrivals on the subcontinent.

A few thousand years after humans reached Australia, we entered western Eurasia, probably from Turkey across the Bosphorus. An easy route on into western Europe would then have been up the Danube. Some archeological finds indicate that the migrating populations also took a more northerly route, from Turkey up through Georgia on the eastern edge of the Caspian Sea, and then west. Matching that route are linguistic and genetic studies, including that of the Eriksson team, hinting at a route north of the Caspian Sea, through Ukraine, perhaps using the Dnieper River for part of the way.

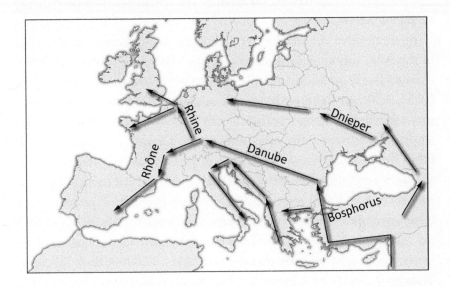

Some potential routes of earliest humans into Europe, emphasizing coastal and riverine routes. *Credit: John Darwent.*

If the earliest Europeans took that northern route, they did so in addition to using one or both of the southern routes. We know this because some of the earliest signs of humans in Europe are in a 44,000-year-old site in southern Italy reported in 2011 by Stefano Benazzi, Katerina Douka, and twelve others. The signs are teeth, and shells with holes bored in them to make beads. Previously, the teeth had been described as Neanderthal, but the new analysis places them firmly as modern humans.

Nevertheless, Neanderthals still lived in Europe forty-four thousand years ago, and Neanderthal genes in modern humans indicate that they and we might have mated with each other, unless these genes turn out to be inherited by both of us from a common ancestor. In chapter 11, I go further into the love-hate relationship of these two hominin species.

Regarding our routes from the Middle East into eastern Asia, largely China, we can only speculate at the moment. We have too few archeological sites and too few archeologists in the region. The Eriksson study indicates the potential, once again, for both a northern and a southern route. However, the strongest genetic and environmental signs get us to about Pakistan in the south, and eastern Kazakhstan in the north, but then become too vague too follow.

By forty-three thousand years ago at the latest, humans had settled in the Altai region of south-central Russia. Sometimes abundant flaked stone tools, the occasional ornament, and a variety of animal bones are the evidence. Where did the Altai people originate? An answer comes from a surprise genetic finding reported in early 2014 by Maanasa Raghavan and a team of over twenty-five other researchers. They decided to analyze the DNA from a 24,000-year-old skeleton of a young boy discovered over half a century previously. The site of the find is in Mal'ta, just north of the west end of Lake Baikal in south-central Siberia, and maybe a thousand kilometers northeast of the Altai. The surprise was that the Mal'ta boy's DNA indicated a strong western European heritage of perhaps one quarter of his genes. The Raghavan team also reported that a 17,000-year-old skeleton of another male from approximately six hundred kilometers northwest of Mal'ta showed similar western European origins. The implication is that a population to which both these skeletons belonged persisted in south-central Russia throughout the last ice age.

Mal'ta is effectively on the Silk Road, which then passes through Mongolia into northeast China. At one point, geneticists thought that northeastern and southeastern Asians had different origins. If so, two routes into Asia would have been a possible explanation. More recently, though, new evidence and new interpretations of the genetic analyses indicate a single origin, almost certainly a southeastern origin. That being the case, the northern route, the Silk Road, to Beijing in northeast China, is presumably a later route, even though the earliest signs of humans near this route are from nearly forty-five thousand years ago. Maybe they reached there but did not go any farther then—too cold to the north, too dry to the east?

Moving farther east in Asia, humans got to Japan roughly forty thousand years ago, it seems. Back then, they could have reached southern Japan overland. That is because the Sea of Japan off the coast of Korea is shallow, less than a hundred meters deep. The last ice age's drop in sea level would therefore probably have exposed the sea floor. To get directly from the Asian mainland to northern Japan at that time, though, people would have had to use boats. Whether they did so, we do not know. We do know, though, that by twenty-five thousand years ago, humans must have been using boats. That is because at that time they reached some of the Ryukyu Islands between the south end of Japan and Taiwan, and to do so they had to cross up to seventy-five kilometers of sea.

Asia so far in my account does not include northeastern Siberia, the necessary overland route from Eurasia to the Americas. I have already mentioned in chapter 2 the earliest finds of humans north of the Arctic Circle at the Yana Rhinoceros Horn site in northeast Russia close to the coast of the Arctic Ocean. Earlier in this chapter, I described finds indicating western European genes in the Altai region of central Russia which, along with eastern Asian genes, might have contributed to the northeastern Siberian populations.

Kari Schroeder and others in David G. Smith's laboratory at my university have shown that one gene in Native Americans occurs nowhere outside the Americas, except in the peoples of one tiny area of northeastern Siberia, a few hundred kilometers east of the Yana site. This particular gene is significant because it is rare in Asia, and yet common in both North and South America. The implication is that a single small population from Siberia peopled the entire Americas.

Beware again, though. Just because all Native Americans came from one small region of Siberia, we do not have to suppose a single, one-time migration from one village. A long-time trickle, say for a thousand years, from that one small region would produce the same genetic prevalence in Native American populations.

As in Africa, so with Siberia and North America: movement of peoples went both ways. Yes, the Americas were peopled from Siberia, but the same sort of evidence that allows us to suppose an origin of all Americans from Siberia allows us also to postulate return movement. According to a report co-authored by Tatiana Karafet, we find in Siberia versions of genes that current research indicates arose in the Americas.

The Bering Strait between Siberia and Alaska is now ocean. Nevertheless, the Siberian ancestors of all Native American peoples would have walked across the Bering Strait. As I wrote in the last chapter, the sea there is less than fifty meters deep, and was dry land up to around twelve thousand years ago.

Returning to the Mal'ta boy with his western European heritage that I described a few paragraphs ago, his DNA showed obvious links to North Americans, as might be expected given all the evidence that Native Americans of both continents originated in northeastern Siberia. The unfolding scenario seems to be, then, that some western Europeans moved through eastern Europe, through Russia, and on to northeastern Siberia. There or thereabouts, they met people from east Asia who had moved north. They had children with each other. And so we find that Native Americans, who all originated from northeastern Siberia, have both European and east Asian genes.

Some western European genetic ancestry in Native Americans could explain why some Native American skeletons show apparently western European features, for example a V-shaped jaw with a pronounced chin. Previously, that connection suggested to some researchers an early overseas colonization of the Americas in addition to the Siberian colonization. However, the Mal'ta finding now explains how European features fit a Siberian origin for all Native Americans.

From northeast Siberia, the founders must have initially moved down the west coast of North America, because the massive Cordilleran and Laurentide ice sheets covered all of Canada and much of the northern USA: northern Washington, Wyoming, Nebraska, Illinois, Ohio, and all of New York and eastern states north of it. Crossing that ice sheet then would have been as difficult as crossing the Greenland ice cap today—in other words, impossible back then.

The coast of an ice sheet is also extremely inhospitable, especially if we take account of calving bergs and the massive waves that they can create. However, a variety of mammals, such as bears and foxes and also sea mammals, were living among the islands off the coast of northwest Canada something like fourteen and a half thousand years ago. Many of these islands would then have been part of the mainland, given sea levels at least a hundred meters lower than now. That implies plenty of food for people tired of a diet of the extremely abundant shellfish that must have lined the coast as they do even now, despite centuries of harvesting.

Once south of the ice sheets, the pre-Clovis now-American people spread inland from the coast, as we know from the finds of stone tools in Texas dated to fifteen and a half thousand years ago.

Most accounts assume a dispersal down the west coast of the Americas, and that is how I have presented the movement. However, Dennis O'Rourke and Jennifer Raff speculate about an eastern route too. It might explain early appearances of people closer to the east than the west coast. However, an eastern route would have entailed seven thousand kilometers of travel along the frigid northern coastline of North America. A movement down the west coast requires getting past only three thousand kilometers or so of glacier.

By perhaps thirteen thousand years ago, the two North American ice sheets had melted sufficiently to allow entry into the region of Montana, North Dakota, and Minnesota. This date is roughly the time that the famous Clovis culture entered the Americas. I described the Clovis culture in the previous chapter, its elegant stone spear points, and the moribund debate about whether Clovis peoples were the first into the Americas.

With a corridor between the ice sheets open, did the Clovis-culture peoples come down the coast (or coasts) like their predecessors, or did they use the ice-free corridor? We do not know, because the culture spread so fast throughout North America, in the extraordinarily short time span of no more than four hundred years, that we cannot separate dates of sites near the southern opening of the corridor from those to the west. Claims that early Clovis sites near the end of the ice-free corridor mean that the Clovis peoples moved through the corridor need to be assessed with this speed of movement in mind.

I need here to mention Dennis Stanford and Bruce Bradley's hypothesis of occupation of eastern North America from Europe some time between twenty-two thousand and perhaps sixteen thousand years ago. This contention of a European connection rests largely on findings in America of advanced stone tools that they say are similar to those of the Solutrean culture of western Europe, which in turn is similar to the Clovis culture found all over the Americas. The Solutrean culture is named after the commune in which such tools were first found, Solutré-Pouilly. It lies in the

Macon district of France, where the delicious Pouilly-Fuissé chardonnay wine is grown and made.

Were the Americas originally peopled by western Europeans, as well as Siberians? Almost certainly not. Indeed, I will follow others and stick my neck out in saying certainly not. The genetic evidence simply will not allow it. If Europeans landed in the Americas thousands of years ago, and the evidence is that they did not, they did not lead to a peopling of the Americas, for their genes are not in Native American peoples of either North or South America, except via immigrants from Siberia.

Archeologically also, the hypothesis runs into serious trouble. For instance, several of the similarities between the eastern American tools and those of the Solutrean culture match features that exist in the tool cultures of many other regions. After all, only so many ways exist to make useful stone tools. Perhaps more importantly, many differences exist between the American and Solutrean cultures. The Solutrean peoples made many other objects, utilitarian and artistic, than just stone tools. So, if the Solutrean culture peopled the Americas, why are all these other aspects of their culture missing in America? I will mention just one more problem with cultural evidence for the Solutrean hypothesis: almost all the dates from the claimed Solutrean sites are from thousands of years after the Solutrean culture ended in Europe. They are, though, within the well-established dates for the non-Solutrean peoples' presence in the Americas.

Anatomically, one of the arguments for an early European entry into the continent was the European features of early American skulls. However, just this year as I write, 2014, we have a report of a skeleton from approximately twelve and a half thousand years ago in the Yucatan Peninsula with unquestionable Siberian DNA, but also European features. The explanation must be that the European features evolved subsequent to the arrival of people on the American continent. Currently, we cannot know whether the changes are a result of genetic drift (the almost random change over time that I will explain in chapter 6), or whether they are somehow an adaptation to the American continents' environment.

Forgetting the Solutrean in America, and returning to substantiated findings and ideas, only a few hundred years after they reached Texas, the first American foremothers and -fathers reached southern Chile, ten thousand kilometers farther on down the western coast of Central and South America from North America. A coastal route seems the best explanation for how they traveled so many thousands of kilometers so quickly.

Nevertheless, David Anderson and Christopher Gillam suggest an inland route as more likely in South America, based on nature of the environment, demography of the immigrants, and archeological findings. I find their suggested route implausible. According to their interpretations, the first Americans managed to get east of the Andes, having entered South America after crossing over the Isthmus of Panama. They then moved south across the rugged terrain and the many rivers flowing east from the Andes, and were finally lucky enough to hit the south-flowing Paraguay River, which could have taken them more or less easily down to where Buenos Aires now sits, three-quarters of the way through the continent. On from there to reach Monte Verde in southern Chile, they would have had to cross another series of rivers, and then get back over the southern end of the Andes.

The many rivers and the wide extent of forest that this route would have had to cross are problematic to me. Up on the eastern slopes of the Andes, the rivers are fast-flowing. Down in the plains below the eastern slopes, the rivers are wide. And these rivers probably did not disappear during the last ice age, for the Amazon shows no sign of the drying that Eurasia and Africa experienced then.

Not only that, but the Amazon forest east of the Andes must have severely slowed any southern route. Tropical forest is not easy to move through. Furthermore, unlike the central African Congo forest, the Amazon forest apparently did not retreat to tiny remnants during the aridity of the last ice age. Its extent was probably reduced by comparison to the present day, but even so, hundreds of kilometers of forest would have been in the way. A stroll down the coastal beaches of South America would have been a far more pleasant route. I admit, though, humans are nothing if not exploratory.

A recent genetic study by Michelle de Saint Pierre and her colleagues indicates that the indigenous people of northern Chile are quite different from those of southern Chile, who in turn are similar to those of southern Argentina. That pattern, along with a calculated age of two of the southern genes of fifteen thousand years, suggests not a movement down the eastern Andes and then across to the west but the opposite, namely rapidly down the coast to Monte Verde, and then east across the Andes. The different genetic profiles of the northern and southern Chilean populations could be explained by genetic mutation and small founding populations.

A problem with the hypothesis of a rapid migration down through the Americas, whether down the coast or inland, is that archeologists have found so little evidence yet for it. Hardly any pre-Clovis sites exist down the west coast of the Americas. But as I mentioned when discussing the coastal route out of Africa all the way to Australia, the hundred-plus-meter rise of sea level since the last glacial maximum at about twenty thousand years ago will have covered most signs of humans passing that way before then. Hardly any sites exist east of the Andes, in large part because excavation in tropical forest is so difficult and because changing river courses over the millennia eradicate sites, as Anderson and Gillam point out.

The final frontiers to our prehistoric spread across the world were its vast oceans. Both linguistics and genetics indicate that the peoples of the Pacific, as well as some New Guineans, came initially from eastern Indonesia. Take linguistics. All the Pacific peoples speak Austronesian, as do the eastern Indonesians and the peoples of the Philippines. The Austronesian speakers must have come originally from Asia, but no trace of Austronesian now exists in mainland Asia. Instead, we find the greatest variety of Austronesian languages spoken in Taiwan, off the east coast of China. Separated from mainland Asia, the native peoples of Taiwan had time to diversify and, happily, escaped replacement by the mainland populations.

The linguistic evidence indicates that from Taiwan, people moved into the Philippines and New Guinea, and on out into the Pacific. Genetic studies both support this route and contradict it, depending on which genes the scientists look at. On balance, though, and given some genes in eastern Indonesia and Polynesia that do not occur in Taiwan, it looks as if our expansion across the Pacific originated in eastern Indonesia, not as might be expected in New Guinea or Melanesia as a whole, given that the region lies in the western Pacific. To refresh memories, Melanesia is New Guinea and the islands immediately east and southeast of it, out to Fiji. (See map, p. 39.)

Geneticists can quite easily distinguish between Indonesia and Melanesia as a source of the Pacific expansion, because the genetic makeup of the people in the two regions is so different. Back in the middle of the nineteenth century, the great biogeographer, Alfred Russel Wallace, noticed that west of Bali and Sulawesi the animals were Indonesian, in effect of Asian origin, but eastward from there and from south of the Philippines they were New Guinean and Australian. The line separating the two biogeographic regions became known as "Wallace's Line." Others have drawn other lines farther east and south in the region, mostly depending on what sort of animal they were studying. The line farthest to the east is Lydekker's Line. The significance here is that people east of Wallace's Line, including the peoples of the Philippines, are genetically Asian. Across the various other lines farther east, the Asian proportion of the genetic makeup drops, until we get to Lydekker's Line, east of which the people are nearly pure New Guinean genetically, like the fauna.

Some of the differences between the genetic findings concerning the origins of the Pacific islanders that I briefly described two paragraphs ago arise from the fact that some of the studies were of mitochondrial genes from females, and others of Y-chromosome genes from males. That is significant, because male and female populations can move differently. I already mentioned in the previous chapter the cases of the Mayflower, the Uto-Aztecans, and Iceland. Sex is relevant in the Pacific too. One can find fast movement across the region by looking at Y-chromosome DNA, but slow movement if one analyzes mitochondrial DNA. Might this difference between the sexes reflect the fact that human males, not needed for care of infants, were the expendable explorers, while the necessarily more sensible mothers followed later?

In one case, just one individual and his descendants seem to have been responsible for the spread of a region's Y-chromosome genes, the male sex chromosome genes. This Y-gene line began in Mongolia about a thousand years ago, and is now, as Tatiana Zerjal and her co-authors point out, present in an extraordinary eight percent of almost exactly the region that Genghis Khan conquered two hundred years later. He and his descendants ruled the region for centuries afterwards. The Mongol Khans are not known for gentleness toward their conquered peoples. They left such a huge genetic mark on subsequent generations in part because they slaughtered so many of their competitors. They would have been especially homicidal to those likely to leave their own genetic mark, namely the resident chiefs.

But the Y-chromosome genes are not always the most widespread. In some places and among some peoples, the mitochondrial genes are. We then see more regional differences among Y-chromosome genes than among mitochondrial genes. That configuration would fit a society in which men were more likely to stay in the place of their birth, but women to move to wherever their husband's place of birth is. Meryanne Tumonggor, Tatiana Karafet, and co-authors showed that this pattern too occurred in the past during the peopling of Indonesia, and of the Pacific from Indonesia. It is the norm nowadays for agricultural societies across the world.

I described in the last chapter how our expansion into the Pacific stalled in western Melanesia, in New Guinea and neighboring islands, for over twenty-five thousand years. Quite a bit of the genetic and archeological data indicate that once humans developed ocean-going canoes, they continued rapidly out to Micronesia and Polynesia. (Micronesia is mostly north of the equator, and includes Guam and the Marshall Islands. Polynesia is mostly south of the equator, and on down to New Zealand in the southwest and Easter Island in the southeast, but includes Hawaii.) However, genetic findings, including differences between the sexes in their movements, indicate that again eastern Indonesians might have been largely responsible for the expansion, not the Melanesians as might be expected, given that they already lived farther east.

The evidence for Indonesian more than Melanesian origins includes the fact that nearly eighty percent of Polynesian nuclear DNA (i.e., DNA carried by both sexes) is of east Asian origin, with only the remaining twenty percent or so of Melanesian origin. As far as differences between the sexes is concerned, it looks as if ninety-five percent of Polynesian mitochondrial DNA is Indonesian, compared to only sixty-five percent from Melanesia.

Here we come to the debate about exactly how the Pacific was peopled. Did they move as an expanding population slowly out from Indonesia through Melanesia, and onward, breeding as they went? Or did they race out from Indonesia again, fast through Melanesia, not pausing much on the way? The extremes have been characterized as the "express train" and "slow boat" models.

Directed migration is one way in which humankind spread across the world. As the King James version of the bible puts it, "And I am come down to deliver them . . . and to bring them up out of that land unto a good land and a large, unto a land flowing with milk and honey" (Exodus 3:8). Within the USA, this sort of movement is exemplified by the two-and-a-half-month, 1,400-kilometer trek of the Mormons from the Iowa-Nebraska border to Salt Lake City, led by the infamously polygynous Brigham Young.

The second way humans spread is by so-called demographic expansion. To quote again the glorious English of the King James bible, "And the children of Israel were fruitful, and increased abundantly, and multiplied, and waxed exceeding mighty; and the land was filled with them" (Exodus 1:7). In other words, people reproduce, the population increases, and those around the edge move on out. Such spread might be exemplified by the slow westward move of European settlement from the east coast of the USA. More than a hundred and fifty years after the first permanent English colony there, settlers had not even reached the Mississippi.

However, with the phrases "slow boat" and "fast train" applied to the peopling of the Pacific, let me quickly admit that we have a bit of a straw person. How slow is slow? How fast is fast? Sixty-five percent Melanesian mitochondrial DNA in Polynesia is a lot of Melanesian contribution. Humans dispersed not just by racing from Indonesia, bypassing Melanesia, nor only by gradual expansion of the population from Indonesia to Melanesia to Micronesia and Polynesia. Rather, humans did both. However, it seems that it was the eastern Indonesians who were the main explorers, first into Melanesia, and subsequently rapidly through Melanesia into Polynesia. Nevertheless, given that the peoples married, so to speak, along the way, might "slow boat" with its connotations of shipboard romance be the better description?

The Pacific islanders, magnificent sailors that they were, did not stop their roaming across the ocean once they had populated most of Polynesia. As a result, through Melanesia, Micronesia, and Polynesia we find quite a mixture of genes from each region. And the same west of the Pacific, with apparent New Guinea genes in Indonesia, for example, as well as the reverse. The story is obviously complex, with lots still to be worked out.

What we can be certain of, though, is that Polynesia was not originally peopled by anyone from South America, Thor Heyerdahl and the Kon-Tiki notwithstanding. For those who did not see the recent Norwegian film *Kon-Tiki*, the original Kon-Tiki was a balsa-log raft that Heyerdahl and a crew sailed in 1947 from South America to French Polynesia in the middle of the Pacific. They undertook the journey to demonstrate the possibility of South American origins of the Polynesian peoples. Genetics would have saved them the trouble, but how could any resultant book be as exciting as Heyerdahl's *The Kon-Tiki Expedition: By Raft Across the South Seas*?

But it was the Polynesians, not the South Americans, who were the outstanding sailors. To a non-Antipodean, like me, New Zealanders and Australians sound the same, which is perhaps not surprising, given that their current ruling peoples largely came from the UK. But New Zealand's original colonists, the Māoris, came not from Australia but from the Cook Islands, two and a half thousand kilometers to the northeast, in the opposite direction to Australia. The small number of women among the founders, maybe no more than a hundred, implies that only a few boats survived the journey, not surprisingly.

Nevertheless, by the time Pacific islanders reached New Zealand, just seven hundred or so years ago, they might have been traveling to and from South America for at least three hundred years. The suggestion comes from study of the genetics of sweet potatoes.

Sweet potatoes are of South American origin. According to Caroline Roullier, Laure Benoit, and their co-authors, eastern Pacific sweet potatoes are closely related to South American potatoes, as if they had been brought to the Pacific islands direct from South America. Although they gave no date for the plant's arrival in the eastern Pacific, others suggested it occurred a thousand years ago. For anyone especially interested in sweet potatoes, the culture around them, and their movement across the Pacific, they will find much to interest them in *The Sweet Potato in Oceania*.

The genetics of chickens and bottle gourds also might indicate transport between Polynesia and South America on the order of a thousand years ago. However, for the moment, that conclusion has various problems. The bottle gourds might have made the crossing by floating on ocean currents instead of being carried by humans. And Vicki Thomson, Ophélie Lebrasseur, and co-authors suggest that the genetic samples of the chickens might have been accidentally contaminated by DNA from modern chickens.

Eight hundred years before the Māori peoples reached New Zealand, eastern Polynesians might have reached southern California. And why not? If by then they had crossed the vast reaches of the eastern Pacific, why stop? The evidence for this Californian landfall is the sudden appearance there about fifteen hundred years ago of eastern Pacific modes of boat construction, specifically the use of sewn planks, along with fish hooks and harpoons also with eastern Pacific design hallmarks. Another sudden appearance of Hawaiian-type barbed fish hooks seven hundred years ago indicates further contact. Some words with seemingly no North American roots, but obvious Polynesian roots, also indicate that the Pacific islanders did not stop going east until they reached the Americas. For instance, the coastal southern Californian Chumash use the unusual word "tomolo" for their unusual sewn-plank canoes, which linguists say is not dissimilar to the Hawaiian word of the time "tumura'aakau," meaning the tree that is useful for making canoes.

Chapter 2 ended with our arrival in Madagascar two thousand years ago. As I wrote there, it looks as if the first humans to arrive in Madagascar might have come not from Africa, but from Indonesia. Whichever group was first, the Indonesians came not across the four hundred and fifty kilometers of the Mozambique Channel that the Africans needed to traverse, but across the three thousand kilometers of the Indian Ocean. Archeology, genetics, and linguistics all lead to this conclusion. For instance, Malagasy, the name for the language of Madagascar as well as the people, is clearly Austronesian. It is closest to one of the languages of Borneo, indeed closest to a language group of southeastern Borneo, Barito.

Three thousand kilometers of open ocean to reach Madagascar. As was the case for New Zealand, that distance might explain why genetics indicates a very small founding population of especially women, just thirty of them. We are talking not only extraordinary boating and navigation skills of these Indonesian sailors, but extraordinary endurance too. I elaborate further on the endurance of long-distance sailors, and women, and the physiological explanations for it in chapter 5.

Once arrived in Madagascar, the initial Bornean population expanded quickly, but soon the peoples in the different parts of the island must have become isolated one from the other. The wonderful language resource on the web, Ethnologue, lists eleven Malagasy dialects/languages. All are related to a language family in Borneo. In other words, Malagasy is still largely a Bornean language, despite quite a large influx of Africans to the island starting not long after the Borneans arrived there to settle, perhaps fifteen hundred years ago.

If Malagasy is related most strongly to a Bornean language, other languages from the region, such as those spoken in Sumatra, Indonesia, and Malaya, have several similarities with the Bornean language, and Malagasy has several similarities with them too. One can imagine that in a region of islands and peninsulas such as southeast Asia, any one community, or boat crew with passengers, would consist of people from a variety of the islands, and hence a variety of languages.

I started this chapter with humans' exodus from Africa. But this was not the only diaspora in our prehistory. Humans migrated within Africa too, and even back into Africa from Europe. I have already described the migrations of the Bantu peoples from their origin in probably west-central Africa, along with the surprising finding that a quarter of the genes of some eastern African hunter-gatherers are originally from Europe from just about three thousand years ago. Some of these peoples must then have migrated to southern Africa, because, extraordinarily enough, approximately ten percent of genes of a Khoi-San population there are European in origin. The genetic findings indicate that these European genes in the East Africans arrived in southern Africa about fifteen hundred years ago.

Did the eastern African genes reach southern Africa via a single migration, or via gradual movement of individuals over generations, reproducing as they went? This is the express train/slow boat—or should I say slow cart—dichotomy that we came across before with the peopling of Polynesia. Demographic expansion seems likely, given the distances involved—at least twice as far as the Mormon migration—and the lack of obvious cause of a sudden migration, such as an extreme dry period.

A distinction between the two types of dispersal is the speed of migration. In the case of our dispersal out of Africa and across the world, some have described our movement to Australia as "rapid." But how fast is "rapid"? One estimate is that it took us on the order of three thousand years to cover the twelve thousand kilometers from the Middle East to Australia. Twelve thousand kilometers in three thousand years is four kilometers per year—or just eleven meters a day, eleven long paces a day. The figure of three thousand years for the journey was arrived at from estimates of dates of the African exodus, and arrival time in Australia. Neither of those dates is at all certain. They could be wrong by thousands of years. But whatever more-or-less reasonable dates we pick, the result is that the journey seems to have been accomplished at a speed of just a few meters a day.

The journey from eastern Siberia to southern Chile has narrower estimates of departure and arrival. Humans might have left Siberia sixteen and a half thousand years ago and arrived in Monte Verde, sixteen thousand kilometers away, no less than two thousand years later. Two thousand years for sixteen thousand kilometers is twenty-two meters a day. We also have good dates of fifteen and a half thousand years for the arrival of people in Texas, as I have previously described. Texas is roughly six and a half thousand kilometers from Beringia and ten thousand kilometers from southern Chile. The rates then are six to seven kilometers per year from Beringia to Texas, or seventeen meters a day, and they are ten kilometers per year or a little over twenty-seven meters per day for the southern half of our journey from Siberia. The Clovis people followed the first Americans, and dates of archeological sites indicate a population front moving at approximately six and a half kilometers per year. That is nearly eighteen meters per day.

Whichever estimate of daily distance we use, eleven meters, twenty-two meters, seventeen meters, twenty-seven meters, eighteen meters, the population is barely moving. A baby could crawl twenty-seven meters in an hour, as some mothers must discover to their alarm. So, for both our movement from Africa to Australia, and Beringia to southern South America, we are talking a rate concomitant with a wave front resulting from population expansion, not a direct migration.

The same is true of the expansion of pastoralists from Turkey to western Europe around sixty-five hundred years ago. Judging from archeological analyses, it took two thousand years to get from Anatolia to eastern Spain, a distance of thirty-five hundred kilometers. The pastoralists expanded, then, at a little less than two kilometers a year, or maybe five meters a day. Pastoralists moving more slowly than hunter-gatherers is easy to imagine, but even so, five meters a day seems extremely slow for flocks of hungry cattle, sheep, and goats.

We have to recognize that all these rates are based on the archeological record, a record that keeps changing. The Texas finds of the earliest indications of people in North America were reported in just 2011. Before that, the earliest reliable dates of any archeological site in North America were from fourteen to thirteen thousand years ago, in central Alaska. The arrival of humans in Texas before they got to central Alaska is not impossible. After all, a large ice sheet covered much of Alaska then. Clearly the humans that were in Texas fifteen and a half thousand years ago bypassed Alaska and left no trace of their passage, or at least no trace that archeologists have yet found.

What other information, then, can we use to estimate rates of movement across the globe by the early humans? Back then, all humans were hunter-gatherers. So how far do present-day hunter-gatherers migrate in a year? Luckily, anthropologists have studied hunter-gatherer peoples for decades, and we have a lot of data on their nomadic movements. Information from studies of over two hundred hunter-gatherer peoples from across the globe indicates approximately eight camp shifts per year and thirteen kilometers per move. Those are averages across the whole world. An example from my state in the USA would be the Serrano peoples of southern California just northeast of Los Angeles. They averaged seven moves a year, and twenty-one kilometers a move. Eight moves per year of thirteen kilometers per move is a little over one hundred kilometers per year. Seven moves of twenty-one kilometers is a hundred and fifty kilometers a year. At those rates, humans would have reached Australia from Africa, or Chile from Beringia, in less than two hundred years.

Stuart Fiedel makes a similar calculation. He points out that in almost historical times, about a thousand years ago, peoples migrated from Alaska to Greenland, a distance of twenty-eight hundred kilometers or so via the Queen Elizabeth Islands in the winter when the Arctic Ocean was frozen and they could walk, in less than a hundred and fifty years. That is nineteen kilometers a year, or sixty-five meters a day with Sundays off. Alaska to

Tierra del Fuego at that speed would have taken them a little under nine hundred years. Certainly slower than the two hundred years of the previous paragraph, but, as Fiedel points out, perhaps these later North American immigrants faced competition from previous residents. If they did, though, it cannot have been serious competition, for signs of any earlier residents disappear in an archeological instant concomitant with the later arrivals.

The range of dates for our earliest arrival in Chile, from the Monte Verde site, is six hundred years, depending on what item was dated, and what method used to estimate its age. That means that even if the earliest Americans took nine hundred years to make the journey, the archeological record could show that they traveled there almost instantaneously.

In sum, then, we have a ten-fold range for rates of movement of the early humans, from a few thousand years to a few hundred years to travel the twelve thousand kilometers from Africa to Australia, or the sixteen and a half thousand kilometers from Siberia to the bottom of South America.

Others have used estimates of population growth rates to calculate how long it would have taken us to fill the American continents. Forty years ago, Paul Martin used the population growth rate of the Pitcairn mutineers and their female companions, a density of hunter-gatherers of one person per square kilometer, and an originating population of one hundred to calculate that the people would have filled North America in only three hundred and forty years. With slightly different but still reasonable values, another calculation concluded one hundred and sixty years as sufficient. The people behind the advancing front would move more slowly, he suggested, because they would find less food. The "Pitcairn mutineers"? This is the Captain Bligh, Lieutenant Fletcher Christian infamous *Mutiny on the Bounty* story. After setting Bligh and a few loyal seamen free in the middle of the Pacific on an overloaded boat with a week's provisions, the mutineers settled with several Tahitian men and women on Pitcairn Island.

Thirty years after Martin, Todd Surovell produced a new calculation. He combined a variety of measures that might indicate rate of a population's spread, including number of camp movements and their distance, reproductive rates, the rate of population increase, and the costs of carrying young children. He came up with figures that span the tenfold range of values calculated from the known camp movements of hunter-gatherers.

A little later, Surovell modeled whether a coastal migration south in response to population pressure from behind would get humans from south of the Cordilleran Ice Sheet to Monte Verde in the roughly thousand years between the archeological dates for people in the two regions, and before any signs of humans inland. This time, he found that they could not have done so. Despite some detailed quantitative modeling, then, we seem to have gotten no closer to a precise estimate of how quickly the first Americans moved through the continents.

However, since Surovell's calculations, we have the Buttermilk Creek site in Texas that is a thousand years older than Monte Verde, and not only inland but far closer to Monte Verde than was the Cordilleran Ice Sheet. Also, Surovell assumed that humans would move on down the coast only once they reached a certain high population density where they currently were. However, it is not necessary to assume that only population pressure would have pushed the first Americans on down the coast.

All that is needed is that the people could harvest food at a higher rate farther on than where they were currently living. Such is likely to be true whatever the density of harvesters. Imagine a density of ten oysters per ten meters along the shore; I take one; only nine oysters left in the current ten meters is less than the ten in the next ten meters, and so I move on to the next ten meters. Ten oysters per ten meters along the shore; a family of ten takes the ten (in the same time it takes me to take one); no oysters left means the family moves on to the next ten meters. I move at the same rate as the family of ten. The advance across the landscape, in this case the shoreline, can occur without population pressure.

That is not to say that population pressure is unimportant. The family of ten has no choice but to move on, however costly the move; whereas if the move is energetic or dangerous, or I need to recover from eating a bad oyster, I can stay where I am because I still have nine oysters left. But I, alone, no population pressure, will move on with nine oysters still remaining if the move costs less than the energy from the extra oyster I will get by moving. This is standard theory in our understanding of organisms foraging for food. It even has a name, the "marginal value theorem."

Experiments with birds, and observations of blackberry-picking parties of humans, confirm it. They show, for example, that the speed with which people move to the next patch depends on how fast they find berries in the current patch. Lots of berries in this patch, and they move on sooner to the next patch. They behave as if they assume that the next patch will be as rich as this one. They take—as they rapidly move on—only the fattest, easiest-to-reach berries. Teachers, parents, guardians, this is an experiment you can perform at home. Try it with Easter eggs. Put a dense supply of eggs in one part of the garden or house, and a thin supply in another part. If the "marginal value theorem" works, participants should rush from potential hiding place to hiding place faster in the first part of the property.

The first humans with their superior tool kit and division of labor between the sexes will have been moving into land effectively empty of serious competitors. It will therefore have been full of fat, easy-to-catch food, oysters and large-bodied animals, neither of which needs much local knowledge to harvest.

David Meltzer is not so sure. He argues that it would pay to maximize the time in any one place, in order to get to know it well, and so harvest it more efficiently. Obviously, then, movement will be slowed. I disagree. The last thing that a moving band of hunters wants to do is allow their prey to learn that they are dangerous. The longer the stay, the more difficult the hunting. And I would not mind betting that we could be talking a matter of days rather than weeks, let alone months.

We are obviously not talking about oysters as prey here. Yet another advantage of them as food, another reason to postulate that a coastal movement is likeliest for the earliest humans spreading through an unknown continent, is that the oysters neither flee nor learn, unlike mammals and birds.

A few individuals moving along a blackberry patch in England is certainly different from a population moving across a continent. Nevertheless, over twenty years ago, John Beaton suggested a two-speed movement to and through Australia, and through the Americas. The transient explorers, as he called them, headed fast through the virgin land, followed by the slower, more residential estate settlers. That was only hypothesis. He had no data other than the observation that the earliest sites discovered by archeologists in both continents were all of about the same age. It looked as though both Australia and the Americas had been filled almost instantaneously.

If we do not have data on the speeds at which the first colonizing humans moved, we do have data on the rate at which another invading species is filling a continent. Let me use an analysis of the speed of the cane toad invasion of Australia as analogous to what might have happened with the first human expansion into the virgin world outside Africa. Humans are not toads, I know, and in another context I probably would not use toads to support a suggestion about human biogeography. However, the toads provide data—actual information—about how an advancing front of animals in a new land behaves. Previously we have had just estimates, and as Meltzer points out, for the initial immigrants into the Americas, we are never going to have anything but estimates, even guesstimates.

In 1935, the Australian Bureau of Sugar Experiment Stations introduced cane toads, native to Central and South America, into the northeast of the continent to control a beetle infestation of the Australian sugar cane crop. These toads, like the first humans, faced few competitors or predators, in part because the toads are poisonous. The study's finding was that the advancing front of toads traveling into territory empty of other toads advanced twice as fast as did the followers, because the animals at the front traveled in straighter lines. They were also fatter and healthier, because they had the first go at the unwary beetles, and because of the lack of indigenous competitors.

Returning to humans, a study of people spreading through eastern Canada from the mid-1600s to the mid-1900s by Claudia Moreau and co-workers found that women at the front of the expanding population reproduced twenty percent faster than did followers, as judged by analysis of genealogies. They did not know why, but suggested easier access to resources—as with the toads.

One of Todd Surovell's findings about humans in the Americas was that fast movement did not necessarily mean slow reproduction. Indeed, the opposite might occur. Women at the front of the advance would need to forage only short distances daily to find enough food for themselves and their offspring. Those coming behind, into land already harvested, would have to go farther each day for their needs. Surovell's calculations indicated that females in the moving front might travel only a quarter of the daily distance of females in a slowly moving population. That was theoretical calculation. What about real data, real information? It turns out that on all three major continents, the hunter-gatherer peoples who moved camp most often reproduced more than those who moved less often.

We have, then, a winning combination of fast migration and fast demographic expansion—in both toads and humans. So I favor a few hundred years for the trip from Africa to Australia, rather than a few thousand years, and the same for Beringia to Chile.

∽

Equating the movement of toads with that of humans is not as extreme, as off-the-wall, as irrelevant as it might initially seem. In the next few chapters I will be making more comparisons between animals and humans, or I should strictly say between non-human animals and humans, because humans are animals too. Consequently, we can expect that the same biogeographic effects that influence where we find large-bodied, warm-blooded animals—which is what we are—will affect where we are also. In fact, we know that some of the biogeographic effects apply likewise to small-bodied and cold-blooded animals. That generality is certainly interesting. But things can become especially intriguing when different sorts of animals do different biogeographic things, such as cover areas of hugely different sizes. Then we have the opportunity to precisely relate the differences in biology to the differences in geography.

CHAPTER 4

HOW DO WE KNOW
WHAT WE THINK WE KNOW?

The science behind the "facts"

I n the last two chapters, I have presented a string of statements about
human origins in Africa, our exodus from that continent, and our subsequent spread across the world, including our routes, and the dates of
various important events in our long journey. Along the way, I have given
some general indications of how we think we know any of the facts and
inferences that I have described. I have mentioned genetic similarities
and differences, for example. But that is only scratching the surface of the
huge amount of "detective work" that scientists have had to perform in order
for me to construct the last two chapters. In this chapter, I will provide a bit
more detail about these detective methods, and how they provide answers
to the many questions about our biogeographic history.

First, we have to find the facts, or what scientists often refer to as "data."
We have to find the bones, the stone tools and other cultural artifacts,

including language (written or spoken), and, of course, the DNA. The bones and other physical artifacts, we find in the ground. We listen to and record the language, or read it. The DNA we can get from cells in buried teeth or bones in the case of ancient DNA, or from blood or cells from the inside of the cheek if we want to examine modern DNA.

Where we find these artifacts and other data can strongly influence the story of our evolution. When the earliest hominin finds were in Eurasia, that is where we thought that our ancestors had originated. When, subsequently, Eugène Dubois discovered Java Man, *Homo erectus*, on Java in 1891, we had to rethink the narrative, and we placed our origins in Indonesia. Nowadays, as I have shown in the previous two chapters, we are certain that Africa is the cradle of hominins and the human species—based on the evidence collected to date. But as I have said, new finds will come, and new finds can change previous interpretations. William Matthew, a famous paleontologist, suggested in 1915 that Eurasia was the cradle of mankind. In doing so, he was replacing previous ideas of the tropics as the cradle. But an editor added a note to a reprinting in 1939 telling the reader of Robert Broom's finding of "very primitive hominids in South Africa," so sending our origins back to Africa. The note was presumably referring to Broom's findings subsequent to Raymond Dart's of *Australopithecus* and *Paranthropus*.

We of course need experts both for the finding and for the identification. A few years ago, the police turned up in my anthropology department. They were looking for Henry McHenry, the department's paleontologist. Someone had reported a human-looking skull in a trash dump. The police, sensibly, decided to find out before they began a murder enquiry whether the skull was really human. Henry McHenry could immediately tell them not to worry—it was an orangutan's skull. Though what an orangutan skull was doing in the trash of a small town in California is another matter.

With the bones, artifacts, lexicons, and DNA in hand—the facts—we then start our analysis. What do all the facts mean? What story do they tell us? We construct family trees out of the bones, tools, languages, and DNA. We make the trees in the same way as if we had only our own family photographs available. We look for similarities and differences, taking account of the age of the evidence. A fossil hominin skull from something like fifty thousand years ago that looks very similar to a modern human skull, and another fossil from three hundred thousand years ago that is less similar to a modern human, are likely to be, for example, Neanderthal and Erectus, respectively.

Ancient bones and artifacts, and these days even ancient DNA, can be picked up from the ground, so to speak. They exist as tangible objects. Past languages do not exist, unless they happen to have been written down. But for the most part, languages are current. Past languages have to be estimated, therefore. Except very rarely, as when a linguist has recorded on a wax cylinder or a tape a very recently dead language, we cannot hear them now. The Ancient Egyptian of the Rosetta Stone was unusual in that the hieroglyphs in which the language was written are phonetic symbols.

Another difficulty with languages as an indication of past biogeographic relationships is that by comparison to tools, bones, and genes, they can change quite rapidly, as parents listening uncomprehendingly to the latest school or pop jargon of their children often discover. The original immigrants to America, coming from a small population in northeastern Siberia perhaps fifteen thousand years ago as they did, presumably all spoke the same language. Later immigrants from the same region might have spoken different languages. Even so, the number coming from that part of the world cannot have been anywhere near the number of indigenous languages now spoken in South America—one hundred and eighty-eight languages in Brazil alone.

With such rapid linguistic evolution, ten thousand years ago might be close to the limit for using words of different languages to discern relationships between languages and therefore any pattern in human movements. Beyond this time span, words have simply changed too much to bear any resemblance to the original language. In fact, for most words, the limit is less than five thousand years. However, some very frequently used words, such as numbers and pronouns, can show little change for perhaps more than twenty thousand years.

With the right choice of words to compare, the match in the patterns derived from archeology, genetics, and linguistics is often satisfyingly close. The expansion of agricultural peoples from the Middle East into western Europe is an example. A main reason for the match is easy to see. As I described in chapter 2, people (genes) tend to move with their culture (artifacts and language). I live now in California, nine thousand kilometers from my home country, England. Although I (genes) have lived in California for a quarter of a century, I still have my afternoon tea at four o'clock (artifact), and no American listening to me speak begins to think that I am a native Californian (language).

When constructing a family tree, whether we are using bones, tools, genes, or languages, we need to know who is the ancestor and who the descendant. If all we are interested in is the shape of the tree, we can often get that shape with no information about age. For instance, one finding supporting the supposition that Africa is the region of origin of all humankind is that the greater the geographic distance of a population of people from Africa, the greater is their genetic difference from Africans. The two patterns cannot be matched for any other region as a source. That fact and its inference are true whether humans left Africa to spread around the world a hundred and twenty-five thousand years ago or only fifty-five thousand years ago. Of course, dates are crucial too. South America peopled by Siberians is pattern. However, South America peopled as early as was most of North America—fifteen thousand years ago—is almost a new picture.

With respect to judging the age of whatever it is we have found in the ground, a simple rule is that the deeper underground the stones and bones lie, the older their age. But the rule is far from inviolate. A burrowing animal, a tree root, a river could have carried bones and stones down into levels below where they first fell. Scientists have rejected several of the oldest dates of humans in Australia and elsewhere for just these reasons.

Another way to age artifacts is to assume that the less sophisticated tools are older than the more sophisticated. The stone tools of what most people know as Olduvai Gorge in northern Tanzania are some of the oldest known. A little museum sits at the edge of the Gorge. I have visited it, and I have to say that some of the oldest stone tools from two million years ago are so primitive that I could not distinguish them from naturally broken cobblestones. But then I am no expert on stone tools. Experts can detect that someone produced the jagged side of the cobble by striking it repeatedly with another stone to produce the cutting edge.

Contrast these Olduwan tools, as they are called, with the sophis-ticated Aurignacian flint spear heads, arrow heads, awls, and so on in another museum that I have visited, the National Museum of Prehis-tory at Les Eysies in the Dordogne of France, and nobody could doubt that the Les Eysies tools came from a much later period than did the Olduvai Gorge tools.

A comment on the name Olduvai. Most of us still term the site Olduvai Gorge. However, go to Tanzanian websites and it is, properly, termed Oldupai Gorge. That is, it is spelled with a "p," not a "v." Olduvai was a mishearing of the Maasai word for the gorge.

Legendary as the Gorge is, it does not look impressive. More a small, shallow valley than a gorge, it lies in the middle of flat, dry, bushy desert scape at the end of a dirt road. Expecting on my first visit something far grander, I felt almost cheated when I saw it. "This is it?" I thought, with a hint of derision. Yet knowing its history, knowing that human-like crea-tures lived there between two and one million years ago, knowing that their remains and their tools lay there for those two to one million years before being discovered by the Leakeys, I shivered as I stepped down into the gorge, though it was midday and shadeless.

The Oldupai Gorge Museum contains, in addition to stone tools, replicas of the 3½-million-year-old footprints from Laetoli, forty-five kilometers from Oldupai. I have not seen the original footprints. However, knowing how excited my wife and I were when we identified gorilla footprints and dung in a Nigerian forest for which no official reports of gorillas existed, I can begin to imagine what it must have been like to come upon incon-trovertible evidence of human-like creatures walking upright so long ago.

Back to the estimation of dates. We mostly want absolute dates, not just relative ones based on what is older than what. To give a date to bones, stones, and other artifacts, we rely on the fact that certain chemical ele-ments in them, or in the ground in which they lie, change form over time, converting from one so-called isotope into another. ("Isotope" is Greek for "same place," and simply means the chemical form of an element. That being the case, why did whoever coined the term not use Greek for "same thing" or "same substance"? Why "same place"? The answer is that the word refers to where the element is in the periodic table.)

The best-known transformation of an element is probably radioactive decay, a phenomenon in which a less stable form changes over time into a more stable form. If we know the natural ratio in the environment of the original unstable and the final stable forms, and we know the rate of change from one to the other, then from the current ratio of the two forms in the artifact or in the material of its close surrounds, we can estimate its age. Ironically, the creationist website AnswersinGenesis gives an admirably clear account of this dating method, using carbon as an example, specifically carbon-14's decay to the stable carbon-12, and hence the ratio of the two. The older the once-living tissue, the less C^{14} we find in relation to C^{12}.

A final word or two on dating. Carbon dating can take us back only fifty to seventy-five thousand years ago. Beyond that, all the carbon-14 has disappeared, so we cannot compare its amount with the amount of carbon-12 in the tissue. Instead, we move to other elements with slower rates of decay. An example would be potassium-argon dating, or the comparison of argon-40 to argon-39. Now we are talking about being able to estimate ages up to a few million years ago, but the principle is exactly the same as for carbon. An advantage of carbon dating is that the material assessed can be the organic material of the sample: the bone, for instance. Also, the rate of change of the carbon-14 into carbon-12 is so fast that the result can be very precise, to within a few decades. The fast rate of decay is why the method can be used over only a few tens of thousands of years. Other forms of dating, such as potassium-argon dating, use inorganic samples. That can be an advantage too, for example when we want to know the age of a stone tool.

With some of these other methods, though, we do not for various reasons date the fossil or tool that really interests us, but rather the soil or rock that surrounds it. And here is a pitfall in the method, one that I have already alluded to. As I said when I first started with methods of aging finds, the sample in which we are interested could find itself in sand older than the sample itself is.

Genetical dating works on the same principle as does radioactive dating. Our cells continually renew themselves. At each replication, the chemicals that make up DNA can change places, or themselves change. Compare the DNA of two species, or populations, count the mutations, and knowing (or perhaps I should say "thinking we know") the mutation rate, we can calculate how long ago the species or populations separated. The same process works for languages.

We do not in fact know from studying the genes or languages themselves what their rate of change is. Instead, we need to find independent evidence of ages, and work from there. So, for example, fossil evidence indicates that the earliest *Homo* arose not far off from two million years ago. If we count the number of genetic differences between humans and that hominin, and divide by two million, we have a mutation rate within the hominin line. Knowing the number of differences between present-day populations, we can then use that rate to calculate when the populations separated, for example, when the line leading to Australian aborigines separated from the line leading to Asian peoples.

In the real world, things are somewhat more complex. Different genes change at different rates. New data, or new ways of testing, or new information about the state of the sample come in, and scientists change their minds about the age of crucial artifacts. For instance, a form of dating (optical luminescence) depends on measuring, on the one hand, the amount of radiation resulting from radioactive decay that is stored in minerals since their complete burial with, on the other hand, the amount in the mineral when it was last exposed to sunlight. But the last exposure to sunlight has to be full exposure, and the burial has to be complete. Nevertheless, with care, we can get a best guesstimate.

Estimating the time since two populations separated is the same as estimating how long any population has been in existence, in other words the age of the population. The longer the population has been in existence, the greater the chance of genes changing within the population. The degree of genetic variety within a population is then itself an indication of the population's age. Hence for continents, African populations—the ones from which all the rest of the world evolved—have the greatest genetic variety. Native American populations, in existence for less than twenty thousand years, have the least.

This interpretation comes with a warning, though. Change over time is not the only way to produce genetic variety within a population. A single population can consist of people from different regions who have come together and interbred. When that happens, diversity rockets. Extraordinarily small frequencies of interbreeding will hugely increase the genetic diversity of the population in which the offspring reside and themselves breed.

The San Khomani of southern Africa exemplify this effect. They are indistinguishable in looks from the neighboring San Namibia. And yet the San Khomani are far more genetically varied than are the San Namibia. But that extra diversity is not because the San Khomani are older. No, the difference is that the San Namibia genes are local, whereas the San Khomani have genes from not only Africa, but from Eurasia too. The San Namibia have not married outside of southern Africa, whereas the San Khomani are clearly a melting pot of peoples, like the USA is.

Most of the times that I have given a date for any event, I have been careful to qualify it with "roughly," "about," "around." No date can be absolutely precise. Inaccuracies occur at many stages in the process of estimation. Nevertheless, with enough measures and enough care, the chemists and linguists who do the dating can give quite accurate estimates of how precise (or imprecise) the dates are.

For instance, I have written that the oldest accepted early modern human is around two hundred thousand years old. In fact, the authors estimated 195,000 ± (plus or minus) 5,000 years (using the ratio of argon-40 to argon-39). Now, beware if you go to the original announcement of the find. That "±" sign in scientific writing has a specific meaning in this context of ranges of estimates. It translates to an approximately two-thirds chance that the date is between the extremes given. If you want the near-full range of possible dates, in other words near certainty that you have got the potential range of dates—the full range of potential "error," as it is called—you need 15,000 years either side (in this example).

Do not be misled by that word "error" here, either. It is not anything to do with being wrong. It is just the jargon for "range of estimate." In common parlance, the previous paragraph's 195,000 ± 5,000 years for the earliest human remains so far can be read as 195,000 years, give or take 5,000 years, even 15,000 years.

Another issue is involved in dating. We know that rarely will archeologists find the very earliest of anything. In the first place, not all of the past is preserved. Even if the past were preserved, we miss most of it. Imagine walking through a village, looking for the village's oldest resident. The chance is near zero that any old person you find is in fact the oldest, unless you look carefully in every room of every house. Archeologists cannot do that. They do not have sufficient time or money. Instead, they search the equivalent of just one or two corners of one or two rooms of one or two houses.

And finally, the scientific methods used for dating are a bit like our memories: it gets more imprecise the farther back in time we go. I might remember to the nearest hour what happened this morning, but only to the nearest day for anything more than a week ago.

So whenever through this book I give a date, it is always a rough date. And very often, earlier dates will soon be discovered.

I need to clarify a few more dating-related jargon words. The exodus from Africa sixty thousand years ago that I wrote about in the previous two chapters is fifteen thousand years or so after the origin there of what archeologists call the Late Stone Age culture. "Late Stone Age" is capitalized, because it is the formal name given by archeologists for the sorts of stone tools made in Africa at that time. Yes, it is not a date. It is a description of the tools. That is appropriate, because often no date can be attached to the tools. All that the archeologists know is what the tools look like.

Annoyingly for non-professionals, stone-tool culture in Europe at the same time is not termed Late Stone Age. Instead, it is termed "Upper Paleolithic." Given that "upper" in this context means much the same as "late," and that "paleo" is Greek for old, we have "late old stone age" for Europe. What is the difference from Africa's "late stone age"? The fact is that the stone tools of the regions differ slightly. Still, what would be wrong with African Late Stone Age and European Late Stone Age?

Europe's delayed equivalent to Africa's Late Stone Age culture, the Upper Paleolithic culture, arose around forty-five thousand years ago, with the first indications of the magnificent Aurignacian technology and art. It is named after the Aurignac region in southern France, where some of the first of the artifacts were discovered. Cro-Magnon is the name commonly associated with the humans of this period, again named after a site of first discovery in France. The culture is associated with not only exquisitely executed and fine stone blades, but carved figurines, bone flutes, and later with evidence of textile weaving, and of course the glorious cave paintings of later in this era. Chauvet and Lascaux might be the most famous of these painted caves, but scores of others exist.

The paintings in these caves are clearly magical in some way. The artists sometimes crawled over five hundred meters into the depths of a cave. And I do mean crawled, so low and narrow were some of the caves then. The original artists did not have to do that to show what a good artist they were, or if their only intent was to draw what they hoped to later kill, or to celebrate what they had already killed. Perhaps they were originally exploring. I imagine their flickering lamps casting shifting shadows on the cave walls, making parts of the walls look like moving animals. And so these ancestors of ours expertly added the outlines of the horses, lions, bison that they saw, producing such realistic images that they have not been bettered since.

Archeologists argue about whether the complexity of the Aurignacian culture originated in Africa or Europe. Certainly, many more examples have been found in Europe. But Europe has more archeologists than Africa, and has had them for far longer. Also, Europe has more caves in which early art is well preserved. In France, you find the highest concentrations of painted caves in the central Pyrenees and the Dordogne region of the central west.

From describing the methods available to tell us about our dispersal from Africa across the world, it is time I gave a few examples to illustrate the findings, and to add more detail to some of the so-far rather broad timeline strokes of our history. I will start with archeological finds and some dates, beginning with the oldest.

We used to think that the Late Stone Age culture started in Africa on the order of fifty thousand years ago, but further research has pushed back the date to over seventy thousand years ago, in both south Africa and east Africa. Not only were humans making sophisticated tools then, we were probably making art. In south Africa, a team of archeologists headed by Christopher Henshilwood has found in a site dated to roughly a hundred thousand years ago, using the radioactive decay method with a variety of elements, all the makings of ochre pigment—the ochre, charcoal, powdered bones, grinding stones, and the abalone shells used as bowls with the prepared pigment in them. Pure ochre carried from some distance away, and presumably also used artistically, dates from even earlier than that.

By seventy thousand years ago, humans might even have been washing the ochre off ourselves after we had painted with it. I say this because humans had started to make use of another form of hygiene by then. Lyn Wadley, Christine Sievers, and their co-authors have found in another south African site dated to about seventy-five thousand years ago signs of bedding made from a tree species, the Cape quince, the leaves of which are aromatic when crushed and contain a variety of chemicals that can act as insecticides. Many woody plants occurred in the area, so the fact that the bedding consists of just this one species indicates active choice of the Cape quince. Lyn Wadley's team also found evidence that the bedding was occasionally burned. The dating method was one of the radioactive-decay ones that the Henshilwood team used. I might add that seventy-five thousand years ago is fifty thousand years before any such indications of hygiene appear in European sites.

To illustrate not only the methods of science, but also its politics, and the reporting of science to the public, I will describe in more detail than I did two chapters ago the controversy over the dating of the Monte Verde site in southern South America. Before the Monte Verde discovery and the establishment of its early date, archeologists and anthropologists considered that the first Americans had been of the Clovis culture. Archeologists classified the culture by their beautiful stone spear heads. They had, as I previously described, a characteristic smooth groove at the base that allowed easy and reliable hafting onto a shaft. However, the very earliest dates for Clovis sites in the Americas are from about thirteen thousand years ago, over two thousand years after the earliest known indications of humans in the Americas, and over one thousand years *after* people were at Monte Verde.

The Clovis culture is tightly tied to North America. Its characteristic artifacts do not exist in South America. Different cultures—in other words, different forms of flaking stone blades and projectile heads—preceded it in Eurasia, and followed it in both North and South America. Debate surrounds its exact origins.

Yet despite Monte Verde, dated over a thousand years before the earliest Clovis, and despite several other sites dated to older than any Clovis site, including the site in Texas that I have just mentioned, journalists and, I am afraid, some scientists still breathlessly describe how "the latest finding" kills the "Clovis first" idea. How can it be gotten into the minds of science journalists that the "Clovis first" idea has been moribund for decades, and dead for fifteen years or more? Yes, some of the claims for older ages of various sites have been shown or argued to be wrong. But Monte Verde should long ago have killed any doubt about our pre-Clovis presence in the Americas. Why has it not?

One answer is that journalists need a contest to get readers interested. Without the dramatic overthrow of an established idea, why would anyone be interested in the new finding? What does it matter that people entered the Americas fourteen thousand years ago, not thirteen? Another answer is that most people find it difficult to give up well-known ideas. The theory of plate tectonics, the movement of massive plates of continent over the face of the earth, was initially strongly resisted by some, despite all the evidence in the theory's favor. As one of the founders of plate tectonic theory, Drummond Matthews, said near the end of a talk on the topic that I attended back in the 1960s, we just have to wait for the non-believers to die.

Both Matthews and I were young then. People who do not find new ideas difficult might be the young, in part because they have not absorbed so completely the old ideas. That is an easy claim to make. In fact, evidence for it exists. The famous philosopher of science David Lee Hull and two co-authors found that among sixty-seven scientists over twenty years of age at the time of the publication of *On the Origin of Species* who were famous enough to leave records, those who accepted Darwin's theory of evolution by natural selection within ten years of publication were on average eight years younger than those who did not accept the new theory, forty years of age as against forty-eight. Tongue in cheek, Hull and his co-authors report Thomas Huxley's contention that scientists should be strangled when they reach sixty years of age so that they do not hinder the progress of science.

That having been said, I have to add that I do not know the average age of protagonists for or antagonists against the claims of extraordinarily early arrival of humans in South America. My age, though, matches the Hull study, in that I am over forty-eight years of age, and I do not yet accept the early ages. Examples are Niède Guidon and co-workers' contention that humans were in northeast Brazil twenty thousand years ago or more. Another early date for humans even farther south is thirty thousand years ago in Uruguay, although the authors carefully and honestly admit that that age is suspiciously unusual.

Of course, some new theories need rejection by everybody. Rupert Sheldrake's theory of "morphic resonance," a theory of knowledge at a distance, epitomized by naïve cattle somehow absorbing from distant experienced cattle the knowledge that they should not cross cattle grids, is notorious rubbish. Even so, it is not as bad as Erich von Däniken's suggestion that extraterrestrials built the pyramids. I cannot bring myself to list their books in my bibliography, but type either name into Google, and you will find not only their books but, I am sad to say, the websites of many believers. To deflect attack against my closed mind, I will add that my beloved mother believed both these ideas.

Rebecca Cann and Sarah Tishkoff are two scientists strongly associated with using genetics to understand how the human species came to be distributed around the world in the way we are. Much of the work now on Africans' genetic profiles is conducted by Tishkoff's highly productive laboratory, whereas Cann works these days on conservation of Hawaiian birds.

The media story twenty-five years ago of an African "Eve" originated from Rebecca Cann's paper in 1987 in which she and her co-authors traced the origins of humans to a female in Africa who lived about two hundred thousand years ago. They did it by analyzing differences between regions of the world in the degree of variation in the populations' mitochondrial DNA.

In brief, from the blood of one hundred and forty-seven individuals from five regions of the world—Africa, Asia, Australia, Europe, New Guinea—they progressively lumped similar individuals into more and more inclusive groups. Think lumping into families, then towns, then provinces. Only with Africa as an origin did they have the fewest number of steps, the simplest pattern, the smallest number of migrations between continents. Given information on the mutation rate of mitochondrial genes, they suggested the date of origin.

Anthropologists had to wait twenty years before publication of the report of a human skeleton from that date aged by the method of radioactive decay. Such a match between two completely different methods of estimating ages—genes and radioactive decay—supports the date obtained from each method alone.

In the same paper that produced "Eve," Rebecca Cann and her co-authors also showed that the peoples of Africa are more genetically variable than are the people of any other region. Many subsequent studies have substantiated their claim, some finding twice as much variation among African populations as among non-African ones. An obvious explanation for this finding of African genetic variability is that humans have existed longer in Africa than anywhere else, and so had the longest time to experience the greatest number of changes in our DNA.

Not only are African populations more genetically diverse than others, but the farther human populations are from sub-Saharan Africa, the less genetically diverse each is. How can that pattern arise? Let me quickly say that these distant peoples do not have fewer genes than Africans. Rather, they have fewer forms of any one gene. One reason for the decrease in variety with distance is that the farther human populations are from Africa, the less long they have been in existence, and therefore the less time they have had to accumulate genetic changes.

Another reason for the drop in genetic diversity of humans as we dispersed from Africa is that along the journeys, only a subset of each population moved on. Imagine a pinball machine (or its video game equivalent) with balls of many different colors in the starting gate (Africa) representing the variety of forms of a gene. Fire the machine, and the farther the balls go, the fewer different colors you see, as only some get past the catching cups.

So, only some eastern Africans moved to the Middle East. Only some Middle Easterners moved to Asia. Only a small subset of them moved to Australia. Only a few Siberians moved to the Americas. And so on. Geneticists have detected two places where they can see a major drop in the genetic diversity from one side to the other of an apparent boundary. One lies between Africa and the Middle East, separated by the Red Sea. The other is between Siberia and Alaska, separated nowadays by the Bering Sea.

We can see this latter drop from the fact that although Siberians have all three main blood types, A and B and O, almost all native South Americans have only type O blood. The population that left Beringia and peopled the whole of the Americas was so small and from such a small area that they all happened to have just the O blood group.

We know that over time, DNA mutates. Consequently, the peoples of different regions have new forms of genes specific to their region. So each population has several of its own forms of genes, called alleles. As humans dispersed, the greater the distance between populations, the longer the time apart, and consequently the greater the chance of mutations in each population, and hence for the populations to differ genetically. That pattern of increasing genetic difference with increasing time and geographic distance allows us to work back from all the regions of the world—to find, as I wrote a few pages back when introducing the methods used to place an age on items or events, that Africa and only Africa is the source of the genes or alleles common to all regions. None of us lacks African genes, but plenty of us lack Australian, or Asian, or American genes. No other way exists for such a genetic pattern than an origin in sub-Saharan Africa.

Indeed, so recently have humans spread from Africa, and from such a small population—perhaps a few thousand—that despite all the genetic differences that I have written about that enable us to plot that spread, still, as I mentioned in the prologue, the whole human species contains less variety of genes than does any one of the chimpanzee subspecies in Africa.

As humans move around the world, we take not only our genes with us, but we also carry the organisms that live with us and on us and in us. Noah of biblical myth took into his ark two of "every thing that creepeth upon the earth" (Genesis 7:14-15). In reality, a family fleeing a flood might have deliberately taken just two of each of their own livestock. But they would have inadvertently taken along with them hundreds, indeed thousands, of other species. They would have taken many insects in the animals' bedding and food. And on and inside themselves and on and in their animals, they will have carried along scores of parasites, lice and nematodes of various sorts, as well as hundreds of species of bacteria.

The story of the spread around the world of those hitchhiking organisms is necessarily a story of our spread around the world. We have archeological evidence for our diaspora from Africa. We have genetic evidence. And now we have the evidence of our parasites and the other animals that live with us. The more routes and types of evidence that we have for any one scientific story, the better is that story tested. The story becomes stronger as each subsequent story matches previous stories, and as different types of evidence produce different sorts of information. It so happens that we have excellent information on the global spread of some of the organisms that live with us and on us and in us.

Take the organism that causes the severest form of malaria, the single-cell *Plasmodium falciparum*. Something like two hundred million people a year suffer from malaria, with maybe half a million of them dying from it. Its native region is Africa, where we see almost all cases of malaria caused by it. Falciparum affects only humans. That being the case, if we have the story of the worldwide spread of humans out of Africa correct, we should see that the story of the spread of falciparum as indicated by its genetics matches the story of the spread of humans. It does. Just as with humans, the genetic diversity of falciparum malaria decreases with distance from Africa. No other region of origin produces this pattern. Similarly, and just as with humans, the farther from Africa the falciparum, the more genetically different it is from the African populations.

One region is an exception. South America. Why? Because with the route to South America via the Arctic, where the mosquitoes that carry falciparum cannot survive, falciparum malaria in South America did not travel with the first Americans from Siberia, but came with the slave trade fifteen thousand years later.

Another example. We humans have in our stomach a bacterium, the spiral rod bacterium, whose scientific name is *Helicobacter pylori*. It is the only bacterium that can survive the acidity of our stomach. All our other gut bacteria live in the intestines. We know quite a lot about our stomach bacterium, because it causes stomach ulcers. Importantly for the current story, this bacterium occurs only in humans. That being the case, a bacteriologist's picture of the spiral rod bacterium's global spread should match the archeologists' and geneticists' and even historians' picture of our global diaspora. It does.

Study of the genetic makeup of the bacterium shows that it, like us, indeed originated in Africa. It left Africa, dispersed to New Guinea and

Australia around thirty-five thousand years ago, dispersed to Asia, and dispersed from Asia to the Pacific islands and the Americas. In fact, it looks as if the Pacific dispersal can be far more precisely located than just Asia. Helicobacter dispersed from Taiwan across the Pacific, and did so starting approximately five thousand years ago.

We can also clearly identify forms of the bacterium in American guts that are closely related to the same bacterium in the guts of West Africans. Obviously we have here a signal of the slave trade. And spread across the world we find the spiral rod bacterium of Europeans, matching the spread of Europeans across the world.

Along the way, the bacterium's genetic diversity decreases with distance from Africa, as I described earlier the human genetic diversity doing. Perhaps most usefully, genetic analysis of the bacterium allows an extraordinarily precise estimate of when it left Africa. The answer is the same as we get from human genetics, namely close to sixty thousand years ago, in fact fifty-eight thousand, give or take a few thousand.

The bacterium travels in us. A time-lapse map of the spread across the Pacific of four other species that live with us almost perfectly overlaps archeologists', geneticists', and linguists' time-lapse maps of humans' spread across the Pacific. One of the four species is the domestic pig. The other three are the Pacific rat, the moth skink, which is a type of lizard, and the sweet potato. Elizabeth Matisoo-Smith conducted part of the research on the pig. She led the research on the rat. Caroline Roullier and Laure Benoit were the lead authors on the study that showed with genetics the movement of the sweet potato from South America to Polynesia. Not incidentally, they pointed out that a linguist had previously pointed that the Polynesian term for sweet potato, "kuumala," was very similar to the South American Quechua term, "kumara."

As an extra strut in the framework of evidence about our movement across the Pacific, if we date the first appearance of Pacific rat bones in New Zealand and of seeds there clearly gnawed by rats, then, as Janet Wilmshurst and co-workers have shown, we find that the dates are the same as we get for first appearance of human bones in New Zealand—seven hundred years ago.

The coincidence between human movement across the Pacific and the movement of the pig, rat, and lizard cannot be accidental. We are talking dispersal over millions of square kilometers of the Pacific Ocean. These three species did not disperse independently. They were not carried on tsunamis, or washed out to sea on storm-tossed vegetation—two common explanations for the presence of terrestrial species on islands. Nor did these three species swim between scores of islands thousands of kilometers apart. They obviously went with boating humans.

Travel halfway around the world to northwestern Europe, and we find that there too, animals confirm human migrations. Adele Grindon and Angus Davison found that DNA of land snails that first appear in Ireland around eight thousand years ago is the same as that of land snails in the eastern Pyrenees, and distinct from populations elsewhere in Europe, including France and the rest of Britain. Why this pattern? They suggest because the river Garonne that rises there and enters the sea at Bordeaux is an easy and long-established route to the Atlantic, and hence around southern Britain to Ireland. The snails might have been taken deliberately, as people in the region of origin have eaten them there for over ten thousand years. About the same time, we see the DNA of Near Eastern pigs turning up in western Europe, with sites along the Danube indicating one possible route. Somewhat later, DNA of mice in the north and west of Britain similar to the DNA of Norwegian mice indicates that mice accompanied the Vikings on their visits to Britain. Presumably they were not pets.

Carry on traveling west, and migrations of a large mammal, the red deer, confirm, or rather match, human migration. Like humans, it got stuck in northeast Siberia until roughly fifteen thousand years ago. Then it moved into Alaska. If it could, so could humans. Indeed, it might have enabled humans to move in by being an extra source of meat.

*

If the appearance of tools and languages and human DNA and, finally, species from elsewhere in new places confirm the story of the human diaspora, so too does the disappearance of native species as humans arrive. A plot of the almost instantaneous extinction of species across the world after our exodus from Africa is a plot of our global dispersal. I come later to this topic of how humans have altered the biogeography of other species.

CHAPTER 5

VARIETY IS THE SPICE OF LIFE

Where we are affects what we are

W e are all African" is the title of the second chapter of this book. All the best information and interpretation that we have indicates that modern humans arose in Africa perhaps two hundred thousand years ago, and that the ancestors of all the rest of us throughout the world left Africa perhaps sixty thousand years ago. But if we are all African, why are we not all the same? Why can someone looking at me know that my parents' ancestors came from Europe, not from Africa, or eastern Asia, or Polynesia? The answer, of course, is that our physical traits, our skin, our hair, the shape of our body, all vary from region to region.

In this chapter, I will be writing about differences between peoples arising because we adapt to different environments. In different environments, different sorts of people survive and reproduce successfully. Where we are affects what we are. Where we are affects what we look like, for example. That is the third of the three linked arguments in Charles Darwin's theory of evolution by natural selection. Although this adaptation to the environment was part of being "fit," the great Charles Darwin did *not* accept that "the conditions of life"—in other words, the environment—affected what we humans looked like, even "after exposure to them for an enormous period of time." Darwin could not see how, for example, skin color correlated with the environment. The "distribution of the variously colored races, most of whom must have long inhabited their present homes, does not coincide with corresponding differences of climate," he wrote.

He was wrong, a very rare event for him. In fact, the "distribution of the variously colored races," or rather, the distribution of skin color in the human species, corresponds strongly with the environment, as does the distribution of variously shaped peoples, and a diversity of other differences between peoples from different regions and different environments. The color of our skin is the most visible instance of regional differences in the biology of humans, so I will start with the topic of skin color.

In brief, people from tropical latitudes are darker than people from outside the tropics. We see a similar pattern among some birds and mammals too, including among non-human primates, as Jason Kamilar and Brenda Bradley showed. Animals in hotter, more humid environments had darker backs than did those in cooler or less humid environments. Chimpanzees are black; baboons are grayish. Latitude is, of course, a proxy measure of the actual environmental influence. The Kamilar-Bradley analysis more precisely indicated that primates are darker where the climate is warmer and wetter. That correlation in animals, specifically birds, had been noticed back in the 1830s by Constantin Gloger. Nobody really knows why the correlation exists.

It is certainly not a hard-and-fast association. Sometimes darker animals live at higher altitudes and latitudes—in other words, where the climate is colder, if not necessarily drier. Maasai herdsmen have twice the proportion of dark cattle in their herds at 2250 meters altitude as they do at sea level, as a very nice study by Virginia Finch and David Western found. Dark cattle absorb more heat, an advantage at high altitudes, and pale cattle suffer less heat stress, an advantage at low latitudes in the dry north of Kenya.

For animals, the variation in color could well also have something to do with hiding from predators and prey. Dark species are sometimes found inside forest rather than outside forest, which might be part of the reason why some tropical species are darker than ones from higher latitudes. The white winter color of the Arctic hare, fox, and ptarmigan would be an extreme example of the camouflage effect. Regulation of temperature probably has something to do with it too.

However, neither camouflage nor regulation of temperature work as explanations for why the color of human skin changes across the world.

We are all aware of the harmful effects of sunburn. But damaged skin is not the only danger from too much sun. Nina Jablonski and George Chaplin argue that the more serious problem for people who live and work in the tropics, as opposed to sunbathing a few days of the year, is destruction by UV light of vitamin B9 in the blood as it circulates near the surface of the skin.

Vitamin B9, also known as folate, or folic acid, or folacin, is responsible for a wide array of normal bodily functions, many of them having to do with production of new cells. Too much UV light and consequently too little B9, and production of red blood cells is slowed, leading to anemia. Probably the most serious consequence is for a fetus. If a newly pregnant pale-skinned mother spends too much time in the sun, the brain and spinal cord of the fetus do not develop properly.

So, people in the tropics need protection from UV light if they are to produce healthy babies. Before sun cream and life indoors, the dark pigment in human skin, melanin, protected us. The pigment cells of tropical peoples produce more melanin than do those of northerners. The melanin acts essentially like a sunshade, absorbing and scattering light rays, so preventing them from penetrating deep into the skin.

The melanin acts in the same way as our earth's atmosphere. The atmosphere, like melanin, can absorb and scatter light rays. The more atmosphere through which the sun's rays have to pass, and the more ozone in the atmosphere, the less the UV rays penetrate. The sun is less directly overhead for less long in temperate regions, and the atmosphere of temperate regions has a higher concentration of UV-scattering ozone. By comparison to people in temperate regions, then, tropical peoples experience more UV exposure.

If UV light is so bad for us, though, why does not everybody have dark skin? Why do Europeans' melanocytes produce less melanin than do, for example, Africans' melanocytes? In brief, why are people from high latitudes pale? We are back to the concept of everything having benefits and costs. Too much sun and we suffer from destruction of vitamin B9; too little sun, and we suffer from non-production of another crucial vitamin.

Sunlight, UV light, has beneficial as well as detrimental effects. UV light might destroy vitamin B9, but it also stimulates the production of vitamin D. Vitamin D is vital for bone formation through its effect on our ability to take in calcium from the diet. Too little vitamin D, especially when we are growing, and our bones are weak. That is why a common manifestation of vitamin D deficiency used to be rickets, observable as exaggeratedly bowed legs.

Although, our bodies evolved to use sunlight as our primary source for vitamin D, with diet being a minor source, we nevertheless now see far fewer cases of rickets in the West than we used to, because manufacturers add vitamin D to several common foods, such as cereals, bread, and milk. Also, we need remarkably little sunlight to get sufficient vitamin D, perhaps just fifteen minutes or so of full-body exposure a day. And we store vitamin D in our body for weeks.

Nevertheless, the fact that humans evolved to get our vitamin D primarily from the sun explains why even now if we hide too much from the sun, we suffer from various ailments associated with lack of calcium. Almost one fifth of people of African origin who live in European and North American cities suffer from calcium deficiency, even rickets, compared to just one fiftieth of people of European descent. Strictly dressed Muslim women in western Europe or other northern temperate regions can suffer too, so covered are they when they go outside.

When I was a child, my family lived in Edinburgh, Scotland, at the same latitude as central Canada, central Russia, and southern Scandinavia. Luckily, my parents knew about the dangers of lack of vitamin D in the winter. My sisters and I therefore suffered each winter morning from being fed by our mother a spoonful of cod-liver oil, rich in vitamin D. The taste was revolting, but none of us suffered from calcium deficiency.

Wait, though: If northern humans need pale skin to get vitamin D from the sun, how can some of the northernmost humans, Arctic peoples, the reindeer-herding Saami of Russia, and nomadic Mongolians, be so dark-skinned? Why do these people not suffer from rickets, especially since, that far north, they are for so much of the time as completely covered as are some Muslim women?

Let me digress for a moment, to comment on my use of "Arctic peoples" instead of the term most of us know them as, Eskimos. Although Eskimo is a description acceptable to some Arctic peoples of Alaska, the Arctic peoples of Canada prefer Inuit. To call the latter Eskimo would be the same as terming Scots or Welsh peoples English. The mistake is not appreciated by the Scots or the Welsh. So, in the same way that I use the term Brits, or British, to mean anyone from the British Isles, which includes Ireland proper, so I use Arctic peoples for all those peoples in the North American tundra, and the Eurasian Arctic peoples too, such as the Saami, if that is where the data that I am referring to come from.

But back to why these northern peoples are unexpectedly dark-skinned. The answer is, in effect, the disgusting cod-liver oil that my mother fed each winter morning in Scotland to my sisters and me when we were children. The dark northerners all had, and some still do have, a diet unusually high in vitamin D, high enough to compensate for the lack of sun. North American Arctic peoples had until recently diets full of the meat of sea mammals and sea fish. The oilier sea fish, such as mackerel, have some of the highest concentrations of vitamin D known in wild animals. The same applies in the Southern Hemisphere, where coastal native Australians, including Tasmanians, are darker-skinned than expected were UV light their only source of vitamin D. The Saami and the nomadic Mongolians survive on the meat of their livestock. Mammalian meat is far lower in vitamin D than oily fish, but it nevertheless has enough to allow the protective effects of darker skin.

Native northwestern Europeans, it turns out, are unusually pale-skinned for the latitude of the region, indeed some of the palest on the planet. Diet explains that extra paleness too, according to Razib Khan. Western Europeans nowadays eat a lot of meat, but until very recently they depended heavily on cereals, such as oats and barley. The significance of cereals to skin color is that they are unusually low in vitamin D. That means that people who depend on them need extra-pale skin in order to absorb sufficient vitamin D from the sun.

This story of diet, vitamin D, and color of skin would be tighter, more believable, if we could find out if the date of the arrival of wheat cultivation in western Europe coincided with when the genes conferring pale skin first appeared. We can.

Though farming began in the Middle East and China on the order of ten thousand years ago, farming in western Europe had to wait for the climate to warm sufficiently for crops to grow, as I described in chapter 3. Agriculture there began around six thousand years ago. And when did a study by Sandra Wilde and ten others (these archeological genetics studies often have many authors) find that a gene that confers pale skin appeared in European populations, but at approximately five thousand years ago. They inferred this date from examination of the DNA of old bones, mostly from Ukraine. They recorded a steep increase in the representation of genes for pale skin between six thousand and four thousand years ago.

Five thousand years ago is almost within historical time. Humans in Africa were erecting stone circles then, apparently for astronomical predictions. I am referring to the Nabta Playa, roughly eighty kilometers west of Abu Simbel on the River Nile. Give us another thousand years, and we have invented writing. The color of our skin changing through evolution by natural selection in historical time means humans evolving now, in effect. We are not talking a process that ceased in the distant mists of time. If Europeans had continued with their cave paintings, and had accurately painted humans, we would see the change in the paintings.

Dark skin in the tropics keeps harmful UV radiation out. Light skin at high latitudes allows useful UV radiation in. The biology, the geography, the chemistry all seem to fit well. But another hypothesis is out there. Nothing wrong with alternative hypotheses. We always want alternative ideas to press us to produce the best evidence possible for our own. Sometimes, though, the other hypothesis is wrong enough that we do not have to do anything with our own. We simply show the problems with the other one. That is what Nina Jablonski did with this other hypothesis.

The claim was that dark skin is better at preventing water loss through the skin than is pale skin. If that sounds weird to you, keep that thought. Nina Jablonski showed that the suggested waterproofing chemistry and biology simply do not fit the facts, indeed are opposite to the facts. Black skin keeps in no more water than does light skin, and, as I will come to later, people of African descent sweat no less than do people of Caucasian descent. Moreover, skin color correlates not with aridity, as one would expect under the waterproofing hypothesis, but with intensity of UV radiation.

The story of human skin color has a wonderful twist to it that illustrates the serendipity that is so often part of scientific discovery. Keith Cheng was looking for genes associated with cancer in an animal often used for genetic studies, the zebrafish. This is a small, black-and-white-striped fish, maybe four centimeters long. The stripes run along the body, though, not across it as in zebras.

Because of the association in humans between pale skin and skin cancer from sunbathing, Cheng became interested in the genetics behind a pale form of the fish, the so-called golden form. Rebecca Lemason led the large team from Cheng's lab that pinpointed the gene responsible for the pale color. It turned out to be the same gene that human geneticists already knew was associated with skin color in humans.

Exactly how the human gene worked, nobody knew. But with the stimulus of the finding with zebrafish, the Lemason-Cheng crew decided to search the huge database available on human genetics, the International HapMap, to see if the gene differed in its form between Africans and Europeans, and if so, how. They found a clear difference, but an amazingly small one—a difference in just one chemical, just one amino acid. A single mutation caused the production of a slightly different chemical, and so changed the color of the skin of a fish and of a human. A study of a tiny aquarium fish helped us elucidate the genetics of human skin color.

The color of our skin is affected by more than one gene, though. I have so far concentrated to some extent on paleness in Europeans by comparison to Africans. However, Asians are also pale by comparison to Africans. Heather Norton and a team of nine others describe how eastern Asians are pale as a result of changes in genes that are different from those that produced pale Europeans. I come back to this issue of different genes for the same useful trait in chapter 9, in relation to the ability to digest milk as an adult.

If we have a geography of skin color, so also do we have a geography of our shape and size. I remember an urban legend from decades ago that people born in Africa were such good athletes because they had an extra tendon in their legs. Urban legends are by definition not true. Nevertheless, as runners, people of sub-Saharan African origin do seem to be extraordinary athletes.

In the 2012 Olympic Women's Marathon, for example, Africans were first and second, and all three winners of the men's event were African. Fourteen athletes share the ten fastest men's Olympic 100-meter times. All are of African ancestry. As of 2014, the last twelve male winners of the London Marathon are all African. With two exceptions, men of African origin have won the last twenty-five Boston men's marathons. Women of African origin, mostly Kenyan as is the case for the men, have won sixteen of the last twenty Boston marathons.

Of course, a bunch of social factors are at play here with the Africans' running superiority. But biology could be playing a part too. People of African origin have, I will go on to describe, an anatomical advantage over people of European or northern Asian origin. The evolved benefit that led to African athletic superiority had nothing to do with running, though.

Let us leave the human species for a moment and turn to animals. What do we see when we compare tropical mammals with their Arctic relatives, the jackal compared to the Arctic fox, say? The tropical mammals are not only smaller, but they have longer limbs and ears than do their Arctic relatives. We see the effect in all sorts of species, including our closest relatives, the primates. The typical tropical primate species living within five degrees of the equator, say the African blue monkey, is—at five kilograms—on the order of half the size of a non-tropical species, such as the nine-kilogram Japanese macaque.

That is not to say that no large primates exist near the equator. Primates of all sizes live there, from the 150-kilogram male gorilla to the mouse lemur, the size of—a mouse. At higher latitudes, by contrast, no very large or very small primates live.

For the small primates, the reason has to do with heat loss in relation to body size. The smaller the animal, the greater its surface area in relation to its mass. More surface area means more loss of heat. What keeps animals warm is fuel: in other words, food. The problem for small animals is that their small bodies come with small mouths and small guts. Small animals in cold climates simply cannot eat enough to keep warm.

Conversely, large-bodied animals have no trouble staying warm. On a freezing winter morning, frost on the ground, the horses in my neighbor's paddock across the road are not just not shivering, they are hot enough that steam is coming off their bodies. The problem for large animals is that they need a lot of food to fuel their mass of muscle. Gorillas, for example, spend nearly half the day feeding. In a temperate winter, they would not be able to find enough food to keep themselves fueled.

But, you are saying, what about lemmings and musk ox? One is just as small as a mouse lemur; the other is larger than a gorilla. They live far north. Yes, and so do quite a few other small and large mammals. The small ones survive the winter by hibernating, at which time they do not need fuel. If they do not hibernate, they live in tunnels under the snow, where they are protected from the worst of the cold. The large ones are ruminants, with a different sort of digestive system from that of the apes and monkeys.

It is not only the size of the body that changes with latitude among non-human animals. So does length of appendages. The antelope jackrabbit of Mexico and the fennec fox of the Sahara have ears twice as long as their respective Arctic relatives, the Arctic hare and Arctic fox. Indonesia's long-tailed macaque has a tail easily as long as its body, whereas macaque monkeys that live outside the tropics, such as the snow monkeys of Japan, have a stubby tail less than one fifth of their body length. What is going on here? Long thin forms lose heat faster than do fat short forms, for the same reason that small bodies lose heat faster than large bodies.

In short, if you live in a warm climate, be long and thin, and also light-weight. If you live in a cold climate, be squat and large.

Of course, scientists have technical terms for these relationships of body size and shape with temperature and latitude. Smaller, more slender bodies at lower latitudes and warmer temperatures is the Bergmann effect. Longer extremities, such as ears and limbs, at lower latitudes is the Allen affect. The effects' names recognize the work of a German, Carl Bergmann, and an American, Joel Allen, who first published their findings of the relation-ships back in the 1800s.

And now we get back to the athletic prowess of peoples from tropical regions, especially Africans.

Humans are mammals, and follow the same rules regarding shape and size of body in relation to temperature as do other animals. So we find that women living near the equator average fifty kilograms in weight; those living in or near the Arctic average sixty kilograms. For men, the comparison is fifty-five kilograms or so vs. seventy. We see the same sort of differences between equatorial and Arctic peoples if we look at weight for height, although now women and men are very similar. Both weigh on the order of three kilograms per centimeter at the equator, but four kilograms per centimeter in the Arctic.

Pygmy peoples are equatorial, and they are the smallest people on earth—an average of on the order of one hundred and forty centimeters. All pygmies on all three major continents live in tropical forest. The point about tropical forest is that it is not just warm: it is humid, too. Shedding heat is therefore particularly difficult.

A quick reminder in passing: the description of peoples as "pygmy" denotes merely their small size by comparison to most other people in the world. It does not denote any similarity in origin. Even the western and eastern pygmy peoples of Africa have been separate for maybe twenty thousand years.

We are not just lighter in weight near the equator, we are thinner. Scientists interested in our size in different climates often gauge body width by the width of our pelvis, measured as the distance between the sharp bony ridges that we can feel at our sides at the bottom of our waist. Tropical peoples have a pelvis nearly twenty percent narrower than do Arctic peoples.

The narrow-bodied tropical peoples also have relatively longer limbs than do people from outside the tropics. Equatorial peoples' legs are around ninety-five percent of the length of their sitting height compared to about eighty-five percent for Arctic peoples. Here, then, could be an explanation for Africans' athletic advantage. They have longer legs with which to propel their lighter trunks.

(In case "sitting height" is not self-explanatory, it is the distance from our seat, the bottom of our pelvis [strictly the ischial bone's base], to the top of our head.)

Note that I say "an" explanation for Africans' athletic advantage. Certainly, personal, social, and cultural factors play a part. With a large proportion of African-American children denied access to the best schools, how else could they and can they get ahead, but in sports? Early very public role models, public precisely because of segregation, such as Jackie Robinson and Jesse Owens, could well have been important in this cultural way too, and presumably still are.

Jesse Owens competed in the so-called Nazi Olympics of 1936, and won more gold medals there than any other athlete. One was in the 200-meter sprint. Second in that event was Mack Robinson, the older brother of Jackie Robinson. Jackie Robinson is famous as the first African-American to play in Major League Baseball. In both these cases and many others, the mere fact of segregation and disparity in opportunity in almost all walks of life could have led to African-Americans focusing more on sports, where superior ability is maybe more immediately evident and less subject to discriminatory assessment of quality.

And it need hardly be said that people of African origin excelled at more than sports. Percy Julian became a highly successful chemist in research and development in the first half of the twentieth century, and was the first African-American chemist inducted into the National Academy of Sciences. In these three cases and so many more, only the most extraordinary ability, will power, and persistence could have gotten the African-Americans to the top. Julian probably became a research and development chemist rather than a university researcher because universities declined to hire him, given his perceived race. Worse, when he was finally successful enough to buy a house in an upscale neighborhood of Chicago in the early 1950s, his house was fire-bombed.

Also, if we are talking a general African anatomical advantage in athletics, why is it that such a preponderance of the African Olympic medalists come from such a tiny area of Kenya? According to the maligned Jon Entine, the Nandi people of western Kenya are the supreme African athletes in the distance events. Yes, the Nandi region lies at relatively high altitude, at roughly two thousand meters, but so do other areas of Kenya, and east Africa in general. Arguably, Kipchoge Keino's amazing performance in the 1968 Mexico City Olympics and then the 1972 Munich

Olympics fired Kenyans into athletics, especially those from Keino's birth region, the Nandi Hills of western Kenya.

But however determined Keino was—and he continued to excel in other areas of his life—could he and the other Africa-born athletes have done it without their long legs?

It need hardly be said that Nandi Kenyans are not the only renowned athletes from a small region of the world. Others are the Mexico Copper Canyon long-distance runners made famous by Christopher McDougall's *Born to Run* with its aggrandizing sub-title, and also, by comparison, the unheralded Fell Runners of Britain's Lake District. A test of the putative anatomical advantage of Africans is obvious.

The lower leg and arm compared to the upper show the same direction of relationship as limb to body. Tropical peoples have long lower sections compared to upper sections, whereas Arctic peoples have short lower legs and arms compared to the upper sections. Given that the lower sections are thinner than the upper, and so have a larger surface area for their volume, they should shed heat more quickly. If so, it makes sense that we find that the heat-losing sections are relatively long in hot regions of the world, but short in cold regions.

However, one experiment indicates a complication with the argument. It illustrates why science has to be so precise in exactly what it measures and in exactly how it expresses its findings. The study found that modern University of Wisconsin students of both sexes lost more heat from their upper leg than their lower leg. That is not surprising, given how much more muscle is in the thigh than in the lower leg. In other words, when I said in the previous paragraph that we should lose more heat from lower than upper sections of our limbs, I should have added "in relation to their mass." The reason that the upper sections of the limbs are not longer than they are in the tropics as an adaptation to shed more heat presumably has something to do with the mechanics of movement.

Now, we all know that well-fed, healthy individuals, whether people or other animals, are larger than poorly fed, unhealthy individuals. Also, tropical peoples have less fat in their diet and live in an environment with more diseases than do Arctic peoples. It should come as no surprise, then, that Africans weigh less than do Arctic peoples, and have different bodily proportions. So how do we know that these bodily differences are not just a result of diet, rather than adaptations to regulate temperature?

In fact, we can distinguish the effects of nutrition and temperature. Good nutrition and health produce heavier bodies and longer legs. We know, for example, that the well-fed second-generation offspring of Guatemalan immigrants into the USA weigh more and are taller than their parents. It turns out that the children are taller because of longer legs, not longer trunks. The contrast, then, between lightweight, long-limbed tropical peoples and heavy, short-limbed Arctic peoples is not a result of only nutrition. If it were, the heavier Arctic people would have longer legs than the tropical peoples. But they do not.

Temperatures change with altitude as well as with latitude. The upper slopes of high mountains, wherever they are, even those on the equator, can experience Arctic temperatures. The native Indians of the Ecuadorean or Peruvian Andes, to pick an example, live in a far colder climate than do the Ecuadorean or Peruvian coastal Indians. I know that from experience. Up in the mountains, I sleep under as many blankets as the hotels provide. On the coast, just a sheet is too hot for me.

As expected from all that I have said to now about latitude and the Bergmann effect, the Andean peoples are heavier than are the coastal native Indians. We know this from studies in Peru. However, the Andeans are also better fed, healthier, and fatter. Once again, how do we separate nutrition from high-altitude cold as an influence on the difference? The answer is that the Andean peoples, by comparison to the coastal peoples, show the Allen effect. The Andean Indians have shorter limbs relative to their body than do the coastal Indians. Good nutrition cannot explain shorter limbs. Only temperature can.

Over time, the correlations between body shape and temperature have lessened, as better nutrition and better overall lifestyle buffer people against the effects of the environment, and hence the benefit of adapting physically or physiologically to it. Even so, by various measures, the correlation still exists.

Another adverse effect of a poor diet is slow growth. A page or two back, I presented the suggestion that pygmy peoples are so small because of the particularly hot and humid environment of the tropical forest. However, are they in fact small because of poor nutrition as a result of the demonstrably poor soils of tropical forests?

No, they are not. They grow just as fast as do people from other regions. The reason that they are so small is that they stop growing at a younger age than do people from other regions. The high temperature and humidity of tropical forest could explain that stop in growth. Grow any larger than an otherwise minimum size for their lifestyle, and they start having difficulty keeping cool.

Besides being hot and humid, tropical forest is infested with diseases and parasites. Pygmy peoples, perhaps as a consequence, are one third less likely than their non-forest neighbors to survive to fifteen years of age. Many studies indicate, at least for animals, that if early death is likely, then the animals had better reproduce early. Reproduction takes a lot of energy, especially for mammalian females, including humans. If pygmy women are going to have the energy for early reproduction, they had better stop growing early.

They do, and by comparison to equivalent non-pygmy populations they indeed start reproducing early, most at sixteen years of age, rather than eighteen or nineteen as in other otherwise-comparable communities, as Andrea Migliano and co-workers showed. Pygmies are small, then, not because of any particular advantage of being small, but as a side effect of the benefits of reproducing early, which entails ceasing to grow early.

Genes are probably involved in the pygmy peoples' small size. The pygmies have, for example, a gene that turns off genes implicated in, or known to affect, the production of growth-stimulating hormone. Several of the pygmy peoples' other genes are associated with immunity against diseases of various sorts, and geneticists have detected at least one connected to early reproduction.

As I wrote at the start of this section, other animals, including non-human primates, show the Bergmann and Allen effects. A non-human species that I have not so far mentioned in this context is Neanderthal. Neanderthals lived across Europe and southwest Asia. Less culturally advanced than modern humans, the Neanderthals must have been more influenced by the climate than the humans that eventually replaced them, especially as the Neanderthals survived well into the arrival of the last ice age. That being the case, we should be able to predict from knowledge of the Bergmann and Allen effects what the Neanderthal body form was like compared to the form of modern humans, especially those from the tropics. Heavier? Squatter? Shorter-limbed?

Indeed. For instance, Neanderthals could have weighed at least twenty percent more than a modern human, in part because of a broader trunk. Neanderthal legs were short, something like eighty percent the length of their trunk, compared to over ninety percent for modern Africans. The length of their arms compared to their trunk was approximately the same as for modern Europeans, which is shorter than for modern Africans. And like Arctic peoples today, Neanderthals had shorter lower legs relative to upper than do living tropical peoples.

But let me here introduce a little bit of scientific disagreement in the form of an alternative hypothesis for Neanderthal bodily proportions. Clive Finlayson, Director of the Gibraltar Museum, suggested that the heavy, squat bodies of Neanderthals and their short robust limbs are not adaptations to cold, but instead that they indicate strength and a highly active daily life. In effect, with less-advanced tools than modern humans eventually had, Neanderthals had to use brawn.

As I have said, alternative hypotheses in science are always useful, because they can stimulate further investigation and refinement of ideas. They are also quite often fun, at least to the onlookers if one enjoys a good argument. The two sides can get quite heated as they expound and defend their own ideas. That is the case with the debate over the small brain size of the miniature Flores "hobbit" in the chapter on island adaptations. The Neanderthal disagreements, though, have so far been objective and polite.

Support for Finlayson's argument that strength and activity, not cold, explain Neanderthal limb proportions comes from the finding that the Neanderthals of warm southeastern Spain had the same bodily proportions as more northerly populations. However, the suggestion that southeastern Spain was not cold could be wrong.

After all, the middle of Spain, less than a hundred and fifty kilometers east of Madrid, had glaciers near the start of the last glacial maximum, when Neanderthal finally died out. Even today, Mt. Mulhacen in southeast Spain has snow on it. As I write, at the end of May 2014, the forecast is for below-freezing temperatures there with wind chill taken into account, and possible snow. If the winters of southern Spain were freezing, or even if only some of the winters were freezing, Neanderthals with Arctic body proportions could have survived better than others.

They could also have led an advantageously active life. In other words, Finlayson in advancing his idea about Neanderthal anatomy and its relation to their active lifestyle did not have to reject the temperature argument for Neanderthals' squat, heavy body. Neanderthals could have gotten two benefits from their size and shape, heat retention and active strength, not just either one or the other.

Tim Weaver and Karen Steudel-Numbers, though, disagreed with Finlayson. They argued that for energetic, active hunting, long legs, especially lower legs, are more efficient than are short ones, as I described when discussing the athletic prowess of people with African ancestry. That being the case, Neanderthals actively hunting but with short lower legs would have become exhausted before the modern humans that replaced them. The Neanderthals were surely strong, but if they had a highly active life, their legs should have been long. Weaver and Steudel-Numbers therefore prefer the keeping-warm argument to the active-lifestyle argument.

The debate continues, though. Might the Neanderthal's more rigid tibia by comparison to modern human distance runners indicate a highly active lifestyle, at the same time the shorter limbs indicate heat saving? Or were the Weaver and Steudel-Numbers measurements not sufficiently detailed to show inefficient locomotion, as a study published just this year of my writing, 2014, argues?

Ryan Higgins and Christopher Ruff have questioned the premise of Neanderthals' inefficient locomotion from a different angle. They point out that the supposition that short lower legs mean inefficient locomotion is based largely on theoretical and experimental analyses of people walking and running on flat terrain. Yet Neanderthals did not live on flat terrain. Neanderthals lived to a large extent in hilly, even mountainous country. There, short legs, especially lower legs, are not a disadvantage, and might even be an advantage.

Evidence that they could be an advantage is the fact that mountainous bovids (cloven-hoofed ruminants such as sheep, cows, deer, antelope) also have short lower limbs by comparison to their upper limb. The two-benefits argument remains valid. Neanderthals' short lower legs both allowed easier movement in the hills, and decreased heat loss.

The possibility that lifestyle might explain Neanderthal body proportions raises the issue of differences between the sexes. We know that in most human societies, including hunter-gatherer societies, the sexes obtain food in largely different ways. I described in chapter 3 with an example from the Torres Strait the common finding that females often go after smaller, less dangerous prey than do males. In the case of Neanderthals, the anatomy of the two sexes is fundamentally the same. Finlayson would therefore need to argue that both led an active life. Indeed, as Stephen Kuhn and Mary Stiner noticed, the paucity in Neanderthal archeological sites of what we typically think of as females' prey indicates that Neanderthal females were doing the same as males. Both sexes were going after the mainly middle-sized to large game, the remains of which are common in the sites.

We need to be careful of that conclusion, though. Rebecca and Douglas Bird describe how Australian aboriginal women hunt small-bodied, numerous game, so providing consistent nutrition for their families and friends. That is in comparison to the men, who go after larger game, in fact dangerous game, in order to achieve status. The distinction is common to many hunting societies. And there lies the archeological problem. Elizabeth Chilton points out that this typical female's prey of small animals such as rabbits or lizards can preserve less well, or be less easily noticed, than the remains of larger animals. Also, are people more likely to cook and eat small animals off-site than large animals? If so, their remains would be more scattered over the landscape, and so more difficult for archeologists to detect than the remains of large animals brought back to a central living site.

In brief, we modern humans show the Bergmann and Allen effects, of heavier bodies and shorter limbs, respectively, at higher latitudes than in the tropics. Neanderthals, by comparison to modern humans, show the Bergmann and Allen effects, even if in addition Neanderthals' leg proportions helped them move through hilly terrain. Monkeys show the Bergmann and Allen effects. And rabbits, foxes, and many other mammals as well as birds show the Bergmann and Allen effects. A reasonable conclusion is that humans have adapted in size and shape to the environment in exactly the same way as have the bodies of the other warm-blooded creatures. If a god created us humans, it was following evolutionary, ecological rules in doing so.

The color of our skin, the proportions of our body differ from region to region—for good biological reasons. Our physiology too differs. The airplane lands at Libreville, Gabon's capital. I step out of the plane's door onto the top of the stairs leading to the tarmac below. And the air is so hot and humid, I feel as if I have walked into a solid wall. I wilt immediately, and for the rest of my time in town I can move when outside only as if slowly sleepwalking. And yet the resident Africans seem unaffected by the heat.

Nevertheless, scientists have not found clear-cut physiological differences between tropical peoples and temperate or Arctic peoples in how they cope with heat. How can that be? Think about it a little, though, and maybe it is not so odd after all. Certainly, the tropics are hot, but even Scotland can reach tropical heat in the summer. The Scots record so far is over 30°C. So, a Scots laborer could well sometimes experience the same temperatures as the Libreville laborer. If so, then over evolutionary time the high-latitude Scots should have evolved the same physiological adaptations to coping with heat as the equatorial Gabonese.

And for the most part, this does seem to have happened. For instance, hot Europeans sweat about the same volume as do hot Africans. That is not to say that no differences associated with perspiring distinguish high- and low-latitude peoples.

What is sweat but salt and water, two of life's essentials? Sweat, and we lose both. And herein lies the problem. Salt and water are often in short supply in tropical climates. The soils of Africa are generally more lacking in minerals than are those of Europe, and drought hits Africa harder than it does Europe. If the soils lack salts, so do the plants in the Africans' diet, making retention of any salt consumed of great importance for survival. If Africans sweat no less than do Europeans, but are short of water, they will be more prone to dehydration than Europeans usually are, making maintenance of blood pressure a high priority for survival.

And what do we find when we compare sweat composition and blood pressure of hot Africans and Europeans? Yes, indeed, Africans are better than Europeans at retaining salt when they do sweat, because their sweat is less salty. Also, they can constrict their blood vessels more in response to loss of water and thereby maintain a higher blood pressure when seriously dehydrated.

The contrast between Africans and Europeans is in fact a general contrast between peoples from tropical and non-tropical regions. When we identify genes that influence ability to retain salt and to maintain blood pressure when water is in short supply, we find, as J. Hunter Young has shown, quite a tight relation between latitude of a human population and the prevalence of five of those genes (strictly, alleles) in the population. The values for prevalence, or the proportion of measured individuals that have the genes, range from roughly seventy percent in peoples from tropical Africa, the Americas, and Asia to around forty-five percent in peoples from non-tropical Europe and Asia. Peoples from the non-tropical but low latitudes of the Middle East and Pakistan are intermediate.

If most temperate peoples are like me, they sweat the most in hot, humid weather. Visitors from the east coast of the USA appreciate California summers because it is so dry here, even if hot. Temperatures in the Central Valley where I live are upwards of thirty degrees centigrade through much of the summer, even above thirty-five degrees. The joke among us when anyone comments on the heat is the inevitable reply "Yes, but it's a dry heat." The aridity makes a huge difference to comfort.

I have already commented on the humidity of tropical forest in relation to pygmy peoples' small stature. The humidity makes shedding heat more difficult, and so makes the forest effectively hotter than savannah. It turns out that if one compares the Mbuti and Biati pygmy peoples of central African forests with the San peoples, who live in wooded savannah, the genes associated with the ability to retain salt and to maintain blood pressure despite loss of water in sweat by constricting the blood vessels are more prevalent among the Mbuti and Biati than among the San.

Unfortunately, the improved ability of tropical people by comparison to temperate people to retain salt and maintain blood pressure can come with a cost. If a person of tropical origin moves to the rich, Western-style diet, packed with salt, they can get into trouble. We find that African-Americans and middle-class Africans in Nairobi are more prone to hypertension and its associated kidney problems than are Europeans. Yes, of course socioeconomic factors and lifestyle might play a part, indeed probably do play a part in African-Americans' susceptibility to hypertension. But a source of that susceptibility is likely to be also the environment in their region of origin, and their evolved adaptation to that environment, in this case to retain salt.

Here then is another of many cases where it could really help doctors to know of regional differences in physiology. The variation could mean that people from other parts of the world are more or less likely than the ones the doctor usually sees to suffer from various medical problems. They might be especially likely to suffer when they leave the environment in which their ancestors lived.

American Africans might on average be even more prone to hypertension than African Africans. In addition to possible social reasons for such a difference, slavery might have induced a biological difference between the two peoples. Many, even almost all, of the manacled, poorly fed, and poorly watered captives on slave ships transporting them to the New World probably suffered from the "flux," or diarrhea. With diarrhea, we lose not just water, but also salt. The loss can be lethal, and we know of appalling death rates on slaving ships. Any Africans that survived such trips would have been even better than the general population of Africans at retaining salt. And so in the current environment of an overabundance of salt, the descendants of those Africans are now more prone to hypertension than is the general population.

If adaptations to prevent heat stress (as opposed to water and salt loss) do not distinguish tropical from non-tropical peoples, adaptations to cold do appear to differ geographically. An example is frostbite among American soldiers in the Korean war of the early 1950s. Many suffered, but more African-Americans than European-Americans did so. What is the difference between heat and cold, then, and latitudinal differences in susceptibility to them? The answer is that temperatures in temperate summers can reach tropical temperatures, but except at the highest altitudes, winter temperatures are lower, often far lower, than in the tropics.

More generally and more physiologically speaking, it appears that peoples from higher latitudes have developed the ability to react to cold by raising their metabolic rates more than can tropical peoples. Consequently, when outside temperatures are near freezing, Arctic peoples have higher internal and skin temperatures than do tropical peoples. The same happens with birds and mammals from the Arctic compared to species from the tropics. In the better studies showing this effect, the researchers controlled for amount of body fat and surface area of the body, both of which vary with latitude. Do that, and the peoples from high latitudes have basic metabolic rates around fifteen percent higher than the rates of peoples from low latitudes.

Similarly, people born and living at high altitudes have higher metabolic rates than those living at low altitudes. The data available so far indicate that metabolic rate climbs roughly four or five kilocalories per day with every 1°C drop in average annual temperature. That is the case whether the drop in temperature comes from living at higher latitudes or higher altitudes.

As far as I can tell from reading the literature, we are mostly ignorant of how much these differences in response to cold develop during the lifetime of the individual, as opposed to being genetically-influenced evolved abilities.

A seeming exception to the generalization that people from cold environments have higher metabolic rates than do the peoples of hot environments are the native peoples of the Kalahari Desert region in southern Africa—at nighttime. Temperatures in the desert at night can reach close to freezing. Instead of keeping themselves warm by raising their temperature at night, the nighttime metabolic rate and hence body temperature of the Kalahari peoples is lower than that of companions of European origin.

If the Kalahari peoples can conserve energy by letting their bodies get cold, why do high-latitude and -altitude peoples not do the same? Indeed, why do they do the opposite—burn extra energy? The answer could be that they cannot afford to do so, because temperatures are often so much below freezing for so much longer that they have to maintain a relatively high metabolic rate to prevent frostbite.

In addition to these apparently evolved overall regional differences between populations, individual humans in all populations can change their metabolic rate according to the temperature that they are currently experiencing. Most of us raise our metabolic rate in colder temperatures.

So, if we want to lose weight, we should exercise in a freezing gym. We will burn off our calories faster because we lose more heat, and because we are producing more heat to lose. That is, of course, as long as we do not react to the lower temperature and higher metabolic rate by eating more. I know that in winter I eat more than I do in summer, presumably because in winter my body is burning through calories faster. So I, like my cat, gain weight in the winter.

An unpleasant experiment for most non-Arctic peoples is to see how long they can hold their fingers in a bowl of iced water. Non-Arctic people take their fingers out long before Arctic people do. An obvious inference is that the Arctic peoples' higher metabolic rate does indeed enable them to function at lower temperatures than people from lower latitudes.

However, we need to be careful in our interpretation of this experiment. We know that expectations and acclimatization can affect behavior. Non-Arctic people are not brought up experiencing severe cold almost daily. When we meet it, we are not used to it, and so cannot withstand it. Lack of alternatives can also influence what we put up with. Just look at the appalling conditions that the poorly outfitted early British Antarctic explorers lived through. "If you had to get out of the tent during the seven hours spent in our sleeping-bags you must tie a string [to close the tent door] as stiff as a poker, and re-thaw your way into a bag already as hard as a board," wrote Apsley Cherry-Garrard describing, in *The Worst Journey in the World*, his, Ed Wilson's, and Henry "Birdie" Bowers's winter journey to collect emperor penguin eggs. Death was preferable to the suffering: "I for one had come to that point of suffering at which I did not really care if only I could die without much pain." Nevertheless, all three of them lived, and Ed Wilson later went with Scott to the South Pole and, yes, died on the way back.

In fact, a main adaptation in relation to the ice-water test is the precise local response of the fingers to the cold. Initially, an Arctic person's fingers and a low-latitude European's fingers drop rapidly in temperature. But then the Arctic person's fingers start warming sooner than do the European's, and reach a higher temperature. Peruvian Andean peoples living at high altitudes have the same response to the ice-water test as do the Arctic peoples. If these cold-adapted peoples did not have that response, then for much of the time they would not be able to function outside.

I have not yet mentioned behavior as a means to alleviate cold. Humans escape cold with behavior in many obvious ways which, except in the materials they use, probably does not differ between cultures. Native Tasmanian peoples, though, had an apparently anomalous reaction to cold. They moved into the mountains as it got colder. In fact, this counter-intuitive movement probably had little to do with a regionally adapted response. Rather, the suggestion is that they were searching for caves in which to shelter, perhaps especially from wind, and caves were at higher altitudes.

As I described, people at higher latitudes are on average heavier than those at lower latitudes. The other part of the world where we find that the native peoples, the original inhabitants, are relatively large is on Pacific islands, at the opposite extreme of the geographic world from the poles. Some of the heaviest people in the world are Samoans.

Rather than having Arctic peoples' solid body build, though, the large Pacific islanders are simply fat. Recent figures from the World Health Organization indicate that nearly three quarters of Samoans are obese. In other words, they have a body mass index (often abbreviated as BMI) of more than thirty, the threshold for classification as obese. For other Pacific islands, the proportions of the population with such high body mass indices range from forty to seventy-five percent of the population. By comparison, the proportions for two countries often considered overweight, the UK and USA, are just twenty-five percent and thirty-five percent respectively.

Body mass index is calculated as mass in kilograms divided by height in meters squared. It so happens that those two easily measured values, weight and height, correspond more or less to the percentage of fat in a normal human body when combined in the equation that I have just described. Of course, the values vary depending on sex, age, and the sort of exercise we take, given that bone and muscle also affect our total weight. But as a near approximation, the equation adequately indicates body fat.

Why are Pacific islanders so prone to fat? Look at a globe and think about the original journeys to those islands. The Pacific Ocean covers nearly half the globe. The initial trips to the Pacific islands covered thousands of kilometers in open boats. The travelers would have been soaked much of the time, and probably half starved by the time they got to an island, let alone by the time they produced their first crop. It is easy to imagine that the people who survived best were those with the most efficient metabolism, a metabolism that was best at converting food to fat.

Once established on an island, if harvests failed, survivors could not easily move elsewhere. They would have to sit out the bad period. Again, those individuals with the most efficient metabolism would have been the ones most likely to survive. In the words of people who study the phenomenon, the islanders had and still have "thrifty" genes, genes that make the body extremely efficient at converting food calories to stored calories—or, in other words, fat. A case of survival of the fattest?

Nowadays, the islanders no longer face starvation. Indeed, quite the opposite. They find themselves in a feast of carbohydrate, sugar, and fat. The consequence? Body mass indices of over thirty. The story is similar to the one of the Africans finding themselves for the first time in their evolutionary history in a perpetual high-salt diet. What was a useful, adaptive beneficial ability in the past environment can be a cost in a new setting.

The thrifty-gene hypothesis has its critics, as Elizabeth Genné-Bacon describes in her review of the various ideas to explain obesity. However, it appears that island animals might also have a thrifty metabolism. In their case, they often have a low metabolic rate, instead of or in addition to efficiently using food. They can afford a low metabolic rate and lethargy, Brian McNab points out, because large predators are rare on small islands.

By now, all this discussion of large islanders might have brought to mind the miniature "hobbit" of the island of Flores in Indonesia, reported to the world in 2001. If Pacific islanders are so large, why is the hobbit so small? Island animals, plants, and people have their own special biogeographical phenomena. Island phenomena are unusual enough to have a chapter of their own later on in the book, in which I talk about metabolic rate and body size.

From Pacific islands, I now move to the heights of the Himalayas and Andes. The Sherpa peoples of Himalayan Nepal are renowned for their high-altitude feats, especially on Mt. Everest. The last time I checked the records, Apa Sherpa had climbed the mountain the most times, twenty-one of them. Babu Chiri Sherpa had spent the most time on the summit without oxygen, twenty-one hours, and once held the record for the fastest ascent from Base Camp, just under seventeen hours to climb thirty-three hundred meters. Sherpas have consistently held the fastest ascent records from Base Camp, with Pemba Dorjie currently the fastest, at eight hours and ten minutes.

The first teenager to the summit was Temba Tsheri Sherpa in 2001, aged then sixteen. The youngest female to have gotten to the top is Malavath Poorna from southern India. She got there on 25 May 2014, aged thirteen years and eleven months, just one month older than the youngest male. I cannot find any records for the heaviest load carried the highest, but I would not mind betting that a Sherpa holds that record too. How do they do it?

My wife and I have trekked with yak herders in the Himalayas of northern Bhutan five times since the early 1990s. The airplane to Bhutan lands at Paro, at an altitude of twenty-one hundred meters. We begin each visit to Bhutan breathless from walking up the Hotel Olathang's stairs, and with a bad headache through the night. By the time of our return there three weeks later, after several nights at over forty-five hundred meters, and passes of over five thousand meters, we can run up the stairs with no trouble and do not need any aspirin. But we still could not chase a yak up a mountain at five thousand meters, as our trek herders do each morning as they round up the yaks freed to forage the evening before.

Brought up in the lowlands of California and Britain respectively, we do not have the blood or lung capacity of the Himalayan Sherpas or Bhutanese, or indeed of the yaks. Nor do we have the blood or lung capacity of native high-altitude Andean Indians and their llamas and vicuñas. All of these high-altitude people and animals have capacities in the thin air that far surpass those of lowlanders.

Some of the differences between lowland people and their Himalayan or Andean counterparts arise simply from the latter living at high altitude and experiencing from childhood the shortage of oxygen there. People born in the lowlands who move as youngsters to the highlands soon become nearly as energetic as do the highland-born. The younger they are when they move, the better they do. Move at two years of age, and a Peruvian Andean, and probably you and I, can move nearly fifty percent more air through our lungs per minute than the Andean could if they move at fourteen years of age.

The information that we have on how and why mountain people do better at high altitude than do lowland people comes largely from Tibetans and Andeans living between thirty-five hundred and four thousand meters, depending on the study, and Ethiopians living at around thirty-three hundred meters. And it comes to quite a large extent from the research of groups led by Cynthia Beall and Lorna Moore.

High-altitude Tibetans by comparison to lowland mostly Han Chinese have a greater volume of blood flowing through their lungs, because they do not constrict lung arteries as much as do lowlanders. The highlanders ventilate their lungs more, and they have more small blood vessels (known as capillaries) in their muscle. Perhaps surprisingly, they have lower concentrations of hemoglobin in their blood. However, it turns out that high concentrations could be associated with mountain sickness. During exercise, the Tibetans have more blood flowing through their brain. And pregnant Tibetan women have more blood flowing through the arteries of their uterus, delivering more oxygen to the uterus and the placenta.

High-altitude Andeans, by comparison to Europeans and South American lowlanders, have a somewhat different list of differences than do the Tibetans compared to lowland Han Chinese. The Andeans have a very high concentration of hemoglobin in the blood, and well-oxygenated arterial blood. Pregnant women get more oxygen to the uterus and placenta by increasing ventilation of their lungs, having more blood flowing in the uterine artery, and having more oxygenated blood.

Until recently, nobody had found that high-altitude Ethiopians had clear adaptations to high altitude, despite the fact that people had lived there at altitudes of more than twenty-three hundred meters for at least seventy thousand years. However, studies this decade led by Laura Scheinfeldt of Sarah Tishkoff's laboratory reported that high-altitude Ethiopians had higher hemoglobin levels in their blood. Gorka Alkorta-Aranburu and a team from Cynthia Beall's laboratory found the same.

Many studies claiming differences between mountain peoples and lowlanders had not controlled properly for what the statisticians term confounding effects. For instance, they did not control for age, or sex, or time spent at altitude. Or if they did, they did not say so.

With those controls, we find less difference between Tibetans and Andeans than without them. We also see fewer differences than had been thought between people born at high altitude and lowlanders who moved up the mountains in each region. For instance, as Charles Weitz and his supporting team show, Han Chinese and Tibetans born at high altitude show exactly the same increase in their blood's hemoglobin concentration over childhood and adolescence. Consequently, if someone did not account for age, and compared young high-altitude Han with older high-altitude Tibetans, they would find that the Tibetans had higher concentrations of hemoglobin.

The human body can adapt to high altitude if raised there from an early age. Are we seeing, with these findings from the Andes, Tibet, and Ethiopia, an effect of adaptation within individuals in their lifetime? Or are we seeing the results of evolved genetic adaptation to what is commonly known as the "thin air" of high altitude? We could expect genetic adaptations. After all, high-altitude Tibetans, Andeans, and Ethiopians have been up there for thousands of years.

So far, the evidence for genetic adaptation to high altitude is not good. Indeed, just last year, one of the longstanding researchers in the field, Roberto Frisancho, stated that nobody had yet conclusively identified genes associated with high-altitude ability among Andeans. I am therefore going to lead up to indications of genetic adaptation by starting with the sort of biological findings that indicate its presence.

I am going to pick the newborn's birth weight as the attribute to discuss in this context of potential genetic adaptation, because a strong correlate of a newborn's ability to thrive is its weight at birth. In general, lightweight babies have more difficulties than do babies of average weight.

Colleen Julian and a team from Lorna Moore's group compared, with a variety of measures, babies of European or Andean parentage (as judged by surnames) born at high altitude (thirty-six hundred meters or forty-one hundred meters) and low altitude (four hundred meters) in Bolivia. They accounted for a variety of factors that could affect birth weight, such as mother's height, sex of the baby, and income of parents.

Whilst at low altitude, the Andean newborns weighed five percent less than did the European ones, at high altitude they weighed nearly ten percent more. The extra ten percent was despite the fact that the Andean mothers weighed less than the European mothers at both altitudes. The high-altitude European newborns were lighter than those at low altitude, whereas the Andean newborns were heavier than at low altitude. Newborns of mixed parentage were halfway between the two. All weighed near three kilograms, but the ones below three kilograms—in other words, of low average birth weight—were the newborns of Europeans at high altitude.

A study in Tibet directed by Lorna Moore found similar results. The drop in newborn babies' weight with altitude was less among parents of populations who had lived at high altitude for probably millennia, Tibetans, than among Han Chinese and Europeans who might have lived at high altitude for only centuries, even though sea-level birth weights were similar in all four groups. Putting some figures to these differences, Tibetan newborns from populations around four thousand meters in altitude were approximately three hundred grams lighter than those born to parents at sea level, whereas European and Han high-altitude newborns were more than four hundred grams lighter than ones born to sea-level parents.

These findings from the Andes and Tibet are the sorts of results that gave the earliest geneticists an idea of the existence of what subsequently came to be termed genes. The offspring of different sorts of parents are themselves different. Nevertheless, purely environmental influences could have produced the differences in the offspring.

That is especially the case with the Bolivian results, because the hospital records did not show how long the parents had been at high altitude there. It seems likely that a greater proportion of the European than Andean parents moved there in adulthood, almost certainly from lower altitudes, given that the hospital in which the babies' measurements were taken was in La Paz, the highest administrative capital in the world.

As another example of a possible influence of evolved gene-based differences, take people born and raised at high altitude, and compare those whose parents are highlanders with those whose parents were lowlanders, and we find that the highlanders' offspring have lungs with greater volume than do the lowlanders' offspring.

Traits evolve when an individual with the trait does better than one without it. Given that birth weight correlates with survival of babies and infants, we can predict that in the absence of modern medicine, a greater proportion of the Andean babies than the European babies would have survived at high altitude in Bolivia, and a greater proportion of the Tibetan than European or Han babies. If any genetic differences contributed to the greater birth weight of the high-altitude babies, the genes involved would spread through the high-altitude populations. The same argument would apply to lung size at altitude.

More directly, in Tibet, when both parents or even just one carried a form of a gene that seems to be associated with ability to carry oxygen in the blood, the children were twice as likely to survive infancy as when neither parent carried that form of the gene.

More directly still, several teams, including ones led by Cynthia Beall and Lorna Moore, have now identified specific genes probably associated with the evolved ability to live at high altitudes. They have done it by searching in available databases for genes that differ between high-altitude and low-altitude populations.

Tibetans, for example, have genes associated with low hemoglobin concentrations and low red blood-cell counts. One would have thought that at high altitude with its lack of oxygen, it would be more advantageous to carry more hemoglobin and more red blood cells in order to be able to carry more oxygen in the blood. In fact, Tibetans do have higher hemoglobin concentrations in their blood than low-altitude Han Chinese, but only a little higher. As I wrote previously, if you do not want to get mountain sickness, you do not want too much hemoglobin, nor too many red blood cells. Too much of both causes the blood to become viscous enough to flow more slowly, and hence not carry enough oxygen to the tissues. Somehow, though, high-altitude Bolivians have a high hemoglobin concentration, and yet manifestly do not suffer from altitude sickness—or else they would not be there.

Abigail Bigham and yet another team including Lorna Moore searched in detail on all chromosomes of individuals of high altitude in Andes and Tibet (a genome-wide scan, in genetics-speak) for differences between their genes and those of lowland peoples. They searched across genes with unknown function, as well as those already known or thought to affect ability at high altitude. The details of the results are effectively incomprehensible to the non-geneticist, written in dense jargon as they are. Suffice it to say that they found several genes in the high-altitude populations that differed from the genes in the low-altitude populations, many of which would have enabled the former to survive better there.

Extraordinarily enough, one of the gene forms that allow Tibetans to live at high altitude apparently came from Denisovans. I will give its name here, as I suspect that it will become famous. It is EPAS1. Emilia Huerta-Sánchez and a very large team reported this finding about the Denisova gene in only July 2014. All the evidence so far indicates that just a handful of other genes show as much change from the common form as this one. The amount of change indicates a strong advantage to individuals that have the altered forms. It seems highly unlikely that Denisovans lived at high altitude. They are far from being modern humans, implying that the skills needed for such an extreme environment must have been beyond them. And yet all the evidence so far indicates that only Denisovans and Tibetans have the high-altitude gene form.

A possible explanation for the apparent anomaly is that the gene form has other benefits at low altitude than the ones it confers at high altitude. The gene is involved in oxygen chemistry in the body. It seems to confer extra ability in endurance athletics, in other words in a sport that probably involves the muscles fighting to get enough oxygen. If running too fast too far produces the same lack of oxygen in the tissues as does high altitude, maybe the same gene form can benefit both activities. Currently, though, we have absolutely no idea of how Denisovans lived.

Through all that I have written so far about adaptations to living at high altitude, I have not explicitly compared the high-altitude Tibetans with the high-altitude Andeans, except with that one mention of Bolivians' surprisingly high hemoglobin concentration. In fact, the ways in which the Andeans and Tibetans cope with altitude differ. In order that nobody has to go back through the past page or two, let me list a few of the differences. While Tibetans can ventilate their lungs more than can Andeans, Andeans simply have larger lungs. Tibetans get more oxygen to their tissues mainly by pumping more blood to them than do Andeans, whereas Andeans carry more oxygen in their blood than do Tibetans, in part by having far higher hemoglobin concentrations.

Correspondingly, most of the genes that differed between the high- and low-altitude populations within the two mountain ranges also differed between the two ranges. The genes that might have conferred high-altitude abilities in Tibet were not the same as those that might have conferred high-altitude abilities in the Andes, which differed again from those in high-altitude Ethiopian populations.

Given that we already knew the physiological means by which the people of the regions differed, and that genes control physiology, this result is exactly what we would expect if evolution of genetically induced traits were occurring.

We thus have more than one physiological and genetic route to the same end, survival and reproduction at high altitude. In other words, the route to the end does not matter, as long as the end is reached, in this case the ability to cope with lack of oxygen.

Darwin's theory of evolution by natural selection has organisms with a useful ability producing more surviving and reproducing offspring than do the individuals without the ability. As I have reported, we now know of several populations of humans where high-altitude people have genes and physiology that enable them to survive and reproduce there better than do people who have moved from low altitude and do not have the requisite genes or physiology. In other words, here we have the results of evolution by natural selection in humans.

People might have moved into the Tibetan Plateau over twenty thousand years ago, maybe even thirty thousand years ago. They might have moved up into the high Andes twelve thousand years ago, as I mentioned in chapter 2. In fact, some of the highland Tibetan subjects of the physiological and genetic studies might have moved into the region only approaching three thousand years ago. As with the story of pale skin in western Europe, then, we have evolution happening in humans in not just archeological time, but in historical time, a point that I will come back to in chapter 9.

As I indicated at the start of the discussion about adaptation to living at high altitudes, humans are not alone in adapting to cope with the resultant low atmospheric pressure and lack of oxygen. Yaks, llamas, and vicuñas do so, and so do other species too, including dogs.

Tibetan mastiffs have some of the same physiological responses to lack of oxygen that high-altitude Tibetans do. The mastiffs have low hemoglobin concentrations in the blood, for instance. However, the genes involved differ from those identified so far in humans, as determined in the case of the mastiffs by comparison with low-altitude dogs in China, and with wolves.

Yaks, as might be expected, have several genes associated with living at high altitude. One is the same as among high-altitude Tibetan peoples, but the others are different.

Similarly, Tibetan mastiffs, high-altitude Andeans and Ethiopians, and, I might add, high-altitude Tibetan antelope and ground tits, each have a gene associated with production of new blood vessels that their lowland counterparts of their regions do not have. However, the gene involved is different for each species.

So far, I have not distinguished between the sexes in their adaptations to the environment, except with the discussion of evidence for a lack of differences in Neanderthals. I have therefore implied that they are effectively the same as regards whatever aspect of physiology or anatomy I have been writing about. However, the sexes do differ, with the result that in some comparisons of peoples between regions, we might need to be explicit about which sex we are discussing.

I easily eat twice as much as my wife, especially if the meals include potatoes or rice. And yet, to my wife's annoyance, I never gain weight. I am the same weight as I was twenty years ago, still within WHO-recommended weight for height. The reason is simple: I have a higher metabolic rate than my wife does, one irrelevant consequence of which is that in winter, our cat sleeps on my warmer side of the bed. So as to avoid unwarranted invidious comparison, let me quickly add that neither has my wife gained weight, and that she too is well within WHO-recommended limits.

In general, males have higher metabolic rates than do females. Of the other traits that I have written about in this chapter, females in general are paler, broader-bodied, shorter-limbed, and fatter and rounder than are males—and in several ways tougher than males. The contrast is as if females came from higher latitudes or altitudes than did males.

Females' paleness by comparison to males would match their potentially greater need for calcium (via vitamin D via the sun) when pregnant and lactating.

Human females' lower metabolic rate, broader body, shorter limbs, and greater proportion of fat all match the fact that a female eats for two, herself and her offspring, and so needs to conserve energy more than does a male. Not only does she carry more fat than a male of the same height or weight, but she carries it closer to her skin than does a male, who packs it more around his internal organs. Females are therefore better insulated than are males, who instead have the fat readily available to metabolize for bursts of energy.

These adaptations of females to conserve energy, like those of Arctic peoples, make the females tough when the going gets tough, in other words when food runs short and the temperatures drop. In the famous Donner Party, stuck for months in unseasonably early and deep snow in the Sierra Mountains of Nevada and California, females were only half as likely to die as were males, Donald Grayson calculated. The same was true, he showed, in another party under similar circumstances, the Willie Handcart Company. I should add that this percentage excludes lone people in the parties, who were especially likely to die, and were all males. The very young were also likely to die, which would be explained by their greater surface area per volume of body, leading to more loss of heat.

Another famous tragedy associated with cold and lack of food is the Dutch Hunger in the winter of 1944/5 near the end of World War II. Germany prevented supplies reaching Holland during this unusually severe winter. Cold, starvation, and associated illnesses killed thousands. The young and old suffered especially, as in the Donner and Handcart parties. And in the worst of the six-month-long Dutch Hunger, working-class men were twice as likely to die as were working-class women.

Less famous, but equally devastating, was the Finnish famine of 1866 to 1868, when harvests of rye and barley, which made up eighty percent of normal energy intake, fell to half their normal level. Death rates quadrupled in the country as a whole by late 1867, early 1868. Males died at a higher rate than did females, as expected from differences in shape of body, amount and distribution of fat, and higher metabolic rate. Additionally, the very young and old died in the highest numbers, as did the very poor, again as might be expected from size of body in the case of infants, and fat in the case of the old and poor.

As always in science, though, we have to wonder about alternative explanations. I have put the resilience of the women in the Sierras and Holland down to a more efficient metabolism and body form, as did the scientists whose studies I am using. However, across the world, societies, and therefore regions, often differ in the tasks that the sexes perform. Did the Donner and Handcart women stay in camp while the men exhausted themselves trudging through deep snow after game? Were the men chivalrous in the Sierras, Holland, and Finland, giving food to their wives and daughters? Instead of the young dying because of a greater surface area per volume of body, was it easier or more practical for parents to give up on younger than older children?

Is chivalry as a cause of differences between the sexes in death during catastrophes too outlandish an idea? It was not in one incident in the last century. A wonderful study on the differences between the sexes in likelihood of dying on the *Titanic* and the *Lusitania* implicates chivalry. In the *Titanic* sinking, but not that of the *Lusitania*, nearly three quarters of the men died, but only one fifth of the women. On the *Lusitania*, death rates were statistically equal.

Why the difference? The *Titanic* hit an iceberg and took two and a half hours to sink, leaving men time to think that they would be saved, and therefore they could afford to act chivalrously. The *Lusitania* was torpedoed and sank in just eighteen minutes. It was every man for himself.

The *Lusitania*, sadly, is more the norm. Another study of sixteen other sinkings since the late 1800s found women twice as likely as men to lose their lives. Were they pushed out of the way by the men, drowned because of their long dresses, or more occupied with saving their children than saving themselves? If they were occupied with saving their children, it did not help, for children were only half as likely to survive as women. That would fit an every-man-for-himself scenario.

Some readers, perhaps especially American ones, might know that more of the Mayflower women than men died in their first months on American shores in the winter of 1620/21. However, we are not talking unchivalrous men pushing women out of the way here, or denying them food. Rather, Cabel Johnson suggests, the reason can be found in the fact that the women stayed crowded in the unhealthful conditions on board ship for the first four months, while the men were on shore in the open air.

These various explanations for differences between the sexes in the likelihood of dying in different situations—physiologically tough women, different jobs, chivalrous men, unchivalrous men, different living conditions—now need to be tested to discern the better explanations in each case for the contrast between the sexes. That is how science works. We make an observation that women have different anatomy and physiology than do men. We come up with an explanation about conservation of energy that works in other contexts, namely contrasts between regions in anatomy and physiology. We see another set of data that seem to confirm the explanation—women survive better than do men when short of energy. But then we think of alternative explanations—differences between the sexes in the sort of work they do, or social chivalry of men.

So now it's back to the drawing board. We need to produce a way of distinguishing the ideas, the hypotheses. But that is for the future. All we can say at present is that the current information on contrast between the sexes in likelihood of death when food is short fits the hypothesis that contrasts in metabolism and anatomy affect how fast we use up our bodily energy, which results over evolutionary time in differences in metabolism and anatomy between peoples from different environments. If the explanations of the contrasts between the sexes turn out to be correct, then the fact that they work in this context strengthens the original biogeographic hypotheses for the Bergmann and Allen effects across regions.

∽

The subtitle of this chapter states where we are affects what we are. People living at high latitude and altitude are different from those at low altitude, because over time the former have evolved physiologies and anatomies that the latter do not have. Once we have adapted to the environment in which we and our ancestors have lived, then what we are affects where we are, because what we are affects where we can comfortably live or successfully reproduce, other things being equal. Or at least it used to. Now it is not so much the case, because we have so many cultural means of countering the effect of adverse environments. The world should be turning not just culturally homogenized, the same jeans everywhere, but also physiologically and genetically homogenized, the same genes everywhere—if I may be allowed the pun. It has not gotten there yet, but is it heading that way?

CHAPTER 6

GENE MAPS AND
ROADS LESS TRAVELED

Barriers to movement maintain diversity

Most Africans are, as I described in the last chapter, darker-skinned than I am, coming as I do from western European ancestry. Africans tend also to have longer limbs relative to the length of their torso than I do. Andean Indian children can play a full game of soccer at altitudes that leave me puffing after a short walk. These differences between peoples from different regions have evolved because people with certain characteristics do better in certain environments than do others. Africans' long limbs shed heat better than do my short ones. The Andean Indians have larger lungs than I do.

But what about the large proportion of red-haired, freckled people that I noticed during my one trip to Ireland two decades ago? Is it really the case that in Ireland, red hair and freckles helped individuals survive, or find mates, or rear their children more successfully than people with other colors of hair or skin? It seems unlikely.

Charles Darwin's answer to the question of why humans differ anatomically from region to region in ways not related to survival was what he termed "sexual selection," as opposed to natural selection. Humans from different regions were not different in so many ways because different sorts of people survive better or worse in different sorts of environments; rather, in choosing a mate, people from different regions happen to prefer different traits. For no particular reason, the peoples of Ireland preferred mates with red hair. Psychological studies often show that we prefer things with which we are familiar. So the children of red-haired parents preferred mates with red hair, and thus the number of red-haired people in Ireland increased.

Maybe. But Darwin did not know anything about genes, which means that he missed another answer to why people of different regions of the world are different. It concerns how, more or less independently of the environment, genetic differences between regions originate and then increase. This is the topic of the first part of this chapter.

Once regional differences exist, they persist only if the populations remain apart. It takes very few individuals to move from one population to another and to reproduce there for the populations to become genetically indistinguishable. So the rest of the chapter is then about the barriers that keep us apart, keep us different, and so contribute to the glorious physical and cultural diversity of the human species.

No two individuals are the same, not even identical twins, and certainly not ordinary siblings, brothers and sisters. So as humans spread through the world, as a few people went north, others east, and a few of those easterners turned south, each little party would have had among themselves a complement of genes that was different from each other little party. Under this scenario, differences between peoples of different regions are, in effect, accidental.

Take the fact that almost all Native Americans or American Indians in South America are blood group O, as I described in chapter 4. No evidence exists to show that O people survive better in the Americas for any reason than do people of blood groups A or B. The sole reason that most native South Americans are O is that the founding population happened to be all O, not blood groups A or B, even though A and B were present in northeast Siberia too.

The same applies, so far as we know, to a number of other traits that distinguish people from different regions. One of the odder ones is the nature of our earwax. It can be wet or dry. As far as anyone can tell, accident of origin is the only reasonable explanation for how wet- and dry-earwax peoples are distributed across the world. Dry earwax is the norm for people from Asia, especially those from eastern Asia, and hence for native Indians of the Americas. Wet earwax characterizes Europeans and Africans. Yet only half the population of Ethiopia is wet. If anyone can see any environmental correlation there, they have better eyesight, so to speak, than any scientists so far. The same lack of correlation with any likely influence from the environment is true of regional differences in the shape of the whorls in our fingerprints, and in the shape of our head.

The possibility that differences between peoples from different regions can be accidents of origin should not be anyone's sole hypothesis, however. For a start, it closes investigation of possible evolved functions of the different gene forms in the different regions. Worse, it could mean that inferences about origins and routes are wrong. If in fact people of blood group O survive better in South America than people of blood group A or B, then we cannot conclude that indigenous South American Indian peoples originated from a small O-blood-group population in northeast Siberia. South Americans could have come from anywhere, and only those of blood group O survived.

Geneticists have a term for what I have just described. It is the "founder effect." It simply expresses the idea that the founding families are different, and therefore so are all of the descendants. The successful first immigrants to the Americas happened to be blood group O. Therefore all the Native Americans that arose from them are blood group O. That is nearly all of them, especially in South America.

Moving to reality from abstract description and analogy, the Irish are genetically different from the English because, according to Stephen Oppenheimer in his fascinating *The Origins of the British*, the Irish founders were originally Celts from the Basque region of northern Spain, as was much of the population of western Britain. By contrast, the founders of eastern Britain's peoples, especially southeast Britain's—in other words, England's—came in higher numbers directly across the Channel and the North Sea from northwest Europe.

Of course, the Irish and English have intermarried and have other origins too, including some Viking "blood," as it used to be called, even if Viking presence is most obvious in northeast Britain. Nevertheless, the Irish and English are genetically different now. They are different not just because the Celts or Basques survived better in Ireland than did who we might call Anglo-Saxons, nor because Anglo-Saxons survived better than did Celts in England. They are genetically different largely because their founders were genetically different. The peoples of Britain are more biologically and culturally diverse than if we had all come from one population of founders.

Understanding the genetic origins of regional contrasts between peoples is of more than academic interest. A number of diseases occur at higher frequencies among some populations than among others. If the diseases were more common among these peoples because of some peculiarity in their environment, the cure would be different than if they arose because of the peculiar genetic makeup of the population.

A well-known genetic disease associated with a particular people and place is Tay-Sachs, named after a British eye doctor, Waren Tay, and an American nerve doctor, Bernard Sachs. As with hemophilia, a person with just one copy of the mutated gene does not suffer. However, if a person receives two copies of the mutated gene, one from each parent, they get Tay-Sachs disease. It results in death in infancy from degeneration of the nerves. This and several other lethal diseases are particularly prevalent among Ashkenazi Jews whose families originated in eastern Europe. In the case of Tay-Sachs, the prevalence among the eastern European Ashkenazi Jews is roughly one hundred times that in the general population.

I do not know of any widely accepted explanation for the high prevalence of Tay-Sachs and the other diseases among the eastern European Ashkenazi Jews. One explanation that would fit the facts, though, is that the founding population was small and by chance had a high incidence of the Tay-Sachs genes, perhaps because an unusually large proportion of the founders came from one family. If the founders then preferentially reproduced among themselves, the problematic genes would have increased in frequency in the population.

The knowledge that Tay-Sachs came from the population's founders, and has nothing to do with the life of the patient or the patient's parents, gives doctors and patients confidence in genetic counseling, as opposed to lifestyle counseling.

However, for another condition resulting from a founder effect, lifestyle can mean the difference between life and death. Porphyria variegata, characterized externally by purple lesions on the skin ("porphyra" is Greek for purple) is a genetically caused condition that can make a barbiturate anesthetic lethal to an otherwise healthy individual.

Until a hundred and twenty years ago, nobody experienced barbiturates in the environment, because nobody knew that they were an anesthetic. Carriers of the porphyria variegata gene were therefore at no disadvantage. And so what happened with the Dutch immigration to South Africa in the mid to late 1600s? The large families and massive population increase that followed in this rich land meant that just two carriers of the porphyria variegata gene, Gerrit Jansz and Ariaantje Jacobs, had thousands of descendants with the gene, in fact on the order of thirty thousand by one estimate. When barbiturate anesthetics reached South Africa, an unusually large proportion of white hospital patients experienced a sudden and utterly unforeseeable increase in the probability of death in the hospital.

In this case, as long as everyone recognizes porphyria variegata for what it is, and knows not to give the carrier barbiturates, the person can live a perfectly healthy life. If one person in a family has the condition, then many family members probably have it. All need to be counseled to tell anesthesiologists not to use barbiturate anesthetic.

That, then, is a brief account of how different founding populations can give rise to regional differences—not only in disease, of course, but in all sorts of other attributes, including cultural traits.

Another random genetic process leads also to populations diverging for reasons unconnected to any differences between them in their environment. "Genetic drift" is the phrase for it, and it perfectly describes the process. The genetic makeup of populations drifts apart over generations simply because nothing prevents the change.

Think of the game "Telephone"—or "Chinese Whispers," as the politically incorrect British name it. The whispered sentence goes down the line of people, diverging more and more from the original as it goes because each person mishears it, until the final player's sentence bears no resemblance to the sentence that started the game.

This concept of increasing difference over time is the biological equivalent of what happens to cultures over time, for example to languages. As a Brit making my first visit to the USA on an American airline, I was alarmed by the pilot's announcement that "the plane will be landing momentarily." I wondered whether I would have time to "deplane" (mutated from "disembark") before the pilot took off again. That is because in Britain, "momentarily" still means "for a moment," not "in a moment." A quote often attributed to George Bernard Shaw sums up the situation: "England and America are two countries separated by a common language."

Americans and English can still mostly understand one another. But with enough changes of meaning, minor differences between languages become different dialects, and finally mutually incomprehensible languages. The changes have nothing to do with the different environments. Nothing in the environment makes people who use "momentarily" more likely to survive or reproduce than people who say "in a moment." But eventually, the languages will differ enough that their speakers cannot understand one another.

So just one culture of humans, maybe no more than a village, peopled South America, starting about sixteen thousand years ago. I described this immigration and the evidence for it in previous chapters. Yet in just the sixteen thousand years since the arrival of those first humans in South America, the continent's density of languages had risen to over half that of Africa's or Asia's, where people had been speaking for maybe fifty thousand years. South America has a density of seven languages in one hundred thousand square kilometers, compared to Africa's and Asia's approximately eleven. And if the variety of languages that have arisen by the equivalent of genetic drift become mutually incomprehensible, then we have the linguistic equivalent of reproductive sterility between species.

Neither of the genetic phenomena that I have just described, neither the founder effect nor genetic drift, would cause populations to diverge were it not the case that barriers prevent free movement of peoples between regions. If people moved freely, all regions would be a melting pot of genes and languages. The existence of barriers to movement is crucial to explaining geographical variation among humans. Barriers limit the spread of people, and therefore of the traits of people, both biological and cultural. Barriers therefore promote diversity, cultural as well as biological.

Geographical barriers, communication barriers, cultural barriers. Those, and how they separate peoples and so produce regional differences across the human species, are the topics of the rest of this chapter.

Discussions of the biogeography of humans in Africa generally separate Africa south of the Sahara from Africa north of the Sahara. That is because the peoples either side of the Sahara are biologically and culturally different. Indeed the biogeographic discussions tend to omit the Sahara altogether because, to all intents and purposes, nobody lives there.

The edge of the Sahara is a terrifying place to stand if you are not accustomed to the sight of miles and miles and miles of sand. You are on the approach road to the Aswan Dam on the west bank of the Nile. Look west out of the bus window, and you will be looking at five thousand kilometers of sand and rock. Look south from Tripoli, and you will be gazing over fifteen hundred kilometers of sand and rock.

But the Sahara has not always been a desert. Rivers used to run through it. The region was intermittently nicely wooded grassland, teeming with game, as in present-day national parks in East Africa. Around a hundred and twenty thousand years ago, when modern humans apparently first left Africa, the world was at the height of the last warm period before the current one, and the Sahara was probably wet. The humans who left Africa and entered the Middle East then might not have had trouble passing through the eastern Sahara.

Subsequently, at the time of our global diaspora sixty or so thousand years ago, the world was well toward a peak ice age. The Sahara was probably dry at the time of this second exodus. To bypass the Sahara, our ancestors would have had to hug the coast. I wrote in chapter 3 of the possibility, even the probability, that humans hugged the coast as we dispersed across the world.

The Sahara is another sort of barrier than a geographical one: it is a knowledge barrier. Its emptiness and the difficulty of work there effectively prevent us from finding traces of our origins there. If we cannot find archeological sites in the Sahara, we cannot find the earliest humans there. If no people potentially related to our ancestors now live in the Sahara, the genetics of living Africans will not allow us to find an origin in the Sahara. A Sahara empty of information leaves northern Ethiopia as the next closest region from which we can obtain bones or genes, and so almost by default leaves it as the region that archeology and genetics can reveal as the home site of the first humans to leave Africa.

Cold prevents life as much as does aridity. Neanderthals were already well into Eurasia as modern humans first moved out of Africa. However, they were anatomically adapted to the cold, as I described in the previous chapter. Maybe modern humans had to wait a further fifteen thousand years until they developed an advanced culture that could cope with the cold before they could move north out of the Middle East into the rest of Eurasia.

I wrote in chapter 3 how human dispersal into northern Eurasia and the Americas was blocked by a massive ice sheet. Go to Central Park in New York and you can see what are called glacial striations on the rock outcrops there. Those resulted as the ice sheet, with rock, sand, and stone embedded in its base, scraped across the rock. Put some sandpaper under a microscope, and you will get an idea of what the bottom of an ice sheet is like.

You will also see in Central Park the occasional huge boulder, big enough for many people to sit, stand, climb, or play on it. These boulders can weigh hundreds of tonnes. (A tonne, remember, is the metric equivalent of a ton.) The boulders were plucked off mountains as the ice sheet advanced south down the continent, and dropped as it retreated. The rocks can be hundreds of kilometers from their source. Hence the geologists' term "erratics" for them.

Probably the most famous erratic, at least in North America, is Plymouth Rock. Originally, it might have weighed something over nine tonnes and been more or less five meters long and one meter wide. Now, broken as citizens moved it, and chipped as people took souvenirs, it is a lot less impressive.

Still in New York, the highest ground in Brooklyn is Battle Hill. It provides yet more evidence of how far south the ice sheet reached, for Battle

Hill is a so-called terminal moraine. A terminal moraine is the ridge formed at the front of a glacier as it melts in place, shedding all the soil, pebbles, and boulders that it has been carrying.

The ice cap covering the whole of northern North America separated east of the Rockies maybe thirteen thousand years ago according to some estimates, maybe fourteen thousand years ago according to another. The western, smaller, third is the Cordilleran ice sheet. The central and eastern two thirds are the Laurentide ice sheet. It was the latter that reached as far south as where New York now is.

The corridor between the ice sheets is a key to understanding routes of humans into especially North America, and for understanding how they so quickly covered the continent. Bison entered the corridor between the two ice sheets a few hundred years after they separated. They entered from both ends. However, anthropologists still debate if and when humans might have used the passage.

The corridor at first opening had to have been appallingly cold and windy, situated as it was between two of the largest ice sheets outside of Antarctica and Greenland. The winds off ice sheets, termed katabatic winds, can reach more than a hundred kilometers per hour in speed. I have been lifted off the ground in Iceland in such a wind. Add to the cold the fact that for thousands of years the land had been buried under a kilometer or more of ice and scraped bare and clean of all life useful to humans. The ground exposed as the glacier melts and retreats is a rubble-and-lake landscape inundated with floods of meltwater. I have visited the edge of ice caps and glaciers in several countries—Bhutan, Iceland, New Zealand, Norway—and they are all barren wastes.

Consequently, I am prone to believe that the earliest arriving humans in the center of North America, for instance the people in Texas fifteen and a half thousand years ago, got there by moving down the coast and then inland.

Before Moses and the Israelites entered the "wilderness" of the Middle East from Egypt, they were stopped by the Red Sea—until their God temporarily "made the sea dry land, and the waters were divided" (Exodus 14:21). And then the sea rushed back and killed all the Egyptians who were chasing them (Exodus 14:28).

This familiar scenario comes from the King James version of the bible. Later scholars suggest that the King James translators did not get it quite right. They argue that in fact the original Hebrew should be translated not as the Red Sea, but rather, the Sea of Reeds. In other words, we are talking a very shallow body of water, presumably somewhere along the Gulf of Suez. Shallow bodies of water can sometimes experience what meteorologists call a "wind setdown." Exodus 14:21 describes the phenomenon perfectly: "and the Lord caused the sea to go back by a strong east wind all that night, and made the sea dry land, and the waters were divided." When the wind ceases, the banked-up waters flood back—in the case of Moses' exodus, swamping his pursuers.

The point is that barring divine intervention, freak natural phenomena, or boats, water is an effective barrier to humans and most other terrestrial animals. That is why the earliest human remains so far found outside of Africa are in the Middle East, the only region that has a dry connection with Africa. In fact, the first global exodus of humans from Africa might have departed completely dry-shod even if they entered Arabia at the south end of the Red Sea, and even without a wind setdown. That is because, as I described in a previous chapter, sea levels seventy to fifty thousand years ago were tens of meters lower than now because so much water was locked up in ice caps as the world headed to the last ice age.

If the first emigrants from Africa did not have boats, the first immigrants to Australia and New Guinea must have had them. Even when global sea levels were at their lowest, perhaps twenty-five to twenty thousand years ago, at least a hundred kilometers of open water separated Asia from Australia and New Guinea. Forty-five thousand years ago, when humans might have first reached Australia, the distance would have been greater.

The earth's curvature prevents anyone in a small boat from seeing more than ten kilometers out to sea. So the earliest arrivals in Australia or New Guinea must have started to paddle out from one of the eastern Indonesian islands—Timor, for example, if they were heading south to Australia—with no idea what was over the edge of their horizon. We can only wonder what drove them to set out. However, we do know that they would not have been in Timor if they had not already crossed at least two ocean barriers hundreds of meters deep along the Indonesian island archipelago on the way to Timor or New Guinea.

Water is such an effective barrier that it was not until approaching four thousand years ago, after the invention of large outrigger canoes, that humans began to colonize the western Pacific islands. As I wrote in the previous chapter, humans did not get to New Zealand from Australia, but from the region of the Cook Islands, near Tahiti, a distance of over twenty-five hundred kilometers, compared to the roughly twelve hundred kilometers from Australia. Why were the Australians not the first immigrants to New Zealand? Simple. Polynesia had the outrigger culture, Australia did not.

For similar cultural reasons, Madagascar, just four hundred and fifty kilometers from Africa, was not first colonized by Africans, but instead by Indonesians from over six and a half thousand kilometers away across the Indian Ocean. Not only were Indonesians from the same Austronesian culture that had peopled the Pacific, and hence highly proficient sailors, but currents and winds were in the right direction for the Indonesians and the wrong direction for Africans.

The Atlantic was such a barrier between the Old and New Worlds that all reliable evidence indicates that it was northeastern Siberians who peopled America across the then dry-land Bering Strait. The next immigrant was Leif Erikson, son of Erik the Red. Erikson reached Newfoundland nearly five hundred years before Columbus landed in the New World. Even so, Erikson's arrival was *thousands* of years after the first overlanders, so effective is the sea as a barrier to us terrestrial humans.

Indeed, waters far narrower than the Atlantic can be a barrier to the spread of humans. On a clear day, one can see across the English Channel. The French are not about to call it the English Channel; they call it La Manche, or "the sleeve." The Channel is just thirty-four kilometers wide at its narrowest point. Nevertheless, the English and French have stayed on their own sides of it for most of their pre-history and history. Correspondingly, they differ in a myriad of cultural and social ways.

They differ genetically too, even now, despite all the means of movement that now exist across Europe. They differ so clearly that a map of the distribution of gene types in Europe looks like a map of European countries. With different gene types in different colors, and no country borders drawn, there is Portugal to the left of Spain, France and then

Britain and Scandinavia north of there, Romania to the northeast, Cyprus in the southeast, and so on.

Geographic barriers explain this sort of separation. Nineteen of thirty-three boundaries between genetically distinct populations in Europe, such as the French and English, are water. This finding is from a study by Guido Barbujani and Robert Sokal. Other water barriers in addition to the English Channel are the Irish Sea, the North Sea, and parts of the Mediterranean. Even the fifteen-kilometer-wide Pentland Firth between north Scotland and the Orkney Isles is such an efficient barrier that the populations either side of it, politically Scots all of them, are genetically distinct. Celts are in Scotland, and Norse—i.e., Scandinavians—are in the Orkneys and Shetlands.

Although none of the European rivers is wide or fast enough to act as a barrier to movement, as judged by the map of European genetic differences, Betty Meggers showed that in South America, the Amazon River is an obvious barrier to several language families.

Personally, I dislike the ocean. You can drown in it. But my wife and I love hiking and trekking in mountains—"tramping," as the New Zealanders say. Mountains, like seas, have for millennia been effective barriers to the passage of humans. With mountains, we are back to where we started this section, with my treks to Bhutan with my wife. Despite Bhutan's political, military, and economic ties to India, Bhutanese is not an Indian language. Its language is Tibetan in origin. Bhutan is a Himalayan country with its highest peaks at over seventy-five hundred meters. Neither India nor the imperialistic British ever conquered the country. Accounts of the difficulty of travel there indicate why.

Thus, in the Barbujani and Sokal study of Europe that I have just mentioned, four of the thirty-three boundaries between genetically distinct peoples are mountain ranges. The Alps are the best known of these, separating French, German, and Italian speakers. Additionally, the Cordilleras of northwest Spain separate the Basques from the rest of Spain.

Moving to the other mountainous areas of the world, I wrote in chapter 2 about how humans appeared to get stuck in the Arabian Peninsula for several thousand years after they left Africa sixty thousand years ago. Could the many mountain ranges around Arabia have prevented us leaving the

Arabian peninsula? East of there, the Zagros Mountains of Iran are covered in snow in the winter now, and until the early 1900s still had glaciers. Did the Taurus Mountains of Turkey or the Caucasus Mountains of Georgia prevent us moving north?

After humans got to Asia over forty-five thousand years ago, it might have taken them another fifteen to twenty thousand years to get into Tibet. The first human remains up in the Andes date from something over twelve thousand years ago at an altitude of two and a half thousand meters. They did not get there until two thousand years or so after humans first reached South America, and then it took another six hundred years to start living above three thousand meters.

Dan Janzen made the interesting suggestion that a mountain range of a given height could be more of a barrier to the movement of tropical species than to the movement of species at higher latitudes. Put in human terms, to get through the year in the lowland tropics, all I would need would be shorts and a T-shirt. So what happens when I start trekking up a mountain there? Very soon, I get cold. And with no winter clothes, I have to turn back. That is not the case in Scotland. Dressed for an Edinburgh winter, I am easily warm enough to scale in the summer Britain's highest mountain, Ben Nevis. I have not walked up Ben Nevis, but I have strolled up Ben Lawers, only a hundred and thirty meters lower than Ben Nevis. A mountain range as high as Ben Nevis would then be no barrier to my traversing Scotland.

Thus we find that tropical species tend to be more constrained by mountain ranges than are the species of higher latitudes. Consequently, unless the tropical species can move around the mountain, they are confined to smaller geographic ranges wherever there is any substantial topography.

The same might be true of humankind to a small extent. Elizabeth Cashdan investigated why the diversity of cultures varies across the globe, and especially why the tropics have a higher density of cultures than do non-tropical regions. In a detailed analysis, she found that in the tropics but not outside them, steepness of the terrain's slope correlated with diversity of cultures. The next chapter is all about the wonderful cultural diversity of the tropics by comparison to higher latitudes.

Oceans, rivers, mountains make travel between places difficult. Independently of whether an obvious geographic barrier exists, though, difficulty of travel in itself can be a barrier to movement. I will give three examples. The first comes from northern Otago, a province in the South Island of New Zealand. Between 1875 and 1914, three times as many marriages occurred between families living within thirteen kilometers of each other as between those living more than that distance apart. Go to Oxfordshire in southern Britain, eighteen thousand kilometers northwest and two hundred years earlier. Over the two hundred or so years from the later 1600s up until the 1850s, the distance between husbands' and wives' birthplaces recorded in marriage registries was roughly ten kilometers. Subsequently, the distance between the birthplaces increased to about twelve kilometers, with some approaching forty kilometers. Quality of bicycles improved over that time, and perhaps also their affordability. But it was probably the arrival of the railroad in the region in 1851 that was a main cause of the greater mobility. And as a last example, Graham Robb, in his marvelous *The Discovery of France*, reports that in one commune in the southwest of the country in the early eighteenth century, over ninety percent of women married a man from less than eight kilometers away.

I might add that in the mid-1900s, social class strongly affected how far apart were the birthplaces of Oxford-born spouses. Spouses of the highest and lowest classes were born approximately a hundred and forty kilometers apart, whereas middle-class spouses were born half that distance apart. Presumably the highest classes could afford to travel, and the lowest classes were forced to do so, looking for work.

If humans do not move far, cultures will not homogenize. In France, as Graham Robb reports, many dialects were almost mutually incomprehensible well into the eighteenth century. Travelers to Paris used dictionaries to understand the Parisians, and Parisian officials in the provinces needed translators. Even in the nineteenth century, a Parisian would have found three dialects of people in the nation of France completely unintelligible— Basque in the southwest, Breton in the northwest, and Oc in the south. Indeed, unless the Provençal people spoke slowly and clearly, the Parisian probably would not have understood them either. Here is the title of chapter 4 of Robb's *Discovery of France*: "O, Òc, Sí, Bai, Ya, Win, Oui, Oyi, Awè, Jo, Ja, Oua." That is "Yes" in a dozen of the major dialects of France, moving clockwise from Provence.

At the start of this chapter's section on geographic barriers, I mentioned Bhutan's lack of paved roads until recently as one reason for its isolation and hence cultural independence. More generally, I am here going to suggest an economic reason why tropical cultures might still have greater diversity than temperate cultures and live in smaller geographic ranges. It relates to the difficulty of travel in the tropics, even where no deserts, ice, seas, or mountains prevent movement.

Tropical countries are in general economically poorer than are temperate countries, for a myriad of environmental, social, and political reasons. Movement is therefore more difficult in the poor tropical countries than in rich temperate countries. Let me give a specific personal example. Consider the fact that I, a Brit, and my wife, a Californian, first met in Rwanda, in 1973. In other words, we met after several aeroplane flights, the cost of which summed to easily ten times what the average sub-Saharan African would earn in a year. The average sub-Saharan African could not possibly afford to travel to Britain or the USA. Poverty is a barrier to movement and therefore maintains regional differences in both the biology and culture of humans.

Even within Africa, poverty makes travel difficult. Some sub-Saharan African countries have less than one hundredth the number of kilometers of road per square kilometer of their land surface as do European countries, and less than one twentieth the number of vehicles per citizen. As a specific example, Kenya, where I was born, had in 2009 eleven kilometers of road per hundred square kilometers of land. That compares to the United States' sixty-seven kilometers, and Britain's one hundred and seventy-two kilometers.

With the same density of roads or vehicles, people in tropical countries would still probably travel less, given the more than twenty-fold difference in average GDP between Europe and sub-Saharan Africa, and over forty-fold difference between the USA and sub-Saharan Africa. Elizabeth Cashdan's analysis relating ethnic diversity in countries to what she called efficiency of transport supports this supposition. Countries with less efficient transport systems, whether road or river, were more diverse ethnically. Moreover, the smaller the boat, and the less far it could travel, the more diverse were the people. Quality of roads, though, did not correlate with ethnic diversity of countries. Cashdan did not offer any explanations for the contrast between water and road transport, but my experience of the African tropical road is one of amazement at how resourceful drivers are in persuading their vehicles to move along what sometimes are barely recognizable as roads.

This contrast between the tropical and temperate regions in ease of transport could be a largely modern one. Travel in medieval Europe was possibly as difficult for the poor as it is in much of Africa now. However, another biogeographic barrier could have been more prevalent in the tropics than outside for millennia.

That barrier is disease. Pathogens and parasites exemplify tropical biological diversity. More of them in terms of both numbers and variety exist in any one site in the tropics than they do at high latitudes, and the species change more from place to place in the tropics than they do at high latitudes. As a result, humans moving from one region to another would face more different pathogens and parasites in the tropics than they would in temperate countries. Could it be the case, therefore, that disease was more of a barrier to the movement of tropical peoples than it was to the movement of temperate peoples? Would Genghis Khan have conquered as much of the world had he been an African or South American warlord? Alternatively, do tropical parasites and pathogens vary so much from region to region because the species that carry them, including humans, find it more difficult to travel in the tropics? I return later in the chapter to this chicken-and-egg issue, and later in the book to the general topic of how diseases have affected our distribution.

Even with no physical barriers, xenophobia can keep us apart (Greek, "xeno": strange—i.e., fear of the strange). If we fear or despise people from other countries, people who do not look like us, then presumably we do not travel to those other countries, or even have anything to do with them within our own country. "Miscegenation. *n*. The interbreeding, cohabitation, sexual relations or marriage involving persons of different race": American Heritage Dictionary, 2000. Frowned upon for long in much of the USA, miscegenation was outlawed in some Southern states well into the 1900s.

"Just like John Kerry, he speaks French too." This was an advertisement by supporters of Newt Gingrich attacking Mitt Romney during the

Republican presidential campaign that ended with Barack Obama's victory in 2012. Some Republicans so despise non-Americans that an American who speaks another language is unfit to govern—even if that other language is that of the nation that not only has a history as a US ally, but gave—not sold—*gave* to the USA the Statue of Liberty.

Still on the topic of diversity and xenophobia, New Guinea has the highest diversity of languages for its size in the world. It has been described as a land in which every valley has its own culture. The steep topography of the mountainous regions of New Guinea, the high rainfall, and the dense forestation must play a part in keeping cultures separate. But the difficulty of movement amply described by all explorers who venture there cannot be all that separates the cultures, because parts of lowland New Guinea are hardly any less linguistically diverse than the mountains. Also, New Guinean peoples trade with one another across the valleys. We need another explanation in addition to topography.

This other explanation can be found in the accounts of early explorers in New Guinea. They have amply described how highly territorial many New Guinean peoples are. Wars between New Guinea tribes are a staple of anthropological material. A culture of active aggression toward others and the fear that it engenders can keep people apart.

More quantitatively, Malcolm Dow and colleagues found twenty-five years ago that across eight Solomon islands, a group of islands southeast of New Guinea, differences in tooth shape and fingerprints associated more strongly with differences in language than with distance between the islands—as if language, not the sea, kept the islanders apart.

Similarly, the study of thirty-three boundaries between genetically distinct populations of Europe that I have written about before could not detect any obvious geographic barrier at nine of the boundaries between the populations. Nevertheless, the languages either side of the boundaries differed. Travel in Italy south from Florence to Rome on the E35/A1. The land is gently rolling hills, no rivers or mountains in the way. And yet northern Italians differ genetically from southern Italians. What separates them? One answer could be that different dialects hindered intermingling.

Even if the English Channel did not separate the French from the British—and indeed it effectively does not, given how easy travel is between the countries now—if the British and French boyfriend and girlfriend cannot whisper sweet nothings to each other, they are not going to marry and have children. And still in this, the twenty-first century, few British or French are fluent in each other's language. And so they rarely marry—as they rarely married a millennium ago.

Different dialects, different languages, different mutually incomprehensible sounds separating populations? That is a concept, a phenomenon that any bird or insect watcher would be completely familiar with. Birds and insects, like us, are not interested in mating with an individual that makes sounds different from their own species' sounds. The result is the wonderful diversity of life around us.

Different dialects might separate northern from southern Italy, but so also might other aspects of their culture. Northern Italian cooking is richer than southern Italian cuisine. As Marcella Hazan describes it in her *Essentials of Classic Italian Cooking*, northern cuisine is exuberant, southern is austere. Could different diets keep us apart? Why not? Samuel Johnson, the famous English writer and lexicographer, defined oats as "a grain, which in England is generally given to horses, but in Scotland supports the people." What Englishwoman would want to marry an oats-eating Scotsman? The French apparently feed most of their maize to pigs. What Frenchwoman would want to marry a maize-eating Englishman? And what English woman would want to marry a frog-eating Frenchman?

We find the same cultural separation in the Amazon. And, as in Europe, we consequently get biological separation. The Xavánte and Kayapó of the Amazon region of central and northern Brazil probably separated from each other less than two thousand years ago. No obvious environmental barrier separates them. Rather, a variety of cultural practices distinguish them. For instance, among the Xavánte, men can marry a group of sisters, whereas the Kayapó are generally monogamous. Correspondingly, as Tábita Hünemeier and a team of ten others report, the two groups are genetically distinct, and can also be distinguished to some extent by the shape of their face and head. In just two thousand years, they have anatomically and genetically drifted apart, separated by culture, not geography. Similarly, Partha Majumder suggests that social barriers explain some of India's regional diversity in cultures.

Humans are extraordinarily good at picking trivial differences to separate "us" from "them." Take a group of otherwise similar students who do not know one another. Give half of them green pencils and the other half red pencils. Now tell them to sort themselves into two groups. And yes, greens associate with greens, and reds with reds. Take a group of otherwise similar beginner skiers who do not know one another. Give half of them green ribbons and the other half blue ribbons to put around their necks so that the instructor can tell them apart on the slopes, as in an experiment reported by Jacob Rabbie. Now tell the skiers to judge each person's skiing skills. And yes, greens judge greens as better skiers, and blues blues. Somehow, completely subconsciously, the students managed to identify the one clear difference among them, and xenophobically act on it. Frogs' legs taste like chicken, yet chicken-eating Brits still despise the French for eating them. Of course, it is not just Europeans who do this sort of thing. Katie Milton saw a greater difference in diet between Amazonian Indians than she could explain by the difference in their forest environment.

Jacob Rabbie and others stress that just because we distinguish ourselves from others and favor those like us over others not like us, we are not necessarily aggressive to the ones not like us. Yes, the green-ribbon skiers thought that other green-ribbons skied better, but they did not start to beat up the blue-ribbons. More quantitatively, Elizabeth Cashdan showed that while people felt more loyal to members of their own group and more hostile to members of the out-group during war with outsiders, in general, feelings of loyalty to in-group members did not correlate well with feelings of hostility to the out-groups.

Nevertheless, once a self-other dichotomy, a like-me/not-like-me difference, is perceived, xenophobia is easily tripped. I have to admit, I have seen it in myself. I am not X. That individual X is offensive. All X are offensive. All X should be banned. More mildly and more pleasantly, but still a perfect incidence of a self-other dichotomy, I could not have been more surprised when, on my last visit to London, traveling on a crowded Underground train, no seats available, a heavily tattooed youth smooching with his girlfriend got up and offered me his seat. So much for stereotypes. However, much as I appreciated the gesture, I did not enjoy the fact that I apparently looked old enough to need a seat.

Why should people be so ready to be xenophobic? In the case of animals and plants that avoid mating with individuals of other species, the answer is that hybrids between species usually do poorly. Indeed, many are sterile. The mule, a hybrid between horse and donkey, is perhaps the best-known example. However, almost by definition, we rarely talk about hybridization within species. That is because it produces no problems. So closely related are all humans that biological problems of reproducing with people from another region cannot explain our xenophobia.

One hypothesis for our dislike of strangers brings us back to parasites. Corey Fincher and Randy Thornhill have suggested that a beneficial consequence of xenophobia (and its opposite, preference for people like oneself, as apparent in family ties and intensity of religious devotion) is that it keeps us away from people in different environments and with different diets. It therefore keeps us away from pathogens and parasites to which we are not accustomed, and from which we would consequently suffer.

Part of the evidence that Fincher and Thornhill use to support their contention is that across the world, as parasite load increases, so does religiosity and preference for one's own kind. However, Thomas Currie and Ruth Mace showed that the global relation with religiosity is entirely driven by the large number of (developed) European countries. They do not have high parasite loads, and are not very religious. Remove them, and no global relationship exists. Indeed, among the five regions of the world analyzed by Fincher and Thornhill, within only one did load correlate with religiosity.

Also, obvious exceptions exist. Ireland has western Europe's highest score for religiosity, but only a median parasite load; and France has the third lowest score for religiosity, but the fourth highest for parasite load. Take Europe as a whole, not just western Europe, and no relation between religiosity and parasite load exists. Do the single-country exceptions, and the lack of the claimed relationship within a majority of the world's regions, negate Fincher and Thornhill's argument? The answer is yes, if they cannot be explained away.

As importantly, Fincher and Thornhill argued only one of the possible reasons for a correlation between religiosity (or xenophobia) and parasite load. They argued a direct benefit—xenophobia keeps us away from organisms to which we are not adapted. But what if staying with our own culture for benefits *other* than avoiding parasites resulted in xenophobia? We would still see a correlation between xenophobia and parasite load, or contrasts in types of parasites and pathogens, but the two would be unconnected.

Almost all of us tend to stay away from strange things as well as strangers. If my dislike of snails as food kept me away from France, and if French parasites were different from Californian ones, then it would look as if strange parasites produced in me a xenophobic reaction to France. But no, it would be the snails that made me not want to go to France. If I do not go to France, I do not carry my parasites to France, nor bring back French parasites to California. In other words, it is the lack of movement between cultures that causes the parasites and pathogens to differ from culture to culture, not the parasites and pathogens that prevent the movement.

Perhaps more realistically, what if we stayed where we were born because we hunted or gathered more efficiently in an area that we knew? As an example, on the same trip as described elsewhere, when my wife and I were working with hunters in Nigeria, the hunters got completely turned around when we left the small area that the hunters knew. We then all had to rely on the compass that I carried to find our way back to camp. If the hunters do not move outside their known area, neither do their parasites and pathogens.

Let me suggest yet another way in which xenophobia could be apparently related to avoidance of parasites, and yet in fact have a completely different cause. I have already mentioned, indeed given facts (number of cars and kilometers of road per person) to show, that movement is far more difficult in Africa than in Europe. Independently of religion or family ties, or parasites, people in Africa cannot move. People in Europe can. The correlation between staying put in Africa, and so avoiding exposure to the high diversity of parasites there, has nothing to do with the parasites, and everything to do with the level of economic development of the two regions.

It seems to me that if humans actively avoid associating with people from another culture—and they indeed seem to—then they do so in large part for the same reason that species, including plants, avoid hybridization—the offspring can suffer problems. In the case of humans, the problems are social, not biological. The difficulties could be lack of communication between the parents, differences of opinion over diet, over holy days, and over any other of the many petty differences that separate cultures.

∽

Xenophobia has many unfortunate consequences, of course. But could a benefit be that dislike of other cultures can help maintain cultural diversity as effectively as can desert, ice, seas, mountains, lack of roads, and maybe even pathogens and parasites? Tolerance would be a more pleasant attitude than xenophobia, but might mere tolerance lead to adoption of the others' customs, and hence to homogenization?

My wife and I have seen the effect of removal of barriers in Bhutan. Bhutan survived as an independent culture in part because of its geographical isolation behind mountain and forest. To the north, the Himalayas themselves protect it. To the south, difficult terrain and thick forest protect it. And until recently, Bhutan has not gone out of its way to encourage visitors, or even movement within the country. The country was effectively closed to foreigners until the 1970s. Not until the 1960s did it have a paved road. Even in the 1990s, still most Bhutanese in the cities wore national dress, and officials still do so. Even the might of British colonialism in the mid-1800s found subjugation of the country difficult, although the Bhutanese peoples' most advanced weapon was the matchlock gun, last used in Europe in the 1700s.

What, though, will be the effect of television's invasion? When my wife and I first arrived in Bhutan in 1996, hotel televisions had only four channels, none of them American. Now, Bhutan television carries scores of channels, many of them American.

CHAPTER 7

IS MAN MERELY A MONKEY?

*Human cultural diversity varies across the globe in the
same way and for the same reasons as biological diversity*

People in different parts of the world differ genetically, anatomically, and physiologically. Good biological reasons exist for the variations. They result from adaptation to different environments, origination from different founding populations, random genetic change within populations, and barriers to movement between populations. That is a summary of the last two chapters. So far, so good.

But humans are defined by culture. Culture separates us from the other animals. Yes, chimpanzees in different parts of Africa use different tools, or use the same tool in different ways, or to collect different foods. But that minimal difference is such a far cry from what we mean by culture in humans that many consider that applying the word "culture" to the regional differences in behavior that we see among chimpanzees demeans the sense of the word as it applies to humans. So to repeat, culture separates us from other animals.

Yet, as I will show in this chapter, cultures show the same geographical patterns as do species. Not only that, but the same biological reasoning that can explain the geographic distribution of species can also explain the geographic distribution of cultures. In other words, even when we are talking about culture, a phenomenon that supposedly separates us from the animals, humans biogeographically do the same as animals. Biogeographically, man—to phrase it alliteratively—is merely a monkey.

Before I go further, I need to say that anthropologists discuss interminably what they mean by the word "culture." That is no different from biologists who have written hundreds of articles and chapters over the decades on what a "species" is—how to define it, how to distinguish species, how to decide when whatever it is they are looking at is a species, a subspecies, or a race, or indeed whether to drop the term altogether. They are still writing.

Nevertheless, biologists can usually ignore all the argument and with highly sophisticated analyses produce deep insights into the process of evolution and the biology of the distribution of species. In other words, despite the details of definition, we can agree that the general concept usually works.

I used to study gorillas in the forests of Africa. When I started, most people agreed that one species of gorilla lived in Africa, separated into three subspecies. Now, as far as I can gather, most primatologists like to write as if two species exist.

I disagree. I have seen both so-called species, heard both, and smelled both—the gorilla male has a powerful and characteristically smelly sweat. And I say that they are one species. But whether we write or think or analyze data as if two or one species of gorillas exists does not matter for almost any biological purpose. In fact, one of biology's most famous evolutionists, George Gaylord Simpson, said that we should use whatever definition of "species" is most suitable for the study we are undertaking.

As far as I can tell, the situation is much the same with cultures and societies and languages. Yes, important debate occurs about exactly what a "culture" is, what a "society" is, where one stops and another starts, what a dialect is, what a language is, what a language family is. For some, language can define culture. For others, mode of life defines culture. Dress, tools, attitudes, nature of marriages, or perceived origins can define a culture or separate cultures. Certainly, if different studies of supposedly the same culture produced contrasting answers, a first question would be whether different definitions of "culture," or different perceptions of what they meant by it, affected the answer.

On the whole, though, the sensible starting attitude is G. G. Simpson's, it seems to me. And on the whole, that is what authors do. Following them, and particularly Daniel Nettle's *Linguistic Diversity*, that is what I do. As Nettle pointed out, the contrasts between regions in, for example, densities of languages is so great that it beggars belief to think that mere difference in definition could produce those differences. China alone, he wrote, would have two hundred thousand languages if it really had the same density of languages as does New Guinea, the region of the world with the highest density of languages. Consequently, I have no preference for any one definition of culture. I use, without questioning their validity, whatever classification of "culture" the original authors used.

Enough on definitions. Back to biogeography. A major biogeographic pattern that most of us with the slightest interest in natural history are aware of is the amazing diversity of life in the tropics compared to higher latitudes. The biodiversity of the tropics is a byword.

Take Ecuador, where my eldest sister has lived for the last thirty-plus years. Ecuador is the same size as Britain. Britain has fifteen hundred species of plants with probably no more than a couple or so still to be discovered, given the density of botanists in Britain. Ecuador has more than twenty thousand species of plants, with quite possibly hundreds more still to be identified if oil companies and others leave any of its forests for future generations. Britain has less than ten native carnivores. Ecuador has nearly thirty.

It is not just Ecuador that is rich in species, and Britain that is poor. The comparisons work within continents too. My wife and I recently spent two weeks in Ecuador and two weeks in Argentina. I am a bird watcher, so of course brought with me *Birds of Ecuador* and *Birds of Argentina*. Ecuador lies on the equator, as its French name, Equateur, indicates. Argentina lies well south of the equator. Only its farthest northwest corner is even in the tropics. Correspondingly, *Birds of Ecuador* is more than twice as thick as *Birds of Argentina*, despite the fact that Ecuador is less than an eighth the size of Argentina. These sorts of huge differences in the biodiversity of the tropics compared to temperate countries are repeated all over the world.

If many of us know about the extraordinary diversity of species in the tropics, few of us seem to have heard of the extraordinary diversity of cultures in the tropics. A linguist could hear twenty-three indigenous languages if he or she traveled the length and breadth of Ecuador. The United Kingdom has half as many, just twelve. Cornish, Gaelic, Manx, Romani, Scots, and Welsh are a sample.

On a continental scale, the people of tropical South America speak four hundred and eighty-seven indigenous languages. In their neighbor to the north, the USA, linguists have counted just one hundred and seventy-six indigenous languages. Tropical South America has over two and a half times as many languages as does the USA in roughly the same area. We get the same contrast between Africa and its northern neighbor, Europe. Africa covers three times the area that Europe does, but has more than nine times the number of indigenous languages, twenty-one hundred in Africa as against only two hundred and thirty in Europe.

In sum, the extraordinary biodiversity of the tropics is matched by an extraordinary cultural diversity.

Go to any one spot in the tropics, say Yasuni National Park of, once again, Ecuador. Stand in the middle of those forests and look and listen. You will see and hear immediately around you far more species of plant and animal than if you went into a North American or British forest. That is one way that the tropics are more diverse than higher latitudes. At any one spot, you will see and hear more species.

Then set off on a path through the forest for a trip of a hundred kilometers. And again, stop and stare and listen. Whatever direction you take, you will find that you are seeing and hearing new species. But do the same in, say, the Great Smoky Mountains National Park in North Carolina and Tennessee, and even though the Park harbors three main types of forest, you will not notice much change, especially if you still find yourself in the Park's spruce forest.

At an even larger scale, travel fifteen hundred kilometers south along the east slope of the Andes from Yasuni in Ecuador to Manu National Park in Peru, and you will see a whole new complement of species. Travel the same distance anywhere in western Europe, and the wilderness will look and sound much the same at either end of the trip.

For cultures, New Guinea has long been famous for its density of languages. It has more languages for its size than any other area in the world. The phrase "for its size" is crucial here. If we want to compare the diversity of regions, whether biological or cultural, we must compare regions of similar size. That is why I compared Britain's linguistic diversity to the diversity in a similarly sized country a few paragraphs back, and why when comparing linguistic diversity between the Americas, and between Africa and Europe, I stated how large they were.

The reason to take account of area is not simply that larger areas are likely to contain more species or cultures than do smaller areas, other things being equal. No, it also turns out that density of species, cultures, languages—in other words, the number per a given area—in fact changes with the amount of area available.

We might expect a one-to-one ratio of number of species to area. Reduce area by half, and number of species drops by half. But no. A rough rule of thumb for species is that a loss of ninety percent of the area results in a loss of only half the number of species. So, if we had a hundred species in one hundred square kilometers of forest, and the forest was reduced to just ten square kilometers, just ten percent of its former area, we would find fifty species remaining—in other words, fifty percent of them. The forest is reduced to ten percent of its previous area, but the number of species is only reduced to fifty percent of the former count.

Therefore, if we want to compare regions for density of languages or species, we have to ask whether they have more or less than the average for their size. Now indeed, New Guinea takes first prize. It has over ten times the density of languages for its size as do similarly forested tropical countries of approximately the same area, such as the Central African Republic, Colombia, or Thailand. In just Papua New Guinea, a political entity that's only about half of the entire geographic island of New Guinea, we have thirteen percent of the world's languages spoken by 0.1 percent of the world's population in 0.4 percent of the world's land area.

One reason for these differences between tropical and temperate regions is that in the tropics, species and cultures cover a smaller area than they do at higher latitudes. Take African primates as an example of what non-human species do. African primates' geographic ranges, measured as the north-south distance of a genus's range, are on average fourteen hundred kilometers long at the equator, but over five thousand kilometers long ten degrees away from the equator.

Human cultures show the same pattern. The average tropical language in Gabon, say, is spoken over approximately six and a half thousand square kilometers, but the geographic range of the average language above sixty degrees of latitude, for example in Sweden, encompasses thirty thousand square kilometers. We see the effect across the globe, and in my analyses in the two continents with a sample of over ten languages, Africa and Asia. However, in independent analyses, Ruth Mace and colleagues have shown that the same occurs in North America and Australia too. They got the same results as I did for Africa, though they analyzed only Bantu languages.

I quite often in this chapter present information for hunter-gatherer peoples in addition to information for other cultures or for languages. Part of the reason is that the relation between the nature of the environment and the hunter-gatherers' distribution should be closer than with agricultural societies, especially agricultural societies rich enough to trade their products. Also, the mobility of hunter-gatherers provides another aspect of biogeography to investigate. And people leading a hunter-gatherer lifestyle, a lifestyle so very different from that of the readers of this book, is surely intrinsically interesting.

The average tropical hunter-gatherer culture, then, the Mbuti pygmy peoples of Africa, for example, covers on the order of thirty-five hundred square kilometers. Go to above the Arctic circle, and the average hunter-gatherer culture there, the MacKenzie Inuit of the Northwest Territory, say, occupy an area twenty times as large, at seventy thousand square kilometers. We see the effect globally, and within North America, Asia, and Australia. African and South American hunter-gatherers do not show the effect. I do not know why they do not in Africa. As for South America, we know from Adriana Ruggiero's work that the Andes, running north-south, exert a strong longitudinal effect on the geography of mammals. Similarly for humans, that longitudinal influence hides in my analyses any associations of cultural diversity with latitude.

In summary, the facts are that we find higher densities of not only species but also cultures in the tropics compared to higher latitudes. Part of the reason for the latitudinal contrast is that species and cultures exist in smaller geographic ranges in the tropics than they do at higher latitudes. Plants, ungulates, monkeys, and human cultures all show the same biogeographic pattern. How is it possible that cultures, languages—products of the human mind—do the same biogeographically as species?

Before I move on to explanations, let me give a brief description of an interesting distinction between humans and other animals.

Similar as their pattern of distribution is in some ways, cultures and species are different in one biogeographic respect: the tiniest patch of tropical forest is full of animal species. Watch a column of army ants for a couple of hours near any one of the research stations in Yasuni National Park in the Ecuadorean Amazon rain forest, and you are likely to see more than ten species of birds following the column and catching the insects disturbed by the ants—in the same way that cattle egrets follow cattle to get the disturbed insects. Sit still for a day in that forest near the research station, and you might see five species of ungulate and, if you are lucky, ten species of monkey. But in that same patch of forest and for many surrounding square kilometers, you would only ever see one native human culture, the Huarani peoples.

The fact is that humans are notoriously territorial. We rarely find people of different cultures occupying the same bit of land. Read accounts of early travelers in the Americas and Africa, and they are filled with stories of battles between the peoples whose lands the explorers are moving through. Cabeza de Vaca got stranded on the west coast of Florida in the early 1500s. He and an ever-dwindling number of companions finally made it to Mexico. Here is Andrés Reséndez in his *A Land So Strange*, recounting from Cabeza de Vaca's story of his journey across the southern USA: ". . . the Mariames engaged in female infanticide to deny potential wives to the surrounding groups who were their sworn enemies. . . . The Susuolas were at war with other groups in the area . . . these Indians made war on one another, constantly engaging in elaborate ambushes. . . . Since Indian men could not venture west for fear of being killed." Other areas of the world have been, and are, no less belligerent toward their neighbors.

The same pattern occurs, then, in humans and other organisms. Whether the distinction between them in the overlap of their geographic ranges is a biologically significant difference is another matter. With humans, we are talking degree of overlap within a species. With other species, the aforementioned antbirds, for example, we are talking overlap between species, even between genera.

If we examine single species of animals and ask about overlap of divisions within the species—overlap of the subspecies, in other words—we find the same as for humans. That is, little overlap in ranges. An analysis that I did of rivers as barriers to the geographic extent of primates in west and central Africa shows the effect. Three quarters of subspecies had a river as the boundary of their geographic range, meaning that, in effect, three quarters had non-overlapping ranges. That compares with forty percent of species and ten percent of genera.

The jargon term for the observed pattern of decreasing number of taxa from the tropics to higher latitudes is "the latitudinal gradient in biodiversity." Biologists have known about the pattern for over two centuries, and still research it and write about it. One of the earliest writers was Johann Georg Adam Forster, who traveled on the second of Captain Cook's voyages around the world in the late 1700s. The "latitudinal gradient in biodiversity" is rather a mouthful, so in honor of J. G. A. Forster I abbreviate it to "the Forster effect."

For some reason, anthropologists became interested in the scientific analysis of tropical diversity far later than did biologists, even though Johann Forster the Younger was renowned in his time for his descriptions of the cultures that Cook's expedition encountered. Indeed, so renowned was he that he was made a member of Britain's Royal Society in his early twenties. We should probably not read too much into the fact that he died of a stroke in his late thirties. Be that as it may, anthropology's late start means that anthropologists have written far less on cultural diversity and the Forster effect than biologists have written on biodiversity and the Forster effect.

Biogeographers have therefore largely ignored what we know about humans. The bible of biogeography, Mark Lomolino and co-authors' simply titled *Biogeography*, contains a dense three-page table listing details of thirty-one explanations for the Forster effect among animals and plants. Humans are not mentioned once.

Similarly, anthropologists have largely ignored what biogeographers know about the distribution of human cultures. As of the end of 2014, I have not found a single modern textbook on the topic of biological anthropology or physical anthropology that even considers the geographic distribution of cultures, let alone relates that distribution to our knowledge and understanding of the biology behind the geographic distribution of non-human species.

Here, then, I will give a quick summary of the variety of ideas out there to explain the Forster effect among non-human species. Several of these cannot apply to humans. Others can. As is so often the case in biology, the answer probably lies not just in one explanation, but in a combination of several.

A non-biological explanation for the Forster effect will not work for humans. I will call it the pencil-box explanation. Imagine that the north-south extent of a geographic range is represented by a pencil. Imagine a lot of colored pencils of different lengths in a pencil box. Where would we find the most overlap of pencils, in other words the greatest diversity of colors? The answer is at the center of the box, the equivalent of the tropics. Why? Because the long pencils in the box have little choice but to overlap at the center.

The idea does not work for humans. For a start, we get a high cultural diversity in the tropics as a result of a high density of cultures each with a *small* geographic extent, not a large one. The cultures of large geographic extent lie outside the tropics. And therein lies the other reason why the hypothesis and its analogy does not apply to humans. The latitudinal extent of human cultures is far too small for this pencil-box hypothesis to operate. It would be like applying the argument to pencils of normal lengths in a box a meter long.

Some of the biogeographical explanations for the general Forster effect beg the question to some extent. For instance, a diversity of pathogens (disease organisms), parasites, and competitors prevents any one species from dominating. Yes, but then why do the tropics have a high diversity of pathogens and parasites? Could it be that a diversity of the species that the disease organisms infect led to a diversity of those organisms themselves? Nevertheless, I go with this hypothesis as part of the explanation for why within the one species of humans we get a high diversity of cultures in small geographic ranges in the tropics.

An alternative to the idea that the tropical diversity of pathogens and parasites maintains a diversity of hosts is that a diversity of hosts allows a diversity of pathogens and parasites. If there were not both monkeys and gorillas in the Virunga Volcanoes, there would not be separate disease organisms for monkeys and gorillas there. So we are back to the question of why the tropics have a high diversity of hosts. I talked about this sort of chicken-and-egg conundrum in the last chapter when discussing the idea that xenophobia kept people away from pathogens and parasites to which they were not accustomed.

Another explanation for the high density of tropical species is that the tropics have a more patchy environment, or more different sorts of environment, or in other words more ecological niches. That environmental diversity would then allow a diversity of users of the environment. Maybe; but if so, why is the tropical environment more diverse? In any case, temperate and Arctic cultures have ranges smaller than the extent of any patch of the habitat. The boreal zone extends all the way around the world, but no culture living there ranges across the top of both the Americas and Eurasia.

The smallness of geographic ranges of tropical species and cultures allows tighter packing of both. Yes, but the geographic ranges of other species overlap, even if those of human cultures do not. Given that many species with large geographic ranges can overlap at any one spot, the size of species' ranges should not influence the number of species in any one area. Also, we are left with the question of why, for a start, geographic ranges are smaller in the tropics. I give an answer to this question in the next chapter.

One persuasive argument for the Forster effect among animal and plant species is ice caps and glaciers, and their waxing and waning over the earth's history as ice ages have come and gone. Tropical species have had more time than high-latitude species to evolve their extraordinarily varied forms because the tropics have not periodically been scraped bare by massive ice sheets.

The problem with applying that argument to humans is that cultures can evolve extremely fast. Take the example of the number of American Indian languages that I discussed in chapter 4 in another context. As I described there, humans have been in South America for probably no more than fifteen thousand years. Yet South America has an average density of languages only one third less than the density in Africa or Asia, where humans have lived for tens of thousands of years longer. Not only that, but humans have been in at least temperate regions for fifteen thousand years, and in Arctic climates for that long, and yet high-latitude peoples have not become as culturally specialized as have tropical peoples.

The biogeographic explanation for the Forster effect that I prefer begins with the productivity of the soil. Biologists evaluate a region's productivity in various ways. In essence, they weigh the amount of plant and/or animal mass that grows or lives in a given area during a given period of time. A variety of factors affect productivity. Energy (for plants often measured as temperature), water, and soil quality are the main ones. In natural systems, temperature and water are more important, because where soils are poor, many plants have evolved to cope with the poverty, as long as the climate is relatively warm and wet.

I need to add another reason why temperature and water are often found to be the most important factors affecting productivity. Thermometers and rain gauges are cheap and easy to get information from: walk up to them and note down what they show. The process is simple and easy. Soil quality is another matter altogether. Samples need to be taken to the laboratory, and a variety of tests done. Consequently, we lack information on soil quality across large regions, and so it does not appear in explanations of regional differences in productivity.

Accepting that drawback of measures of productivity, let us go with the productivity argument. Except at the height of summer, land outside the tropics is not as productive as land in the tropics. The reason is easy to see. The tropics are relatively warm year-round, and have more or less intense sunlight year-round. In regions that also have water year-round, plants can grow year-round.

Outside the tropics, the short days and freezing temperatures of winter put plants on hold for several months of the year. At sixty degrees latitude, up by Anchorage in Alaska, or the Hudson Strait in Canada, or Oslo in Scandinavia, or the middle of Russia, the land is only one fifth as productive over the year as it is at ten degrees from the equator.

Especially in the lean months, then, many animals outside the tropics, again including humans, need to range farther each day, week, or month than they do in the tropics if they are to find sufficient resources to survive. The density of individuals in temperate and Arctic populations is therefore lower, which means that for a species or culture to persist at a viable size in the long term, the high-latitude species and cultures need a larger geographic range than do the low-latitude ones. Needing a larger geographic range, species and cultures cannot be packed as tightly as in the productive tropics, and hence diversity of species and cultures is lower at high latitudes than in the tropics.

Conversely, if the high-latitude species or culture cannot for some reason expand its range, then the size of its population might not be great enough to survive the proverbial slings and arrows of outrageous fortune. Look at any map of global warming, and you will see that climatologists predict greater increases of temperature at the higher latitudes than lower latitudes. Indeed, higher latitudes tend to have less stable climates than do the lower latitudes. Yes, African forests declined to remnants during the last ice age, but they still survived. They did not disappear under ice caps more than a kilometer deep.

As a consequence of wilder swings of climate, a high-latitude species or culture of the same number of individuals as a persistent low-latitude one might nevertheless still not be safe. The prediction is clear. High-latitude species and cultures should not only have larger ranges than low-latitude ones, but they should have ranges large enough to harbor larger populations.

I do not know whether that prediction is true of animals, but I do know that it is of humans. High-latitude hunter-gatherer cultures live in larger populations than do low-latitude ones, even though the former live at lower densities. Compare Arctic, temperate, and tropical hunter-gatherer cultures, and their average population sizes are, respectively, forty-three thousand, one hundred and forty-eight thousand, and two hundred and twenty-four thousand people.

In general, then, explanations for why species are more diverse in the tropics than at higher latitudes seem to work also for cultures. We should not be surprised at that finding. We are all carbon-based life forms that need energy (ultimately from the sun) to survive and reproduce. With the same biogeographical explanations for why both cultural diversity and biological diversity vary in the same way with latitude, perhaps we should not be surprised that of the top twenty-five most bio-diverse countries, two thirds are also the most linguistically diverse.

If the argument is correct, then measures of productivity itself, as opposed to degrees of latitude, should correlate with the density of individuals in a culture's population. Indeed, latitude must merely be an index of the actual environmental factor that is affecting individuals. Productivity should correlate also with distances moved by individuals, and hence with the geographic range size of cultures. And productivity should correlate also with the diversity of cultures.

Such is indeed what we often find. As productivity goes down, the population densities of cultures drop, distances moved either daily or in camp movements go up, range sizes of cultures increase, and the diversity of cultures within the region decreases.

The correlations are never perfect. Sometimes aspects of the environment other than merely productivity correlate best with diversity of cultures. Length of growing season is one such factor. But in some way, the measures usually indicate how much can be grown. The relationships are especially strong among hunter-gatherers, presumably because fewer extraneous factors, such as intensification of agriculture and trade, affect how many people can live in an area.

To give some real figures, I will compare Alaska in northern North America with Colombia in equatorial South America. Annual productivity in Alaska is less than one twentieth of what it is in Colombia. An average Alaskan hunter-gatherer culture might live at a density of less than three people per one hundred square kilometers, move camps over a total distance of roughly four hundred kilometers in a year, and live in a geographic range of perhaps eighty thousand square kilometers. Go down to Colombia, and the averages are near eighteen people per square kilometer, who move camps over a total distance of approximately two hundred and eighty kilometers in a year while living in a geographic range of three thousand square kilometers. Smaller ranges in Colombia then correlates with greater cultural diversity there than in Alaska.

If winter prevents plant growth at high latitudes, the dry season can prevent plant growth in the tropics. The Sahara, dry all year round, is millions of square kilometers of almost no green at all, and hence has a low diversity of cultures. Daniel Nettle pointed out an exception within the region, an exception that supports the suggestion, that year-round productivity allows a self-sustaining population to live in a small area, and hence contributes to a high diversity of cultures in a region.

He showed that across West Africa, the longer the growing season, the smaller the geographic range of the cultures. A nine-month growing season, and the average culture had a geographic range of two thousand square kilometers. Three-month season, and the range is thirty times larger.

A couple of cultures, though, had unusually small geographic ranges for the length of the growing season. In one case, Nettle could tie the exception to the fact that though on the edge of the Sahara, in desert, the culture lived next to Lake Chad. In other words, in the midst of near-desert, with an otherwise extremely short growing season if one was an agriculturalist, the region could support a sustainable density of people in a small area because they had access to a year-round supply of food, and hence, effectively, a year-long growing season.

Twenty-five years ago, my wife and I spent three months in Nigeria. Along with Ibrahim Inahoro of the Nigerian Conservation Society, our job was to count gorillas in the west south-east of the country. Before publishing our conclusion, we needed to check a rumor of gorillas farther north than where we had worked. So we journeyed up to the center of the eastern side of Nigeria, where forest gave way to woodland. I had never before seen anyone trying to grow crops in such parched land. The sight of the small patches of cassava, each plant in its mound of dry earth, is still vivid in my mind.

The size of a geographic range covered by the average hunter-gatherer society is in general larger than that covered by the average language. The reason is easy to see. By and large, hunter-gatherers do not practice agriculture, by definition. Without the dense supply of food that is the advantage of agriculture, the hunter-gatherer society needs more land to survive.

I have had the pleasure of a field trip to the White Mountains of the California-Nevada border led by Robert Bettinger, an archeologist in my department. The whole trip was fascinating, but I will never forget Bob Bettinger pointing to the flimsy little wild-grass seedheads and telling us that the Native American people used to harvest them. A single maize plant in the corner of a kitchen garden would produce more calories, I reckoned, than all the grass heads in my whole field of vision on that slope in the White Mountains.

James Brown might have published more on the topic of biogeography than anyone else on the planet up to now. He likes an additional suggestion for the paucity of species outside the tropics. It is that metabolic rates are slower where it is colder. Consequently, everything tied to metabolism is slower. And in living things, that is everything, including speciation rates. I can see that this would be true across species. Within the human species, however, we already know from chapter 5 that Arctic peoples have higher metabolic rates than do tropical peoples.

I need to point out an interesting pattern in some of the biogeographical relationships that I have just been writing about. James Brown highlighted it nearly twenty years ago. At low latitudes we find all sizes of geographic range or, for example, size of body of primates. But at high latitudes, we find only large range sizes of animal species or humans, or only large body sizes.

In effect, anything is permitted in the tropics. But at high latitudes, climate restricts what is possible. A culture or species can find enough resources to persist in any size of range in the lush environment of the tropics; but at high latitudes, where resources are not plentiful year-round, species and cultures need a large geographic range to survive. Although the authors do not say anything about it, the pattern is very evident in the analysis of the size of the geographic ranges of the Bantu languages in Africa that I mentioned earlier. Their graph shows a nearly forty-fold spread of the sizes of geographic ranges near the equator, but only a five-fold spread at twenty-five degrees, or near the Tropic of Capricorn.

If, overall, the tropics are more diverse biologically and culturally than higher latitudes, in part because of the denser packing of species and cultures, we also find hotspots of extra-high diversity within the tropics. For monkeys and apes, the Cameroon highlands and the mountains bordering the eastern Congo Basin are two such areas in Africa. In South America, the main hotspot is in the headwaters of the Amazon. Borneo is where to go in Asia to find the densest collection of species. Go to the east coast of Madagascar to see the most species in the shortest time.

My wife and I visited Patricia Wright's study site at Ranomafana National Park on Madagascar's east coast some years ago. On our first day in the Park, we saw six of its seven species of day-time lemur. We saw the seventh the next day. I do not know where in the USA one could see in two days seven species of any one sort of large day-time mammal, let alone seven of one sort. When we go to our neighboring national park on the coast of California, we often see, as we hike, the two species of deer there, but in twenty-five years of going to the park, we can remember only one day during which we saw three of the region's four species of carnivore.

Accepting that an even distribution of anything is extremely unlikely, we will see hotspots in the distribution of almost everything, simply because of random effects. Even I, with my minimal knowledge of computer programming, can get my computer to put spots, literal spots, any color I like, on a virtual map. One does so by entering coordinates for where one wants the spots to appear. Take a list of random numbers as the coordinates, and you will not get an even distribution of spots of color. You will get a patchy image, an image of concentrations of color amidst bare areas, of hotspots and coldspots of color.

So some cultural or linguistic hotspots might have no other explanation than randomness. Randomness, though, is an unlikely explanation for the geographical overlap of so many hotspots of biodiversity and cultural diversity. New Guinea, the Amazon region, the Congo region are hotspots of biodiversity. And they are hotspots of linguistic diversity too. These regions are all in the tropics. Given that some have argued the same environmental explanation for tropical cultural diversity as for tropical biodiversity, perhaps an overlap in the geographic siting of hotspots of the two forms of diversity is not surprising.

High tropical productivity is not necessarily the only influence on diversity, though. Outside the tropics, the Himalayas and Karakorams have some of the highest densities of languages in the world. An explanation for that diversity is the same as one of the explanations for New Guinea's high density of languages, namely the barriers to movement that were the topic of the previous chapter. Much of eastern New Guinea is extremely mountainous and forested, and hence difficult to move through. Not coincidentally, the Himalayan and Karakoram ravines and crests are especially deep and high, and their rivers especially fast-flowing.

A hotspot of cultures used to exist also along the northwest coast of the United States, from southern Washington through Oregon down to northern California. Early records indicate that west of the Sierra Nevada and Cascade mountain ranges, one could hear in the region a total of sixteen language families and forty-four languages. That is a density of nearly twenty languages per quarter of a million square kilometers. Such a high density is what we expect for the tropics. Colombia, for instance, has seventeen languages in the same area. We do not expect such a high density of cultures at forty degrees latitude north, well into the temperate zone.

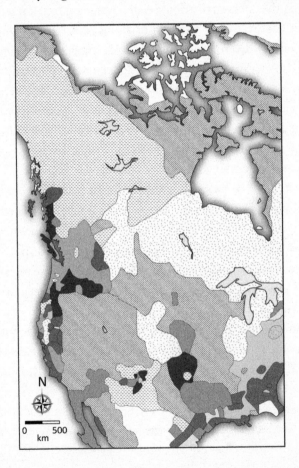

Map of western and central USA Native American cultures. Note high diversity of languages of small geographic area on west coast and Gulf Coast. *Credit: Redrawn by John Darwent from WikiCommons from I. Goddard 1996.*

The author, seated amongst his parents and three sisters around a sign for the town of Harcourt in Normandy, France. He was born in Kenya, moved to Britain, and now lives in California, USA. One small example of the movement of peoples around the world. (Chapter One) *Credit: A.H. Harcourt.*

Adam and Eve, by Hendrik Goltzius. Given that humans originated in Africa, all Adam and Eve pictures should of course show Africans, not white Caucasians. (Chapter Two) *Credit: WikiCommons.*

Clovis spear and maybe arrow points, characteristic of the second main culture that migrated into North America. The longest blade, from Fayette, Illinois, is a little over 14 cm long. Others from South Dakota down to New Mexico. (Chapter Three) *Credit: J. Darwent: Dept. Anthropology, UC Davis.*

Hunter-gatherer family. A family of Inuit people, who could have averaged 50 km or more each camp move for a total of over 500 km per year. (Chapter Three) *Credit: George R. King; WikiCommons.*

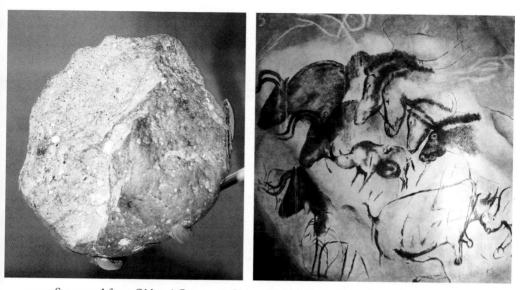

LEFT: Stone tool from Oldupai Gorge, an Olduwan chopper from maybe two million years ago, and barely distinguishable from a broken cobble stone. (Chapter Four) *Credit: British Museum; WikiCommons.* RIGHT: Chauvet cave painting, from about 32,000 years ago. (Chapter Four) *Credit: Anthropos Museum, Brno, Czech Republic; WikiCommons.*

Japanese snow macaque. Lives in northern Japan, where it snows in winter. Note short tail. (Chapter Five) *Credit: A.H. Harcourt.*

LEFT: Long-tailed macaque, Indonesia. Lives in the tropics, where it is warm all year round. Note long tail. (Chapter Five) *Credit: T. Brown; WikiCommons.* MIDDLE: Kalahari peoples drop their temperature in the freezing desert night to conserve energy. (Chapter Five) *Credit: Yanajin33; WikiCommons.*

LEFT: Himalayan resident: Bhutanese yak herder with yaks. (Chapter Five) *Credit: A.H. Harcourt.* RIGHT: Andes resident: Quechua lady and child. (Chapter Five) *Credit: © Alfredo Carrasco Valdivieso.* Both peoples show several, but different, adaptations to living at high altitude where oxygen density is low, as do the yaks.

African savannah—what the Sahara could have been like in wet periods, so enabling passage by early humans. (Chapter Six) *Credit: A.H. Harcourt.*

Coastal glacier of the sort that could have made movement into the Americas from Siberia difficult, given calving bergs from its face (blur at left center), and the impossibility of landing on it in case of need. (Chapter Six) *Credit: M. Clarke; WikiCommons.*

Edge of an ice sheet, a barren waste, with no food, or shelter, or firewood near it, and large melt pools extending out from it, so making migration between ice sheets unlikely. (Chapter Six) *Credit: A.H. Harcourt.*

Outrigger canoe of the sort that might have finally enabled humans to people the whole Pacific. (Chapter Six) *Credit: WikiCommons.*

The English Channel/La Manche separating British culture from French culture. (Chapter Six)
Credit: European Space Agency.

The Orkney Islands, separated from Scotland by the Pentland Firth, and so with a different genetic heritage. (Chapter Six) *Credit: M. Norton; WikiCommons.*

The Zagros mountains of Iran, perhaps a barrier to movement of early humans out of the Middle East. (Chapter Six) *Credit: S.Asadi; WikiCommons.*

LEFT: National dress of Bhutan. An example of a culture maintained through lack of contact with other cultures and, in Bhutan, by decree. (Chapter Six) *Credit: A.H. Harcourt.* RIGHT: Ecuadorian Achuar man (with feather headband). Ecuador, the same size as the United Kingdom, has twice the number of indigenous cultures and several times the number of indigenous species as the UK. (Chapter Seven) *Credit: © Alfredo Carrasco Valdivieso.*

LEFT: New Guineans. For its area, New Guinea has ten times the highest density of languages, or cultures, as do any other similarly forested tropical countries of roughly the same area, such as the Central African Republic, Colombia, or Thailand. (Chapter Seven) *Credit: S. Codrington, Planet Geography, 2005; WikiCommons.*

RIGHT AND BELOW: Tropical (Ecuador) vs. non-tropical (Scotland). Year-round warmth and wet in the tropics allows animals and humans to live at high densities in small areas. High latitudes, unproductive for six months of the year, force animals and humans to range over large areas if they are to survive. (Chapter Seven) *Credit: A.H. Harcourt.*

The cave in which the hobbit remains were found, Liang Bua, Flores, Indonesia. You can see the excavation pit near the cave's entrance. (Chapter Eight) *Credit: Rosino; WikiCommons.*

Otherwise large animals become small on islands. Dwarf elephant (bottom left) in an Egyptian fresco within Rekhmire's tomb, from thirty-five hundred years ago. The long tusks indicate that it is not a young elephant. (Chapter Eight) *Credit: Rockartblog; WikiCommons.*

LEFT: On isolated islands, cultures (and species) unique to the islands often evolve. Carving on a meeting house of the Māori, a distinctive culture of New Zealand. (Chapter Eight) *Credit: Kahuroa; WikiCommons.* RIGHT: Fulani herdsman of west Africa. The Fulani, like western Europeans and some other peoples, have evolved the ability to digest milk in adulthood. (Chapter Nine) *Credit: J.Atherton; WikiCommons.*

LEFT: The Reverend Thomas Savage, a European missionary, who lost two wives to disease in tropical west Africa. (Chapter Ten) *Credit: Special Collections, University of Virginia Library.* RIGHT: Jeffries Wyman, Harvard professor, and with Thomas Savage co-namer of the gorilla to science as *Troglodytes gorilla*, now *Gorilla gorilla*. (Chapter Ten) *Credit: WikiCommons.*

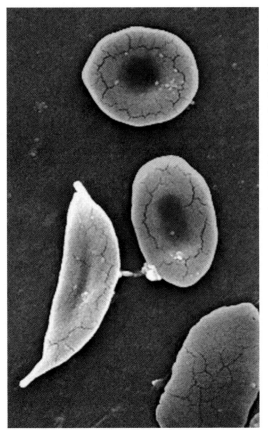

LEFT: The misshaped cells of someone with sickle-cell anemia, a condition that in its mild form protects against malaria, but in its serious form, as here with cell on left, causes illness, even death. Healthy cells would be perfectly round discs with a central indentation, nearly like the top cell. (Chapter Ten) *Credit: OpenStax College; WikiCommons.* BELOW: Workers in a sugar plantation on Trinidad, where the irrigation canals provide ideal breeding ground for mosquitoes. (Chapter Ten) *Credit: Richard Bridgens; WikiCommons.*

Dryas flower, after which the Younger Dryas cold period is named. (Chapter Eleven) *Credit: H. Storch; WikiCommons.*

Giant deer, or Irish elk, extinct around ten and a half thousand years ago, one thousand years before the arrival of humans. The warming climate and lack of food were probably the cause of its demise. (Chapter Eleven) *Credit: Wellcome Library London; WikiCommons.*

Gelada baboon, confined by climate warming and humans to the top of the Simien Mountains, Ethiopia. A male grooming one of his females. Note the fur blowing in the cold wind. (Chapter Eleven) *Credit: R. Waddington; WikiCommons.*

Kudzu, native to Japan and southeast China, is a pest in many other parts of the world now, smothering native species. Here it is all of that dark background, Maui, Hawaii. (Chapter Eleven) *Credit: Johnny Randall-NC Botanical Garden.*

Canyon de Chelly, Navajo Nation Reservation, Arizona, one of the more infertile parts of the USA. (Chapter Twelve) *Credit: katsrcool; WikiCommons.*

LEFT: Shuar man, a member of an Ecuadorean-Peruvian Amazon culture bringing its culture back from the brink. The Shuar founded in 1964 the first ethnic federation in the Ecuadorean Amazon. This federation uses radio programs, a printing press, and other means to maintain their culture. (Chapter Twelve) *Credit: © Alfredo Carrasco Valdivieso.*

By contrast to this coastal density of cultures, go east from coastal northwest USA, east of the Sierra and Cascade mountain ranges, and we find just four language families and ten languages, roughly the number expected for temperate latitudes. Seven times as many cultures occur in California as in an equivalent area of Nevada, where densities are as low as in the Arctic. In California, the Yukian, Pomoan, Wintuan, Maiduan, Palaihnihan, Shastan, Nadene, and Algic cultures and more used to live there. Most of Nevada was just Uto-Aztecan range. On average, the coastal peoples lived in ranges that were less than half the size of the inland Native Americans' ranges.

A window seat on an airplane flight across the region immediately indicates why we get the contrast between the coast and inland in the number of cultures. As one flies inland, say from San Francisco to Las Vegas, the color of the land below changes abruptly from green to brown at the crest of the Cascade Mountains and the Sierra Nevadas. The rain clouds coming from the Pacific Ocean to the west drop most of their content on the Pacific coastal mountains and on the western slopes of the Cascades and Sierras, leaving almost nothing to fall to the east. To the west is what has been called the USA's breadbasket, the fertile and well-watered Central Valley of California. To the east is near-desert. Indeed, the San Francisco-Las Vegas flight goes over Death Valley.

Coastal populations in a green land have another advantage over inland populations in a brown land. The coastal peoples can use the sea, so effectively increasing the productivity of their territory. I described in chapter 3 the enormous mounds of discarded shells found along the coasts of the world. Not only can the sea's bounty be directly consumed, but agricultural populations can increase the productivity of their land by using seaweed as a fertilizer. They still do so in parts of the British Channel Islands and Scotland, and the web has hundreds of sites devoted to its use by gardeners.

If this coastal-productivity argument explains the cultural diversity of the west coast of the United States, it should apply elsewhere. And so it does. The western Gulf Coast had a greater density of Native American cultures than did the Dakotas or Minnesota, or Manitoba or Ontario. In the south we have Tonkawa, Atakapa, Adai, Natchez, Caddoan, Tunica. Up north in Canada, just Ojibwa (also Ojibwe) and Cree.

The Gulf Coast northward is a latitudinal comparison as well as a coast-inland comparison. So are we looking at latitude as the correlate here and climate as the influence, or at coastal high productivity of resources versus inland relative poverty as the influence? These sorts of analyses of human cultural diversity are too new for a definitive answer. Understanding of geographical variation in diversity of cultures is an open field for enquiry, so open that it has not yet made its way into standard textbooks of physical anthropology.

Brian Codding and Terry Jones provide a different explanation for the high density of cultures in California. They relate it to the order of arrival of cultures in the region. The first arrivals established themselves on the productive coast. The newcomers did not supplant the previous ones, because the first arrivals had built up large populations that were not easy to supplant. The later ones therefore settled inland. In effect, two waves of peoples doubled the density of cultures. The latest arrivals of all, the Algic and Athabascan peoples, did however supplant coastal peoples. They could do so because of their superior technology.

A believable hypothesis, maybe, but how do we test it? Journals have limits to how long articles can be, and the argument as published needed more explanation. Brian Codding kindly gave it to me. I asked him why the earlier cultures with their higher population density were unable to prevent any immigration of the later cultures. His answer was that hunter-gatherers, who have to move a lot, find it difficult to defend any sort of territory.

I asked him why the two latest, more technologically advanced cultures did not expand and supplant the residents, and indeed had some of the smallest ranges of all the cultures in the region. His answer was that they depended heavily on salmon runs, which meant that they confined themselves to near the rivers in which the salmon ran.

A problem with testing their order-of-arrival argument as an explanation for concentrations of cultures in other parts of the world, such as the tropics or New Guinea, is that we might never know the exact order of arrival, given how long some of the cultures have existed there. However, the salmon-run idea could well explain the fact that in Australia, coastal aboriginal peoples and those who lived at the mouths of big rivers had smaller ranges than the average.

The general explanation here would be that, as with the fishing cultures on Lake Chad, we have an example of high productivity allowing a people to live in a population of sustainable size in a small territory. Alternatively, is it the case that that specialization (on salmon, in the case of the Algic and Athabaskan groups along the California coast) prevents expansion?

Currently, I prefer the argument that links productivity to diversity, because it seems to me to be better substantiated and more generalizable than arguments that rely on order of arrival and degree of specialization. Nevertheless, the idea that a productive, stable environment allows specialization, and gives more time for it, and hence allows a concentration of specialized species in a small area, is a strong one in biogeography.

However, if contrasting productivity of the land between coast and inland in the western USA explains the contrast in cultural diversity between the two regions, then we have a nice substantiation of the productivity explanation for the Forster effect. Many aspects of the environment in addition to productivity correlate with latitude. But with the same productivity-diversity correlation in the western USA, but now with a longitudinal pattern, those other correlates with latitude are removed as potential alternative causes of the Forster effect. In general, the more situations that a hypothesis can explain, the more faith one has in its solidity.

Thomas Currie and Ruth Mace provide an interesting variation on the idea that high productivity allows high density of cultures via allowing survival in small territories. They propose that in modern times, cities can be considered as highly productive sites, even if they import their food rather than growing it themselves. If cities follow biogeographical patterns, they should also be culturally diverse sites. In London, England, one can hear over three hundred languages in its schools. That is a density of one per five square kilometers. More quantitatively, another study that controlled for all sorts of other potential influences showed that change in wages in cities in the USA between 1970 and 1990 matched change in diversity of languages.

In sum, with high productivity, many cultures can live packed in small ranges; with low productivity, a few cultures each have to live in a large range. The hypothesis works in recent times, as well as during humanity's global hunter-gatherer days.

Through this chapter, I have consistently separated findings about hunter-gatherer societies and findings about all cultures, in part because of the two main sources of raw information about cultures, Murdock's on hunter-gatherer societies and the website Ethnologue, which lists languages. Several hunter-gatherer societies speak essentially the same language as their agricultural neighbors. Nevertheless, hunting and gathering is clearly a different way of life, a different form of culture than farming.

I discussed hotspots of diversity some pages ago. It turns out that eastern Asia might have a hotspot of hunter-gatherer cultures, mostly in the tropics, as expected. We do not know much about them in comparison to our knowledge of African hunter-gatherer societies. However, we do know that some of the hunter-gatherer peoples resemble genetically and linguistically their agricultural neighbors more closely than they resemble other nearby hunter-gatherer peoples. An example are the so-called Negrito peoples of the Philippines.

Genetical studies indicate that the Negrito peoples arrived early in the region—indeed, that they were some of the earliest arrivals. Subsequently, some of them became agriculturalists, while others remained as hunter-gatherers, even if they traded with their agricultural neighbors, as do the pygmy peoples of Africa. By contrast, another group, from northern Thailand, the Mlabri, seemingly reverted from agriculture to hunting and gathering perhaps just five hundred years ago.

In the last chapter, I described Dan Janzen's explanation for the generally small geographic ranges of tropical species, which in turn explains the relatively high density of species in the tropics. Janzen's argument was that tropical species experience mountains as barriers more than do temperate species, because the tropical species rarely experience freezing weather. Consequently, they are not so adapted to withstand cold, with the result that they cannot cross high mountain ranges, and are therefore more often confined in relatively small areas by comparison to species of high latitudes. The high-latitude species, by contrast, by definition must survive winters, which means that they can withstand the cold of high elevations as well as of high latitudes, and therefore their geographic ranges are not constrained by mountains.

The argument might explain high tropical biological and cultural diversity in mountainous areas of the tropics. Its problem as a general argument, though, is that if you want to see examples of high tropical biological and cultural diversity, go to the Amazon or Congo forests. They are a problem for the argument because they are effectively flat for hundreds of kilometers in all directions, across hundreds of thousands of square kilometers. Mountains cannot explain the biological and cultural diversity of these forests.

An explanation for the Forster effect in contemporary or recent human cultures is imperialism. While cultures that still rely quite heavily on hunting and gathering persist in South America, Africa, and Asia, they are gone, of course, from much of North America and Europe. Imperialist agricultural societies have replaced them there. So is it the case that imperialism over the ages in its various forms—the Mongols and the British Empire, for example—is all the explanation that we need for a low diversity of cultures with large geographic ranges in the poorly productive lands of high latitudes?

No, it is not. As I said, hunter-gatherer ranges are largest precisely where one might expect that imperialist cultures would have severely constrained them, namely outside the tropics. Also, we see the Forster effect within regions of the world conquered by northern imperialistic powers.

The human history of the world is an account of non-stop invasion and counter-invasion. Nations and cultures arrive, expand at the expense of others, and leave. Dynasties extend their reach over neighboring peoples. The Mongols erupted out of central Asia and almost conquered western Europe. I remember old atlases showing much of the world as pink, reflecting far-reaching British dominion. If powerful temperate cultures can expand way beyond their original homeland, why have not a few powerful tropical cultures done the same, so producing ranges of the same size as those of nations in higher latitudes?

I do not know what historians might answer. Biogeographers, though, suggest that, among other factors, disease and parasites can prevent cultures, as well as plant and animal species, from expanding. This was the suggested alternative in the last chapter to the Fincher and Thornhill idea that religion prevented movement between regions, and hence movement into areas where the immigrants would not have been adapted to the disease organisms. Thus Jared Diamond suggested that movement of crops and livestock along the length of the Americas and Africa was more difficult than in Eurasia, because in the former the long axis was north-south, and hence across a variety of environments and diseases. By contrast, the east-west axis of Eurasia resulted in a similar environment all the way across, including of parasites and pathogens.

This argument is similar to Robert MacArthur's suggestion from twenty-five years earlier of why fewer temperate species moved south than tropical species moved north. He suggested that a temperate species going south would meet too many competing species, not to speak of a wealth of parasites and pathogens. That would not be the case for a tropical species going into the relatively depauperate temperate regions. So we find that many attempts to introduce temperate crops into tropical regions failed, whereas gardeners can grow several tropical plants in temperate regions.

What about movement within the tropics? After all, I am trying to explain why so many species as well as human cultures have small ranges within the tropics. In the previous chapter, I wrote on the topic of disease as a barrier to movement. As I wrote there, disease organisms show the same biogeographic patterns as do other species. They tend to be more diverse within the tropics than outside. Hence, in part, all the inoculations I had to take before going to work in various tropical African countries.

Correspondingly, and crucially for the argument, each disease species covers a smaller geographic range in the tropics than outside the tropics. That means that a tropical culture would not have to move far before it encountered diseases to which its members had not evolved resistance. And so the culture and its people would be confined to the relatively small geographic range of their ancestors.

Thus Elizabeth Cashdan found that along with various potential climatic influences on cultural diversity across the world, high parasite loads correlated with high diversity. By "load," she meant a measure of both variety and number of parasites. Oddly enough, she found this correlation with parasite load (as well as a number of other environmental correlations) not among what she termed non-complex societies, but among chiefdoms and states. She could not explain that disparity, and I cannot either.

The idea that pathogens and parasites can control where people live is logical; but in chapter 6, I raised the chicken-and-egg conundrum of which came first, pathogens and parasites, or people. Did disease organisms prevent the movement of people, or did people staying put for another reason (xenophobia, maybe) mean that because the disease organisms were not moved, ones from different regions did not mingle, and so the organisms differed between regions?

We have evidence for both possibilities. I go into the topic in more detail in a later chapter, but good reason existed for why west Africa was once termed the white man's grave. White men there died in droves from tropical diseases, even though at the same time they found enough healthy Africans to fuel the slave trade. African diseases might, then, have prevented European hegemony in west Africa.

South and North America illustrates how the movement of people moved diseases. European incursion there took mosquitoes, malaria, yellow fever, measles, and smallpox into the continent and devastated the Native American populations.

I described previously the high cultural diversity of the western coastal United States, and put it down to high productivity of the land. But can diversity of disease explain any of it? The answer is that it cannot. At approximately forty degrees north, the region does not have a great variety of diseases.

Californian newspaper headlines might scream about West Nile virus in the region, but so far this year, mid-September 2014, the Centers for Disease Control report that California has experienced just twelve deaths from the disease. A flea or two, a mosquito or ten, and a tick or two is usually all we have to suffer. And that is number of individual animals I am talking of, not species. That is a far cry indeed from the swarms of biting flies and mosquitoes my wife and I have experienced in Gabon and Nigeria. Although the tundra of the far north also swarms with biting flies, if only in the summer, they are simply unpleasant. They do not carry disease.

Some evidence for the importance of disease affecting where we are and where we are not is the finding that genes connected with resistance to disease vary more across populations from different parts of the world than do genes associated with adaptation to climate or to diet. In fact, the genetically influenced resistance was more specific than the word "disease" indicates. The resistance was largely to parasitic worms, such as hookworm, rather than to the far faster evolving bacteria, single-celled organisms, or viruses. Common diseases of these last three sorts of organism are, in turn, cholera, giardia, and AIDS. Again, though, the parasites might differ between peoples of different regions because for some reason other than the presence of parasites, we do not move between the regions.

Even if disease prevented the expansion of imperial temperate peoples into especially Africa, I do not see that it can have prevented expansion inland of west coast Native American cultures in the United States. That being the case, we are left understanding how those cultures could exist in a small area—high productivity allows a population of viable size—but not why the more bellicose of them did not expand. Maybe we need a historian, not a biogeographer, to answer that question.

✑

Culture, whatever we might mean by it, is to a large extent a product of our mind. Our minds, not the environment, determine our language. Yes, the environment plays a part in our dress—no brightly colored feathers in the Arctic. But the fact that I speak English whereas a Parisian speaks French, the fact that I might wear a peaked cap whereas a Frenchman might be more likely to wear a beret, has little or nothing to do with the environment or our biology.

Nevertheless, as I have described in this chapter, human cultures biogeographically behave in many ways as do many non-human animal species and even plant species. Humans are more culturally diverse in the tropics in the same way that many animal and plant species are more taxonomically diverse in the tropics. Not only is the pattern the same, but the same hypotheses that explain our tropical cultural diversity also explain the tropical biodiversity of non-human animal species. When we are talking about the global distribution of human cultures, we could be talking of the distribution of thousands of other animal species, even plant species. Biogeographically speaking, man is—in a perhaps surprising number of ways—merely a monkey, a mantis, even a myrtle.

CHAPTER 8

ISLANDS ARE SPECIAL

Size and metabolism in a small environment

B iogeographically speaking, islands are in some ways just small bits of mainland. Indeed, sometimes islands are literally small bits of mainland. They have broken away from the edge of a continent and moved out to sea, propelled by plate tectonics. Alternatively, they are still strongly connected to the mainland, but the connection happens for the moment to be flooded by high sea levels. Between ice ages, sea levels are high, as they are today. During ice ages, the water gets converted to ice, and sea levels drop. At the height of the last glaciation, roughly twenty thousand years ago, sea levels were over a hundred meters lower than they are now. If global warming does not prevent the return of the next glacial period, sea levels will then fall as in previous ice ages, and many coastal islands will return to being part of the mainland. Britain and Ireland will no longer be islands but, instead, simply the western edge of Europe, as they were during the last ice age.

That could be a pity, because in many ways, islands are special. Cut off by water, isolated from contact with others of their kind, island animals and plants can go their own way evolutionarily. The result is many species unique to a particular island and found nowhere else. The lemurs of Madagascar are a classic case for primates. The nene, or Hawaiian goose, is classic for birds.

A relevant piece of biogeographic jargon out there is the word "endemism," or "endemic." It means confined to a particular region. The lemurs are endemic to Madagascar. One often thinks that endemism means confined to a small area. However, the size of the region is irrelevant. In Hawaii, some bird species are endemic to just one island, but the kangaroo is endemic to Australia, and the reindeer is endemic to northern Eurasia.

Biogeographers speak of island "rules" to describe and explain a number of general ways in which island fauna and flora differ biologically from mainland fauna and flora. Humans follow some of the rules, but not others. To introduce one of the island rules, I am going to start with "Flo," the Flores island "hobbit." Strictly, Flo is probably not relevant in a book on human biogeography, for reasons that I will come to. However, the story allows me to introduce an island rule that Pacific-island humans break. Understanding why a phenomenon breaks a rule is often just as informative concerning underlying causes as understanding the reason for the rule in the first place.

In 2004, the announcement in the journal *Nature* of a new species of the genus *Homo* from Asia, *Homo floresiensis*, electrified the anthropological world. The species name comes from the discovery of 40,000- to 13,000-year-old bones in a cave on the island of Flores, east of Java and Bali in the Indonesian archipelago.

"Flo," or the "hobbit," as the tiny new species affectionately became known, was the first new hominin species claimed in Asia for over a century. The last one, oddly enough, was the first *Homo erectus* ever discovered. Named Java Man, it was found in 1891 by the Dutch paleoanthropologist Eugène Dubois or, to give him his impressive full name, Marie Eugène François Thomas Dubois.

The hobbit was strange in many ways. Apparently alive on Flores well over ten thousand years after the last non-human hominin died out elsewhere in the world, its skeleton turned out to be weird enough that doubts immediately arose regarding almost everything to do with it. Arguments and counter-arguments flared, burned out, and flared again. And the arguments were not, sad to say, always politely expressed, reasoned differences of opinion. Sarcastic commentary with the word research in quotation marks to denigrate the others' scholarship marked one exchange. Other different opinions were characterized as "unsubstantiated assertions" by a "vocal group." Suggestions of, shall we say, incomplete reporting of measurements, along with accusations of mishandling of the original material and denial of access to it, exemplify just some of the all-too-human nature of the early debate.

If the hobbit had been only slightly different from all human ancestors, probably only paleontologists, a journalist or two, and the occasional informed member of the public would have paid any attention. But the hobbit was extraordinary. It had such a mix of modern, ancient, and its own unique features, such as almost ridiculously long feet, that disagreement was inevitable.

Perhaps the most amazing feature of the hobbit was its small size—hence the nickname. Let me say that only one near-complete skeleton has been found, so I refer to the hobbit in the singular. The hobbit stood just over one meter high. That is thirty centimeters shorter than any human pygmy. I am a normal-sized male for my generation in the UK, at one meter seventy-eight centimeters. The hobbit would reach the bottom of my rib cage. The hobbit was stocky, so at thirty kilograms it weighed more than many modern-day pygmy adults do. Added to its amazingly small stature is an amazingly small brain. The one skull found shows a brain so tiny—four hundred and twenty-five cubic centimeters, one third the size of a modern human's, in fact the size of a chimpanzee's—that we have to go back three million years in human evolution, beyond *Homo* to the australopithecines, to find another hominin with a brain this small.

Is the hobbit a relict australopithecine species? If not, what is it? How do we explain its extremely long feet? How can the hobbit be a hominin, and yet have so small a brain? Why is this hobbit so small? How do we explain its mix of modern and ancient traits?

With no evidence of any australopithecines outside of Africa, one of the early explanations for the hobbit's small body size, and especially its small brain size, was that it suffered from microcephalic dwarfism. The "microcephalic" part of that is just the typical doctor's Greek jargon for "small-brained." I cannot resist a story from my brother-in-law. He had a swollen knee. He went to the doctor. "Aah," said the doctor, "you have patellitis." Patellitis is Latin and Greek for swollen knee!

Other pathologies to explain the hobbit are Laron syndrome, or a form of cretinism. Both of these conditions come with small stature. The first is a genetic condition, while the second can result from mineral deficiency, especially lack of iodine. It characterizes individuals who lack fully functioning thyroid glands, the glands that produce hormones essential for full growth. A recently suggested pathology to explain the hobbit's features is Down syndrome, which can explain the odd mix of modern and apparently ancient traits.

If the hobbit was in fact a modern human suffering from a disease, it would not be relevant to this book on the biogeography of humans, given that none of the suggested diseases to explain its small stature and brain size confine themselves to any one part of the world. However, the hobbit lived on the small island of Flores, in eastern Indonesia, and the biogeography of small islands is highly relevant to the hobbit. Conversely, the hobbit is relevant to the biogeography of small islands, given how intensely scientists have studied it.

Flores covers on the order of thirteen and a half thousand square kilometers, approximately the size of Connecticut in the USA, or Northern Ireland in Britain. A feature of small islands is that species that are large-bodied on the mainland often evolve to become smaller, sometimes far smaller. On the islands of the Mediterranean Sea, such as Cyprus, Malta, Crete, and Sicily, elephants and mammoths shrunk over many generations' time to half the height that they were on the mainland, so they ended just one and a half meters high at the shoulder.

California's Channel Islands too had their own pygmy mammoth, just a little taller than the Mediterranean elephants. Similarly, the mammoth of Wrangel Island off the north coast of Siberia was roughly thirty percent smaller than the average mainland mammoth, as judged from their tooth sizes.

Flores is no exception to this phenomenon of miniaturization. An extinct form of elephant in mainland Asia, stegodons, were some of the largest elephants ever to have lived. The Flores version on the island at the same time as the hobbit was one third smaller than its mainland relative. The adults, at about eight hundred and fifty kilograms, were too big for the hobbit to hunt, but they hunted the young ones. Archeologists have inferred this from cut marks on the bones in the cave.

Another animal that the hobbits almost certainly hunted was the Komodo dragon. It still lives on the island of Flores, as well as on nearby Komodo Island. This dragon is in fact just a lizard. I have introduced it not because the hobbit probably hunted it, but because it illustrates the other aspect of the "island effect" on the size of animals' bodies. In contrast to the miniaturization of large-bodied species, small-bodied species sometimes become large on islands.

Well-fed Komodo dragons can grow to three meters long and weigh up to seventy kilograms. Even at a more normal two and a half meters and fifty kilograms, it is still the largest lizard in the world, and scary enough. It is a meat-eater, like almost all of its kind, the monitor lizards. Big enough to bring down a deer, it could easily have taken a careless or sleeping hobbit. Flores also has a giant rat, with a body length of up to forty-five centimeters.

Another island giant we all know is the now famously extinct dodo of the island of Mauritius in the Indian Ocean. It was a member of the pigeon family, but a pigeon that weighed twenty kilograms. That is ten times larger than the current largest pigeon in the world, the beautiful crowned pigeon. The city pigeons that Americans and Europeans are used to seeing weigh close to a third of a kilogram.

The cross-over point between otherwise large species shrinking and otherwise small species evolving to become larger lies somewhere between one tenth of a kilogram and one kilogram. Starting at more than a kilogram or so, island species evolve to become smaller. Less than one tenth of a kilogram or thereabouts, and the island species become larger.

If Flo is a new species, how big was its ancestor? The ground where the Flores hobbit remains were discovered dates back to nearly one hundred thousand years ago. Assuming that the remains did not somehow sink from higher, younger levels in the cave, and assuming also that the hobbit is not a representative of a human population with a pathological condition, that age means that the most likely potential ancestor of the hobbit is *Homo erectus*. We already knew that the hobbit's ancestor had to be larger than one kilogram. But Erectus is one of the larger hominins. Adult Erectus weighed in the region of fifty kilograms. So the hobbit follows the island miniaturization rule.

Why do large-bodied species evolve to become small on islands, and small ones to become large? Over the decades, biologists have produced a myriad of suggestions. I will summarize what I think might be the simplest and most widely applicable. It seems to me that we mainly have to take account of the opposing influences of food and predators, of eating and avoiding being eaten.

The basis of the food argument is change over time in the amount of food available. The climate varies over decades and centuries, and the food supply fluctuates in step. That happens everywhere. The drawback for animals on islands is that when conditions get bad, the animals find it difficult to leave the islands to search for greener pastures. The animals that we now see on islands are necessarily, then, those that have adapted to survive the poor periods.

Now, large animals need to eat more food in total than do small animals, even if small animals need to take in food at a higher rate. On a small island, large-bodied animals, including hominins, will find themselves short of food more often than will small animals. As an extreme but historic human example, there could hardly be a smaller island than an expedition's sled in the middle of the Antarctic continent. And who was the first person to die on Scott's return voyage from the South Pole but the largest, Edgar Evans.

On a small island, then, with a limited food supply and no opportunity to migrate to richer areas, not only are smaller animals more likely to survive than are larger ones, but more small animals than large animals can live on the food that is there. Small animals are therefore less likely than are large ones to dwindle to such small numbers that they go extinct.

Not only that, but small animals reproduce faster than do large ones, other things being equal. Therefore, populations of small-bodied animals can recover faster than can large-bodied from an event such as a famine, with the result that they can get back to safer numbers before the next environmental disaster.

In sum, small-bodied individuals are more likely to survive on small islands over the long term than are large-bodied individuals, meaning that the species as a whole evolves to be smaller-bodied.

We see the result on Flores. The large-bodied elephants and hominins evolved to become smaller than their mainland relatives. The fifty-kilogram Erectus became the thirty-kilogram hobbit.

Smaller animals in smaller areas is a general rule. Australia's largest plant-eating mammal, a form of wombat, might have weighed two tonnes. Two tonnes is half the weight of the mammoths and mastodons of North America and Eurasia, and of the stegomastodon of South America.

I probably need to say again that a tonne is the modern metric measure used almost everywhere in the world but the USA. It is in fact very close in weight to the USA's "ton." A tonne is one thousand kilograms, which is two thousand two hundred pounds as near as makes no odds, or just two hundred pounds more than the two thousand pounds of the USA's ton.

The explanation of more food for small animals, which are more likely to recover from a crash in numbers on a small island, does not explain why species that are small on the mainland become larger on islands. Large-bodied animals need more food, and recover more slowly from population crashes. We should see only a one-way change in the size of animals' bodies on islands. But we do not.

We need a different explanation now. The answer is danger from predators. Predators need a large area in which to live, larger than do plant-eating animals of the same size. That is because a predator's food, say a rabbit, is rarer on the ground than is the food of rabbits, grass. A general rule is that meat-eaters need in the region of ten times the area that a vegetation-eater needs.

Consequently, small islands that can support plant-eating mammals often cannot support the mammalian carnivore that would eat them. If the hobbit still lived, it would have nothing to fear from Flores's largest mammalian predator, the introduced civet. It weighs maybe three kilograms.

So now it is an advantage for the smallest animals to become too large for an island meat-eater to eat them. Conversely, large herbivores on islands, which might on the mainland have evolved to be too large to be taken by predators, can afford to become smaller. Indeed, the smaller an individual, the less likely it is to starve, as I have already explained.

Another advantage to a small animal of becoming large, and we are now back to a food explanation, is that it can defeat opponents in competition over food. No mainland rat would want to come up against the giant rat of Flores.

Of course the largest species do not evolve to the original size of the smallest, nor the smallest to the original size of the largest. Hominins get as small as chimpanzees but not as small as rats. Island rats become as large as a small dog but not as large as a chimpanzee. A larger individual might win a fight over food, but then it needs more food and might not find enough. A smaller individual might need less food and so be less likely to die of starvation, but then it might be small enough to be eaten by the island's small predators. In brief, the range of bodily sizes on islands becomes less than on the mainland from which the species originally came, as the largest and the smallest do better by being medium-sized.

But not the Komodo dragon. The Komodo dragon is an island anomaly. Its relatives, the monitor lizards on the mainland, are already large, many over five kilograms, some even as large as ten kilograms. If the cross-over point between large animals becoming small and small becoming large on islands is a kilogram or less, why is the sixty- or seventy-kilo Komodo dragon so much larger than its mainland relatives? Not only has it broken the island rule, but it is a meat-eater. Meat is rare everywhere. As I have said, large mammalian carnivores cannot survive on small islands. How, being a large meat-eater, can the Komodo dragon be as large as a mainland panther?

The answer lies in the low metabolic rate of reptiles. Komodo dragons rest almost immobile for many days after a large meal. These days, goats are put out to attract Komodo dragons for tourists. Formerly, the lizards would have eaten the endemic rat of the islands, ground-nesting birds and their eggs, and any of their own species smaller than themselves. The dragons need far less food than does a mammalian predator over the same period of time, and so can live on far smaller islands than can any mammalian predators of the same size. In fact, a general rule of the biogeography of small islands is that we can expect the species there—birds, mammals—to have a lower metabolic rate than their mainland relatives.

And so the Komodo dragon survives on Flores, a danger to any hobbit, and a danger these days to careless or unlucky residents and tourists. Smithsonian.com reports two killings of humans there by Komodo dragons in the last ten years. I cannot resist adding that the website reports also that Sharon Stone's former husband, Phil Bronstein, was mauled by a Komodo dragon in the Los Angeles Zoo.

Flores is not unique in its lack of large mammals, the existence of its giant lizard, and its periodic shortages of resources. Australia as a continent is relatively small. We know that it is a dry continent, and it is a continent with periodic serious droughts. Like Flores, it has always had few large-bodied mammals. Like Flores, too, it has a large lizard, one nearly as large as the Komodo dragon. Indeed, it once had a far larger one. Megalania it is called, Greek for the "great roamer." The largest megalanias might have reached seven meters in length and weighed a thousand kilograms. I stress the "might," because a complete skeleton is yet to be found. Potentially more realistic average weights for megalania are still impressive, at four meters and a hundred and thirty kilograms. That is far larger than the Komodo dragon, but then Australia is far larger than Flores. Again like Flores, Australia's largest carnivorous reptile was far larger than its largest carnivorous mammal. That was the marsupial lion, which at roughly a hundred and thirty kilograms weighed about as much as a present-day female lion, or less than one fifth the weight of a megalania.

In sum, we have a nice match of situations, Flores on the one hand and Australia on the other. On both, we see the same biological pattern, the evolution of giant reptilian carnivores in the absence of competition from mammalian carnivores. We should thank our stars for the asteroid that wiped out the dinosaurs, or we might still have tyrannosaurs roaming the earth.

I have so far left the hobbit in scientific limbo as either a dwarf, microcephalic modern human, suffering from some pathology, or an evolved miniaturized island form of an early human ancestor. And that is where I am going to leave it. The latest hypothesis that it is an individual suffering from Down syndrome was published too recently for a response from those arguing that it is related to Erectus, or is an Erectus. Judging from the debate so far, it seems highly unlikely that either side is ready to cede victory to the other.

However, the hobbit's small body and brain size fit the well-established biogeographic effect of island dwarfing. Nor as a miniaturized island human is it unique. Human fossils on the small island nation of Palau (465 square kilometers) indicate a population of people smaller than present-day pygmy peoples, even if nowhere near as small as the Flores individual.

"The island dwarfing effect in biogeography." That is the rule that Flo the hobbit allowed me to introduce. Big animals evolve to become small on islands. But what did I write regarding Pacific islanders in chapter 5? "Some of the heaviest people in the world are Samoans," I wrote. The Samoans and other Pacific islanders are an exception to the island miniaturization rule, then. The suggested explanation for them has to do with efficient metabolism. Only those who were able to store extra calories as fat could survive the long voyages to the distant islands and the likely periods of starvation on the islands.

Stone tools on Flores indicate that hominins might have been on the island from at least a million years ago. The island holds no signs of the hobbit, though, as of more or less thirteen thousand years ago. What killed them off? A volcanic eruption is one possibility. Indonesia has many active volcanoes, Tambora and Krakatoa among them. A hardened layer of volcanic ash, tuff, lies over parts of Flores. That tuff turns out to be approximately fifteen thousand years old, approaching the same age as the last signs on Flores of both the island's hobbits and its miniaturized stegodons.

An alternative to a volcanic doomsday killing off the hobbit is the arrival of modern humans. As I describe in chapter 11, the arrival of modern humans on small islands has been the death blow of hundreds of species. Is the hobbit yet another of these human-caused extinctions? At the moment, it seems not, for it looks as if humans did not land on Flores until maybe eleven thousand years ago, in other words some thousands of years after the volcano and the last signs of the hobbit.

Another well-established island rule concerns the number of species on an island in relation to its size. Flores is a small island. It had food enough for just one large-bodied herbivorous (plant-eating) mammal to survive on it, the elephant-like stegodon, and one large carnivorous mammal, the hobbit. By comparison, nearby Java, ten times the size of Flores, had at the time the hobbit lived four large mammalian carnivores—tiger, leopard, hyena, bear—and seven large herbivorous mammals. These seven were two elephant species, two rhinoceros species, and three cow species.

I wrote in the last chapter on number of species and cultures in relation to size of a region. Why have I used a paragraph repeating not only what I have said before, but to some extent repeating the obvious? Of *course* larger areas contain more species and languages. How could they not? Why do scientists so often expend so much effort researching the obvious? The answer is that the devil, the interest, is in the details. Here, the interest lies in two sorts of details.

First, science with its accurate and precise data and its statistics can tell us the exact relation between—in this case—area and number of species. Science can tell us exactly how many species we can expect to lose with a certain loss of area, or conversely to gain with an increase in area. If I want to double the number of species in a wildlife reserve, how much larger do I need to make the reserve? That is a very practical question, which needs a concrete response. I know the answer surprised me when I first heard it. I gave details in chapter 7, but I will repeat the finding here: a doubling of the number of species requires a reserve not just twice as large, but something like ten times larger, other things being equal.

Human languages show the same sort of relationship. We find more of them in larger areas. We see this in sub-regions, such as among the Solomon Islands and over large regions, such as the Pacific. Palau, for example, at four hundred and sixty square kilometers, has four native languages. French Polynesia, ten times larger, has nine. As for species, then, so for Polynesian languages: twice as many languages needs ten times the area.

In the last chapter, I discussed how the nature of the environment affects the number of species and languages. By comparison to the last chapter, where I produced quite a bit of evidence to indicate that we find more languages in more productive areas of the world, mainly the tropics, Michael Gavin and Nokuthaba Sibanda could not find any obvious connection between the number of languages on Pacific islands and the nature of the environment. The result is not surprising. Most Pacific islands are in the tropics, and hence warm, wet, and productive the year round. As Gavin and Sibanda wrote, ninety percent of their islands had a twelve-month growing season.

But if some findings are not surprising, others are. And surprise is the second reason to work out the exact details of something as obvious as the fact that a larger area can hold more species or cultures. I have worked in five countries in Africa. In increasing order of size, they are Rwanda, Uganda, Nigeria, Tanzania, and the Democratic Republic of the Congo. The order of number of indigenous languages is almost the same, from three in Rwanda to two hundred and fourteen in D.R. Congo. The odd country out is Nigeria. Nigeria has five hundred and ten indigenous languages, more than five times what we would expect for its size among African tropical nations. Anthropologists and linguists have known for a long time about the country's diversity of languages. However, it is only once we know the general relationship of geographical area to diversity that we can see just how very unusual Nigeria is.

In science, understanding the reasons for exceptions can be as important as understanding the reasons for the general rule. If you want to identify people who really understand their subject, ask them to explain the exceptions.

I am afraid I have to admit that I have not been able to find a confirmed explanation for why Nigeria is so culturally diverse.

I am going to suggest an explanation, though. It comes from non-human biogeography. When two biogeographic zones meet—lowlands and mountains, forest and savannah, western and central Africa, central and eastern Africa—often the border zone has an unusually high diversity of species, because it contains representatives from both zones. The hill and mountain forests of southeast Nigeria and southwest Cameroon, where the Guinean and Congolian biogeographic zones meet, and of the eastern Congo, where the Congo zone meets the Shaba zone, are well known as such highly diverse meeting zones.

Ornithologists flock there to augment their life-list of species. Could it be that Nigeria is so culturally diverse because it is right at the corner of Africa where west Africa meets central Africa? The suggestion should be easy to test, needing only information on the regional origins of Nigeria's tongues.

Returning to the Flores-Java comparison, Flores, at one tenth the size of Java, should have had one half the number of large mammals, if the two were showing the area-by-number-of-species relationship that I have described. Yet Flores had far fewer than that. It has just one fifth Java's number of species. The discrepancy relates to the fact that Flores has always been an island, and Java has not. At the height of the last ice age, maybe twenty thousand years ago, Java was part of mainland southeast Asia because sea level was so much lower than now. It was open to receive any of Asia's mammals that could walk to it. By contrast, mammals on Flores had to have floated there.

If Flores had fewer land mammals than we might expect from its small size by comparison to Java, it has way more languages. It does not just have more than expected for its size: it has more than Java does. Instead of half Java's number of languages, Flores has nearly twice its number, twenty-one as against eleven.

Flores is close to three other islands: Sumba, Sumbawa, and Timor. That proximity does not explain the contrast with Java, though, because none of these islands shares indigenous languages with Flores. I will add that two of them also have more languages than does Java, although they are all far smaller than Java.

Why they should be so linguistically diverse by comparison to Java, I have no idea. However, the fact that they are all close together, and yet share hardly any of their languages, illustrates another fact about island cultural diversity.

Usually, the farther from a source an island is, the fewer species it has. That seems not to be the case for human cultures, in this case languages. The lack of an association between distance of islands from a potential mainland source and linguistic diversity on the islands is true worldwide. I analyzed the relations, taking account area of the island and latitude (given that we know that area and latitude correlate with linguistic diversity). The result was that distant islands had as few or as many languages as did islands close to a mainland.

Michael Gavin and Nokuthaba Sibanda found that among hundreds of Pacific islands, those closer to a mainland hosted a slightly greater number of languages than did distant islands. The effect, though, was minimal, so minimal that an equation connecting area plus distance with number of languages on islands produced no better a prediction of the number than an equation with area alone. In other words, the effect of distance on diversity is negligible by comparison to the effect of area on diversity.

The fact that Gavin and Sibanda found an influence of distance when they analyzed diversity of languages with distances alone does not necessarily conflict with my finding of no effect worldwide. Maybe Pacific islanders are different from people in the rest of the world. Also, our analyses were not exactly the same. For instance, they took account of a number of possible influences that I ignored, such as climate and soil fertility. At the same time, they ignored latitude, which my analysis showed to correlate as strongly as area with number of languages. And finally, our statistical methods differed.

But that is just armchair speculation. Furthermore, Elizabeth Cashdan's findings raise the question of why the potential frequent contact among the peoples of Flores and its surroundings did not lead to a homogenization of language. In chapter 6, I reported that she found that the easier movement by water was, the less difference between regions in their languages. Her sample was the globe, and her data were nations, for many of which rivers, not the sea, would have been the water route. But why would movement by rivers be associated with fewer languages in a region, yet movement by sea not make any difference, or, in Flores's case, be associated with greater linguistic diversity? I have no bright ideas for the answer.

Let me return now to the effect of Sumatra on Java's variety of languages. I mentioned a few paragraphs ago that Java was connected to mainland southeast Asia twenty thousand years ago, and so was open to receive Asia's terrestrial animals. Humans are a terrestrial animal, so at first sight Java should definitely have more languages than Flores, which has always been an island.

However, Java was a narrow end of a narrow peninsula. In the same way that only a few humans got out of Africa north of the Red Sea, or across the Bab el Mandeb Strait, or were able to penetrate the narrow ice-free coast of Alaska into the rest of the Americas, might it be that only a few could move from eastern Sumatra to western Java? And once some cultures had settled in western Java, might others have found it difficult to get past them, given how narrow the whole island is, and remembering from chapter 7 how territorial humans are?

Whether or not isolation affects linguistic diversity, it can affect another aspect of culture. Tasmania is the classic example.

The Tasmanian aboriginal peoples arrived from Australia thirty-five thousand years ago or thereabouts. I described in chapters 2 and 3 how the world was then heading rapidly into its last ice age, and sea levels were dropping as water became locked up as ice. When humans first arrived in Tasmania, the region was the end of a broad peninsula running south from southeast Australia. The northern half of the peninsula was a more or less flat plain just seventy meters or so above sea level, providing easy access to the peninsula. But ten thousand years ago, the ice age was ending, ice caps and glaciers were melting, and sea levels were rising, eventually by over a hundred meters. Tasmania got cut off from Australia. The sea between Tasmania and Australia is now the Bass Strait. It is only about fifty meters deep, but it is two hundred kilometers wide. Australian aboriginal peoples' sea-craft were little more than close-shore, one-man canoes. Two hundred kilometers was therefore a very effective barrier, essentially isolating Tasmania.

The small population of people then stuck on Tasmania gradually lost the skill to make all sorts of useful implements. Nets disappeared. Boomerangs and barbed fishing spears went. The Tasmanians even discarded cold-weather clothing, though the island gets snow in the winter.

This simplification by the Tasmanians of useful artifacts lies in contrast to the finding for historic hunter-gatherer societies. Among them, the complexity of their tool kit increases with latitude, and with decrease in the productivity of the land, and might increase with increasing risk of famine. For instance, temperate and Arctic hunter-gatherers have three times the number of food-gathering tools, especially weapons and traps, that tropical peoples do. In other words, the Tasmanians lost tools, even though they are in an environment in which other such hunter-gatherer societies seem to gain them.

Consequently, the current theory is that the Tasmanian population was so small that without continual interaction with surrounding cultures, continual renewal of skills, and continual input of artisans when local artisans died, the ability to make these artifacts was lost.

The Tasmania effect is not confined to Tasmania. Michelle Kline and Robert Boyd recently found that among the ten islands of the western

Pacific that they studied, the number of types of indigenous tools used to exploit the sea matched the size of the human population on the islands. Islands with few people had a limited variety of tools; islands with many people had a wider array.

For instance, Malekula with a population one tenth the size of the population of Manus had thirteen forms of tool compared to Manus's twenty-eight. Kline and Boyd did not point out that their ratio of population size to types of tools is extraordinarily close to the one-half-to-one-tenth ratio of the number of species on islands in relation to the area of the islands.

Two islands in the Kline-Boyd study were obvious exceptions. One was Tonga. With a population only one third larger than Manus, it had twice as many sorts of tool as did Manus. To have twice as many, it should have had ten times the population of Manus if it had been following the species-area ratio, or four times the population if following the cultures-area ratio. The reason for Tonga's unusually high diversity of tool types is that it had more contact with other islands than did Manus, even though Tonga is farther out in the Pacific than is Manus. The second island was Trobriand. It had too few types for the size of its population, especially as it has a high degree of contact with other islands. Kline and Boyd did not tell us what might explain Trobriand's relative paucity of tool variety.

Kline and Boyd accounted for other factors than size of the population as a possible influence on diversity of tools. We know that the number of scientists who have conducted surveys can affect the number of items reported. We know that just this sort of bias affects counts of bird and plant species. Where do we find the most birds? Where bird watchers spend the most time "twitching," as the Brits call it. And where is that, but where it is easy to get to—along roads. Kline and Boyd checked for this effect by asking whether the variety of tool variations reported correlated with the number of publications concerning the islands and the number of authors of those publications. They found no correlation.

One potentially important factor that they did not report as an effect, though, was size of the islands. What if larger islands had more habitats, the exploitation of which would be more efficient with more types of tool? I tested for an effect of area, and found none. Correspondingly, the exceptional islands of Tonga and Trobriand were neither unusually large nor small.

The Tasmania effect works, then—more people means more ideas, more complexity of tools. Even laboratory experiments show the same result, as Maxime Derex, Marie-Pauline Beugin, and their collaborators found. The larger the group cooperating on a computer simulation to make a stone tool or a fishing net, the more likely the group was to design an effective product.

In human evolution, we can see bursts in the complexity of tools and art. Painting with ochre appears in South Africa a hundred thousand years ago. Elegant stone blades and decorative shells appear in South Africa seventy thousand years ago. The glorious cave paintings of western Europe appear roughly thirty-five thousand years ago. In the dry, difficult Australian continent, we do not see signs of complex culture until going on twenty thousand years ago, which is maybe twenty-five thousand years after humans first got there.

Might it be that at those times in those regions, human populations grew large enough for a synergy of ideas and innovation to blossom? Paul Mellars suggested this idea more than fifteen years ago. An alternative hypothesis promoted by Richard Klein is a reorganization of the human brain at these crucial points in human cultural evolution. Klein's idea is as yet untested, and might be difficult to test. Also, while it might just work for humans a hundred thousand years ago, maybe even fifty thousand years ago, which are the main dates that Klein was talking about, is anybody going to accept the idea of a major brain reorganization among Australian aboriginal peoples just twenty thousand years ago? The population-size argument for a blossoming of cultural innovation seems more compelling, and some mathematical modeling relating projected population sizes in different parts of the world to the date of the appearance of cultural innovation there has shown that the idea could work.

To bring to a close all this discussion of the effect of area, numbers, and distances on diversity, I am going to end this section with a study of species in human homes. Arturo Baz and Victor Monserrat have shown that small Madrid apartments, equivalent to small islands, have fewer species of lice than do large apartments, equivalent to large islands. It is not just small apartments that have fewer lice. So also do new apartments. It takes time to build a full complement of lice species in an apartment.

If islands are sometimes poor in number of cultures and species compared to mainlands, they are often rich in cultures and species found nowhere else in the world. I have already mentioned the lemurs of Madagascar. As described earlier, this phenomenon of species or cultures specific to a particular region is known as "endemism," and the species as "endemics." Earth's species are almost certainly endemic to earth. Other planets are not going to have humans, even if they have intelligent life. Many of Africa's species are endemic to the continent, found nowhere else. Kenya has nine endemic bird species, one of which lives on only Mt. Kenya.

Endemism arises in the same way among cultures as among species. Isolated from the rest of the world, the cultures and species evolve in their own directions, undiluted, unaffected by the wider community of dialects, languages, populations, subspecies, or species.

Of course, island species and peoples got there from somewhere else. We can still see their relation to their origins. The kiwi birds of New Zealand, native to only New Zealand, are related to Australia's emus and cassowaries. The Māori peoples and culture, native to only New Zealand and its nearby islands (or Aotearoa, as New Zealand is termed in Māori), are closely related to where the New Zealand Māoris came from: the Cook Islands of central Polynesia.

Islands are characteristically rich in unique species and cultures because they are isolated. They characteristically have few species and cultures, because they are both isolated and small. But though poor in number of species, islands often have denser populations of the species that *are* there. They have more individuals per given area than among related species on the nearby mainland. Madagascar provides a prime example. Its lemurs live at three times the population density of Africa's monkeys. What is going on?

Predators and competitors are common answers. As I have already explained, small islands have few predators, meaning that populations of the prey, in this case lemurs, can build to higher levels than on large islands or mainlands. Small islands have few species, so any one species faces few other competitors, and therefore populations can build to higher levels than on large islands or mainlands.

Madagascar indeed has fewer sorts of predators than does Africa. For instance, the island has just one family of mammalian carnivore, the euplerids, a mongoose-like animal made up of less than a dozen species; Africa has five families and over sixty species of carnivore. Madagascar's diversity of carnivores is less than we expect from just the contrast of its area with Africa, but then Madagascar is hundreds of kilometers out into the Indian Ocean from Africa, and therefore difficult for terrestrial animals to reach.

Nevertheless, if the prey, the lemurs, can build to high densities, so can the few existing sorts of predators. In this case, then, the small number of species of predator cannot explain the high density of lemurs. Rather, with all the terrestrial predators coming from just one sort of family, maybe a lack of variety of predators can explain the high lemur densities. Maybe the lack of any terrestrial predators larger than a mongoose allows the largest lemurs to do particularly well, because they are too large for a mongoose-sized predator to take them. In fact, the largest lemurs live at the same densities as similarly sized monkeys in Africa.

A lack of competitors might then be a better explanation for the high densities of most of Madagascar's lemurs. The main other arboreal, plant-eating mammal in Africa is squirrels. Madagascar has no squirrels, whereas Africa has more than fifteen species, together competing for many of the same foods that the monkeys eat. An explanation, then, for the high densities of Madagascar's small and medium-sized lemurs, ones roughly the size of squirrels, is that the lemurs face no competition from squirrels. With more food, lemur populations reach higher densities than do Africa's monkeys. In sum, islands with fewer competitors can be good places to live for the species that make it there.

But this book's title is *Humankind*, not *Lemurkind*. I showed in my *Human Biogeography* that neither hunter-gatherers, nor speakers of indigenous languages, lived at higher densities on islands than they did on mainlands. Why not? Why do we not do the same biogeographically as other mammals?

A lack of predators on islands should not be relevant to densities of modern humans. We are so much larger than most predators that can make it to islands. Yes, the Komodo dragon is heavier than a human, but only the most sleepy human would be taken by a lizard. That leaves competitors.

Island humans should face less competition if islands have fewer species than mainlands of the same area. And of course they do. No Indonesian island smaller than Sumatra has elephants, and none smaller than Java has rhinos. Vanuatu in the western Pacific does not have locusts. Hawaii in the central Pacific does, but they were introduced by humans in only the 1960s.

A main difference between humans and other animals is that humans can easily kill off competitors, at least non-human ones. As I mentioned, the hobbit hunted stegodons. Even if we humans did not kill off our non-human competitors, other humans are probably any one human's main competitor. Perhaps humans are better at killing our main competitor than are other animals. Alternatively, humans can probably more readily leave an island when competition becomes intense.

Another important influence on density that we know of is disease. Disease epidemics can cause populations to crash. If competitors find it difficult to reach islands, then so too should diseases. Malaria was originally an Old World disease, so if it is spreading out to Pacific islands, Asia is the likely source. Vanuatu, one thousand kilometers east of New Guinea, is as far as malaria has reached. Vanuatu defines the so-called Buxton line, the line east of which islands are so far from Asia that malaria has not yet reached them. Mosquitoes have gotten farther out, and carried diseases with them, but apparently not yet malaria.

Nevertheless, hardly anyone has studied the biogeography of disease on islands, as far as I can tell from my search of the literature. We have innumerable counts of plants, insects, birds, people on islands compared to similar areas on mainlands, but next to no counts of disease organisms that I know of. This is despite the continual warnings from the Centers for Disease Control regarding various mosquito-borne diseases on Pacific islands. For instance, earlier in 2014, Fiji was suffering from an outbreak of dengue, a viral disease transmitted by mosquitoes.

I did, though, manage to find one biogeographical study of islands and disease in humans. It comes from nearly fifty years ago. Francis Black reported that measles was more prevalent on islands with larger numbers of humans. Hawaii, for instance, with over half a million people at the time of the study, had measles throughout the year. The Cook Islands, sixteen thousand people, had measles just one month a year. Disease prevalence, then, has the same relationship with population size as does variety of tool types. More generally, it matches biogeographical relationships of biodiversity with area. Number of people available to host a disease is equivalent to number of hectares available to host a species.

Two islands in Black's study were unusual. Guam had a higher incidence of measles than expected for the size of its population. French Polynesia, essentially Tahiti, had an unexpectedly low incidence. Remember Tonga, with its unusually high number of tools for the size of its population by comparison to Manus, because Tonga had unusually high contact with other islands? It turns out that contact explains Guam and French Polynesia. Guam, as an American military base, had a regular turnover of inhabitants. In other words, it had effectively a larger, less isolated population than the size of its resident population or its position in the Pacific indicated. French Polynesia, by contrast, way out in the middle of the Pacific, the farthest island out of the thirteen Pacific Ocean islands examined, is unusually isolated.

Statistically, the population size of the islands explained on the order of two thirds of the variation among the islands in number of months with measles. As the exceptions indicate, though, distance east into the Pacific—distance from the nearest main landmasses—also had a little over a ten percent influence.

On the assumption that statements concerning proportion of variation explained might be unintelligible to readers who are not professional scientists, let me try an analogy. You ask Americans whether they prefer Democrats or Republicans. You get a mixed answer. Nowadays, the USA seems split 50:50. However, two thirds of the responses match the dominant political party of the state that the respondent lives in. Coastal states tend to vote Democratic, inland states Republican. In a scientific paper, that result would be written as "Two thirds of the variance (that's the strict statistical term) is explained by state of residence."

But, of course, some coastal residents vote Republican and some inland ones vote Democratic. What else explains the distribution of votes? One tenth is explained by the sex of the respondent—more females than males vote Democratic. So some male coastal voters will buck the trend and vote Republican, and some females from inland States will vote Republican.

After that, all sorts of other factors explain the remaining quarter of the responses. For example, we know that age has an effect—a greater proportion of old people than young people vote Republican. I will end this short explanation by saying that two thirds of variation of a biological system explained by one influence, in this case population size, is an unusually large proportion. Biologists, biogeographers are happy if they can explain twenty percent of the variation observed.

Pacific islands are not peculiar in their association between population size and disease. The same relationship exists on Atlantic islands. Iceland's population in the mid-1900s of one hundred and sixty thousand people had measles during roughly two thirds of the months. The people of St. Helena (where Napoleon was exiled by the British), and the Falklands, with less than five percent of Iceland's population (and less than fifteen percent of the area), got measles less than two weeks each year on average.

These results reflect what happened with the bubonic plague that hit France in the early 1700s. Records indicate that all cities (islands in a sea of countryside) of over ten thousand suffered from the plague, whereas hundreds of villages of less than one hundred inhabitants escaped it. Travel likely also contributed to the difference. Almost certainly, a greater number of people entered and left the large towns than tiny villages.

∽

In sum, if you want to see many species or cultures, do not go to an island, especially a particularly distant island. But if you want to see unusual species or cultures, then an island is the place to go, especially an isolated one. If you want to be healthy, travel to the most distant, least visited island possible. If you want to do novel research on the biogeography of diseases, then as far as I can tell, islands are where you want to go, because we know extraordinarily little regarding disease on islands. However, if you are from a mainland, make sure *you* are healthy before you go, because we know that island species and peoples can be extremely susceptible to mainland diseases. I come back to this susceptibility in the next chapter but one.

CHAPTER 9

WE ARE WHAT WE EAT

*Our diet affects our genes, and different regions
eat different foods*

N early two centuries ago, Jean Anthelme Brillat-Savarin famously told us that we are what we eat. He also famously told us that the "discovery of a new dish confers more happiness on humanity than the discovery of a new star." Brillat-Savarin was French. He lived through the French Revolution of the late 1700s, dying in 1826 at the age of seventy-one. A magistrate by then, he is most famous now for his *Physiology of Taste. Meditations on transcendent gastronomie ... dedicated to Parisian gastronomes.* So important is the book to gastronomy that no less a food writer than M. F. K. Fisher took it upon herself to translate it into English.

Brillat-Savarin knew nothing regarding the biology behind the geography of our diet. But if our external environment (where we are) affects our anatomy, our physiology, and our cultural diversity (in other words, what we are), then it follows that what we put into our body from the external environment also affects what we are. The peoples of different regions of the world have different diets. And those different diets have produced over time some of the differences that we see between peoples from different parts of the world.

This chapter then is about how our diet affects our physiology, about how people from different parts of the world have evolved over thousands of years different physiologies because their diets differed in the past, or differ now. In one case, diet has affected us through inducing changes in our physiology in our lifetime. But on both time scales—evolution over millennia, and change during a lifetime—our diet affects us. Where we are affects our diet, which in turn affects *who* we are because it affects *what* we are.

The only person I have met who had to go to the hospital after an evening of drinking was a Japanese student. And this despite the fact that I was an undergraduate at a British university, one of a set of institutions not known for the sobriety of its junior members. The hospitalization surprised me, because the Japanese student had drunk nowhere near as much as did some of the British university students that I had known, and who suffered only a hangover the next morning. At the time of the Japanese student's hospitalization, though, I did not know about the relative susceptibility to alcohol of about thirty percent of Japanese and other southeast Asian peoples compared to people from the rest of the world.

The problem for susceptible southeast Asian peoples is not the alcohol itself, but rather the chemical that the body converts alcohol into. It is called acetaldehyde. Acetaldehyde is effectively a poison. Whereas much of the rest of the world seems to be able to convert the acetaldehyde into harmless chemicals quite quickly, if sometimes not quickly enough, the susceptible southeastern Asians lack the enzyme that facilitates the conversion.

As a result, many Asians who drink more than a little alcohol suffer more than do, say, Europeans or Africans drinking the same amount. To add insult to injury, the Asians get less of a lift from the alcohol itself, because, it seems, their physiology is extra-fast at breaking down the alcohol into acetaldehyde.

Less buzz from the booze, and a quicker and more intense hangover from the acetaldehyde, and it turns out that, sensibly, the peoples of southeast Asia drink on average not far off from half the amount of alcohol per person that western Europeans drink. Correspondingly, they have lower rates of alcoholism. Note that alcoholism is not the same thing in this context as alcohol poisoning. The former is long-term addiction. The latter is what happens after a bout of drinking. Compare consumption per head in western Europe, and according to WHO's list of eighteen countries by alcohol consumption in 2010, the people of sixteen of those countries drink on average more alcohol per person than do Japanese people. Those two relatively temperate nations are Iceland and Italy. For the interest of native English-speaking readers, the United Kingdom is seventh on the list. Were they added, Australia would be fourth, and the United States and Canada would be sixteenth and seventeenth, respectively, all ahead of Japan.

We have no substantiated explanation for why southeastern Asians have a different alcoholic physiology from westerners. It could be pure chance—the founder effect that I wrote on three chapters ago. Alternatively, the southeast Asians could have evolved sensitivity to aldehyde as a means of reducing alcohol consumption. Yi Peng and co-workers from the Chinese Academy of Sciences in Kunming suggested this idea. Their argument is based on the fact that timing of the evolution of the gene that enables the body to break down alcohol shows the same geographic pattern as does rice cultivation. In essence, then, southern Asians would have cultivated rice and drunk rice wine for longer than westerners have cultivated their cereals and drunk beer. The Asians have therefore had longer to evolve a physiology that leads to their drinking less alcohol, and so avoiding alcohol's other disadvantage, namely the deleterious consequences of drunken behavior.

This idea of increased suffering to prevent disadvantageous drunkenness seems to argue for the evolution of a Puritanical physiology. I don't see that such punishing deterrence could work. My feeling is that, on the one hand, the means of prevention are on the one hand at least as costly as the drunkenness, and on the other hand that the drunkenness itself carries enough cost to prevent overconsumption. Evolving further means of prevention would carry no advantage. But that is merely opinion at present.

I do want to stress, though, that there is more to drinking, alcoholism, and the associated suffering than just physiology. Wealth and culture play a large part. Of the top twenty countries for alcohol consumed in relation to GNP, twelve are eastern European, and six are African. The Muslim country that drinks the most per head is thirty-ninth on the WHO 2010 list, and the ten countries with the lowest level of alcohol consumed per head per year are all predominantly Muslim.

Ecuadorean Andean Indians have the same alcohol-aldehyde physiology as do the Japanese. Yet Ecuadorean Andean Indian males have a high incidence of alcoholism. Drinking to get drunk is a cultural norm in the case of at least one group of the Ecuadorean Andean Indians, as among so many western European undergraduates. But, as Carola Lentz describes it, the drunkenness of at least some Ecuadorean Andean Indian societies is not associated with anti-social behavior. In those societies it is part of a quiet, eventually near-comatose event that seems to have a lot to do with male bonding.

A commonly suggested relief for a hangover is a glass or two of milk. With milk being mostly water and a little fat, the remedy probably works. But it would work for only one third or so of the world's population in only some regions of the world. For the many others in the world, that glass of milk would make the morning after even worse.

The story of the evolution of the ability of adults to digest milk has been told several times and is still being told, because new details are still appearing. Archeology, genetics, medicine, and evolutionary biology are all involved. Pascale Gerbault, Anke Libert, and six others have provided one of the more complete recent reviews of facts and ideas. Here follows a summary.

We all, of course, can drink milk as infants. But in early childhood, most of the world loses the ability to digest milk or, more precisely, to digest milk's sugar, known as lactose. The word comes from the same Latin root as lactation, the act of a female giving milk. At around three years of age, by which age most children are weaned, the body ceases to produce the enzyme that breaks down the lactose.

Our bodies digest the lactose in the small intestine, because that is where we produce the break-down enzyme, called lactase. If a person cannot produce lactase, but continues to consume milk or milk products, the lactose passes undigested through the small intestine. In the large intestine, the lactose, instead of being broken down by lactase and absorbed through the walls of the intestine, is broken down by bacteria. Gas and water are the byproducts—in other words, flatulence and diarrhea. Belly cramps and vomiting can also occur. None of these causes serious trouble in rich households, but in poverty, and especially where clean drinking water is lacking, diarrhea and vomiting can be dangerous because of the associated loss of fluids. Unlike pathogen-induced diarrhea and vomiting, such as we get with cholera, lactose-induced illness is rarely serious enough to kill.

One of my greatest pleasures is cream in my morning coffee, and heavy cream on my desserts. I have freely indulged my creamy pleasures through my adult life because I am a native-born western European. Most people of northwest European origin continue to produce the lactase enzyme into adulthood, and so will not have experienced the symptoms of lactose intolerance.

The same goes for the people of much of west Africa. Populations in Saudi Arabia and Pakistan also can digest milk as adults. Most everywhere else in the world, though, the symptoms of lactose intolerance will be familiar to those there who have made the mistake of trying, say, a western-style ice cream with its high content of milk. Much of southeast Asia, including southern China, and southwest Africa are especially cold spots for the ability to digest lactose.

If you want an easy sample of the worldwide distribution of what the scientists term "lactase persistence," look at the "Got Milk" advertisements on the web. Almost all the people featured holding a glass of milk and with their upper lip smeared in milk are Caucasians. A few are of African origin. Hardly any are Asian. Before the pictures started repeating, I counted twenty-six Caucasians in the advertisements, six people of African origin, and just two who might have had Asian origins.

The family members of Pegasus Books' associate publisher, Jessica Case, also nicely illustrate how region of origin affects diet via lactose tolerance and intolerance. Her father is of European ancestry, and her mother is Chinese. Correspondingly, her father can happily enjoy milk and cheese; her mother cannot. Jessica is one of four siblings, three females and one male. Her brother is lactose-intolerant, but neither she nor her sisters are. That biased sex ratio applies also to her ten cousins on her mother's side of the family, all of them of Chinese and European ancestry. All the males are lactose-intolerant; none of the females is. Science has not shown that lactose intolerance is linked to sex, so the link between sex and the syndrome in Jessica's family is a result of unlikely chance. Somehow the males, but not the females, inherited the control gene that switched off production of lactase.

What is special in the environment of northwest Europe and some parts of Africa that correlates with their populations' ability to drink milk, whereas most people from other regions cannot? The answer is that from at least eight thousand years ago in Europe, and maybe five thousand in Africa, the people have herded livestock. Livestock produce milk. How could the herders not use this potentially important source of nutrients and fluid?

As they did so, more and more of the population evolved the ability to digest milk into adulthood, because they survived better with this extra source of fluid and nutrients than did those who could not digest lactose. Genetically, they ceased to switch off the genes that allowed them to produce lactase. Peoples of other regions who did not herd livestock, and so did not find themselves in a milky environment, continued to shut down lactase production at the end of infancy. There is no benefit to producing a useless enzyme, and indeed probably a cost.

One nice study showed that in western Europe we find not only a concentration of people who can drink milk as adults and of the genes that enable them to do so, but also a high diversity of domestic cattle genes associated with the production of milk proteins, along with a heavy concentration of Neolithic cattle-farming sites. The diversity of the genes implies a long history of milking. I describe it as a "nice" study, not just because of the demonstration of overlap of three phenomena, but because the phenomena belong to such different fields of enquiry: medicine, genetics, archeology, and evolutionary biology.

Scientists interested in the origins of the ability to drink milk in adulthood have debated which came first, the genes that allowed the ability, or dairying. In the former case, the gene would have had another job, in other words another chemical advantage, and dairying would have been more readily adopted by populations that carried the gene than by populations that did not. How do we distinguish the possibilities?

The answer is to go to populations living before dairying developed. In other words, examine the skeletons of people who died thousands of years ago, some before the origins of dairying and some after. A team that did this found no evidence in the pre-dairying Neolithic skeletons of the genes that allow digestion of lactose in adulthood. It looks as though dairying preceded the evolved ability to drink milk as an adult.

The buzz-phrase in scientific circles for what happened here, a change in the environment (milking) to which humans adapt genetically (lactase persistence) is "gene-culture interaction," or "niche construction." Both now have a large literature. "Nicher" is French for "to nest," and indeed part of the niche-construction literature concerns the ways in which birds change their own environment when they nest, and the effects of that change on the birds.

One might think that because the genes that enable the manufacture of lactase into adulthood have the same job in African and European herder peoples, they would be identical in those two populations. But they are not. Although on the same chromosome and close in position to one another, nevertheless the genes are different. Strictly, I should say the "alleles" are different, because we are talking of the same gene in Africans and Europeans, but different forms of it. Groups headed by Catherine Ingram and Sarah Tishkoff demonstrated this. Let me say again that for anyone interested specifically in genetics of African populations, Sarah Tishkoff is the name to look for.

As a specific example, one of the most recent detailed studies of the distribution of the alleles associated with lactase persistence comes from Sarah Tishkoff's group in a paper headed by Alessia Ranciaro. The group identified one allele especially common in eastern Africa, another in the Arabian Peninsula, another in western Europe, and a fourth in Eritrea. In all these regions, herding was one of the most dominant lifestyles. However, so also was hunting and gathering.

Hunter-gatherers are, by definition, not herders. What on earth are they doing with gene variants that allow them to digest lactose as adults? Well, the full chemical name for lactase is lactase-phlorizin hydrolase. The so-called lactase breaks down not only lactose, but also phlorizin. Chew a little bit of cherry bark or apple bark, and that bitter taste is phlorizin. The connection to hunter-gatherers is that at least one group of hunter-gatherers, the Hadza of Tanzania, use phlorizin-containing parts of plants as a medicine.

Here we might have an example of an ability evolved for one benefit, in this case medical, also being useful for another potential benefit, in this case if the Hadza lifestyle changed and they became pastoralists.

People started to herd livestock and drink milk within just a few thousand years of one another in northwest Europe and Africa. So far apart did they live, though, with so little contact, that the two populations evolved different genetic mechanisms to benefit from milk. And it looks as if a third set of genes exists among Saudi Arabians who drink camel milk.

I mention this difference in genes/alleles doing the same job to emphasize an aspect of evolution by natural selection. More than one way exists to build a bridge. To quite a large extent, the outcome is what is important, not the means by which the outcome is produced. Birds fly fine, and bats fly fine. However, they do so using different means. Wings certainly, but trailing feathers from the arm and hand bones of birds, skin stretched between the arm and finger bones of bats.

The fact that the genes involved in the ability of adults to drink milk differ between Africa and Europe has given rise to the idea that some benefits from milk might differ between the two regions. In Africa, the benefit might be a source of clean liquid, an item often in short supply there. In Europe, the greater benefit from milk could have been its vitamin D in a region short of sunlight for several months of the year. We encountered the relation between the sun and vitamin D in chapter 5, where I wrote on skin color and why it differs between regions of the world.

This potential geographic contrast in the benefits of the ability to drink milk beyond infancy is supposition at present. In science, we test suppositions (or "hypotheses," as scientists prefer to call their ideas) by making predictions regarding what we should find were the supposition true.

With regard to milk consumption into adulthood, the predictions have not so far borne fruit. For instance, Clare Holden and Ruth Mace found no link between a lack of clean water and the ability of adults in a population to digest lactose. But, as the phlorizin story earlier indicates, maybe the original benefit of lactase production was not in fact the lactase, but the phlorizin part of the full name of the enzyme, lactase-phlorizin hydrolase.

Whatever the benefit, it is a remarkably strong one, one of the strongest ever demonstrated. Genetic calculations of the rate of spread of the gene through the population indicate that since the gene arose, people with it produced approximately ten percent more living offspring than people without. This is a difference comparable to the benefit from adaptations to malaria or sickle-cell anemia, which I come to in the next chapter.

By now, some readers might be saying: wait a moment, we know of people from regions other than northwest Europe and west Africa who nevertheless milk their animals. There cannot be a tourist who has gone to East Africa who has not seen the large herds of cattle there, and the Maasai who manage them. And yet only around two thirds of adult Maasai appear to have genes that enable them to use the lactose in milk.

Maybe some of the Maasai have the same adaptation as Somali pastoralists of eastern Africa, namely a non-genetic way to digest lactose in adulthood. Adult Somali pastoralists do not produce lactase after infancy, and yet they drink milk with no ill effects. Perhaps they can do so, Catherine Ingram, Charlotte Mulcare, and co-authors suggest, not because they have evolved genetically to continue to produce lactase into adulthood, but because their gut bacteria produce lactase. As far as I know, nobody has yet worked out how or why the bacteria do so, or why not all humans have gut bacteria that can digest lactose. We do know, though, that different populations of humans have very different complements of bacteria in and on them. I come back to this phenomenon when I write about how other species influence where humans live.

Mongolian pastoralists are another example of a people outside the adult-lactose region of northwest Europe and west Africa who milk their livestock. However, the Mongolian pastoralists do not drink the milk. Instead, they turn it into yoghurt before they consume it. Crucially, the fermentation process that produces yoghurt converts the lactose into sugars such as glucose that the adult body can digest with no trouble.

Alternatively, if milking populations let the milk sit, maybe heat it a little, keep stirring, then the liquid portion, the whey, will soon separate from the remainder, which is essentially cheese. The cheese is free of lactose, because the lactose is dissolved in the whey. My wife and I saw Bhutanese herders producing cheese in this way while they kindly allowed us to shelter in their yurt during a monumental rainstorm.

The herders let the cheese dry to the consistency of what we are used to as cheese, cut it into cubes, threaded the cubes on strings, and then dried it to the consistency of slightly rubbery chalk. It lasts a long time like that. To eat it, they suck it as we in the west would suck a hard candy, but their cheese lasts way longer in the mouth than any hard candy that I know.

To separate the cheese from the whey, the Bhutanese ladies scooped the cheese from the huge bowl, maybe a meter wide, that was sitting on the coals. Seven thousand years ago, the people of Poland achieved the separation in a different way, as Mélanie Roffet-Salque and her colleagues showed. The people (was it the women who had the job?) used pottery sieves. The whey drained through the holes, leaving the cheese behind. A thousand years before then, the people of the Middle East, Turkey, and southeast Europe were all using pottery to store milk. How they used the milk, whether the people drank it or converted it to cheese, we do not know. We do know, though, that dairying in Europe probably came from the Middle East. We can guess this from the finding that European cows are more closely related to Middle Eastern cows than they are to the native aurochs. That was the conclusion from an analysis by Ceiridwen Edwards, Ruth Bollingino, and a team of over thirty others of mitochondrial DNA in bones from archeological sites dating to several thousand years ago.

Dairying might go even farther back, to the origins of domestication, ten and a half thousand years ago in the Middle East. The idea is a clever interpretation by Jean-Denis Vigne of skeletal remains of sheep, goats, and cattle. If the people were using the livestock mainly for meat, skeletons would be of near-adults or adults, he reasoned. If they were using them for milk as well, a more-or-less large proportion of the skeletons should be of calves—killed so that the humans rather than calves get the milk. A fairly large proportion of the skeletons were indeed calves.

Lactose is a so-called simple sugar, or, in Greek, saccharide. The staple diet of most of the agricultural world—rice, wheat, tubers, and so on—consists of a form of carbohydrate, starch, which is a polysaccharide (many joined saccharide molecules). As with alcohol and milk, the biology of its digestion varies across the world.

Digestion of starch begins with saliva. Saliva is not there just to wet our mouths. It contains an enzyme, amylase, that begins the breakdown of starch. The breakdown continues in the small intestine, where the amylase comes from the pancreas. Amy1 (that's AmyOne, not AmyL) is a major gene responsible for our ability to produce amylase. Current estimates are that a little over ten percent of our genes come in multiple copies. Amy1 is one of these. Importantly for a story on human variation around the world, people from different regions of the world have different numbers of copies of this gene.

People who eat a lot of starch—for example, people of western European origin, such as myself, with our diet of wheat and potatoes; the Hadza people of Tanzania, who eat wild tubers; and the Japanese, who eat a lot of rice—have an average of six copies of the Amy1 gene. People who do not eat much starch (two pygmy populations and a pastoralist people of Africa, and a fishing people of Eurasia) have an average of four and a half copies of the gene. It is as if the number of copies matches the ability to digest starch.

What about our closest primate relatives? Chimpanzees eat very little starch indeed. If they are following the same physiological path as humans, they should have even fewer Amy1 copies. They do, averaging only two. I have not found data for other animal species on their number of Amy1 copies. However, a study from thirty-some years ago indicates levels in blood serum of amylase in several monkeys and apes, including the chimpanzee, that are similar to human serum levels. One species had a level six times that of humans. Maybe, then, the starch story is more complex than simply the number of copies of the Amy1 gene indicates. Maybe, mirroring Somalis' ability to digest milk, these monkeys and apes had a gut bacterium that could produce amylase.

An advantage for humans of starch from tubers is that getting at tubers is difficult. A lot of work can be involved in digging them out of the ground. That sounds like an oxymoron. However, the benefit of a difficult-to-get-at food is that few other animals use it. One that does is the naked mole rat, an extraordinary animal with a termite-like social system and lifestyle. One female breeds, a few males mate with her, and the rest of the large subterranean colony are workers that look after the young, dig tunnels, and search for tubers. Digging through the rock-hard earth of eastern Africa is difficult enough for humans, especially as grandmothers are one of the main diggers in a Hadza community, according to Kristen Hawkes and her co-authors. But at least the humans can search for the tubers by walking and looking for the stems, whereas the mole rats have to search by digging. Incidentally, the naked mole rat is indeed naked—it has no visible hair.

To digest starch, we really need to cook it. Try eating a raw potato, and you will end up with a bellyache. If we can cook, though, then tubers, especially tubers in dry ground, have a great benefit. Unappetizing as at least the Hadza consider them, they are a year-round food supply, available when foods that the Hadza prefer are in short supply. The Irish grew so much potato because it would grow in the wet ground where wheat would not. A main advantage of cooking in relation to tubers is that it makes more of the carbohydrate's energy easily digestible. A cooked tuber provides more energy to the consumer than does an uncooked one. Richard Wrangham and Rachel Carmody suggest that the invention of cooking could explain the increase in size of that energetically expensive organ, the brain, that began approximately two million years ago.

Man cannot live on tubers alone, though. We need folate, vitamin B9, in our diet. We last came across folate in chapter 5, where I described how vital it is to bodily functions, and how too much sun denatures it. A problem with a diet heavy in tubers is that tubers do not contain much folate. But Angela Hancock and co-authors reported that some genetic evidence indicates that people for whom tubers are a dietary mainstay have a form of a gene that improves production of folate in the body.

If tubers are a reliable source of carbohydrate, they can also be a fairly reliable source of a variety of toxic glycosides. Almond nuts and cassava famously contain glycosides chemically combined with cyanide. Cassava is a staple in the tropics in all three major continents. People reduce the level of cyanide by mashing the cassava and soaking it for many hours in water, or by leaving the mashed paste in the sun for hours, when it gives off hydrogen cyanide. Beware, though: these methods of preparation do not necessarily remove all the cyanide.

Not all forms of glycoside are toxic. Indeed, many have beneficial properties and are, in effect, medicines. Others are sweet, and used as sweeteners, tens—even hundreds—of times sweeter than sucrose.

Like milk for adults in much of the world, seaweed is not a food for most people. A large part of it consists of lignin, which the majority of us cannot digest. Instead, seaweed is merely a convenient wrapping to hold our sushi roll intact. Or for the health-conscious among us, seaweed simply acts as the dietary fiber that the nutritionists encourage us to eat. The Japanese are different, though, as we might expect from the amount of seaweed in their meals. They can digest more of it than the rest of us can. But their ability to do so is not a result of an evolved adaptation, at least in the normal sense. They obtained the ability in an unusual way.

The story of how the Japanese can digest seaweed concerns a gut bacterium, as was the case with the Somali pastoralists' ability to consume milk as adults. In the case of the Japanese and sushi, the story concerns a bacterium with the rather attractive name of zobellia that lives and feeds on the sushi seaweed, the gene that allows that bacterium to digest the seaweed, and another bacterium in the gut of the Japanese. Here is how the story goes.

The gene in the Japanese peoples' bacterium that allows them to digest seaweed did not originate in their gut bacterium. Instead, the gene came directly from zobellia. A laboratory directed by Jan-Hendrik Hehemann discovered this fact. Zobellia cannot survive in the human gut, because our stomach's acidity kills it. The only beneficial bacterium that lives in our stomach is *Bacteroides plebeius*. Somehow, though, the gene that enables zobellia to digest seaweed moved into the stomach bacterium of Japanese people or a Japanese person before the stomach acid killed the zobellia.

The scientific name of sushi seaweed is *Porphyra*. (Do not confuse this name with porphyria, the disease that I wrote on in chapter 6.) Zobellia and Porphyra are rather attractive names. I can imagine naming daughters Zobellia and Porphyra.

Scientists have no idea how the transfer of the zobellia gene happened. All we can say is that such sideways movement of genes across bacteria is not uncommon. But the result is that thanks to the zobellia gene, Japanese people can digest seaweed and benefit from its nutritional content, while the rest of us cannot, even if many of us outside of Japan now eat sushi.

So as with milk, perhaps with alcohol, certainly with seaweed, we get different environments (savannah, coast) producing different behaviors (drinking milk, eating seaweed), which cause different internal environments (lactose, zobellia in our diet), which lead to changes in our physiology (lactase production in adulthood, a new gene in our gut bacteria), which changes our external environment (increased use of livestock and the sea), which leads to further change in our internal environment.

In these accounts of physiological adaptations to consuming milk, alcohol, or seaweed, I have suggested that the ability arose via evolution by natural selection. People with the ability raised more surviving children than did people without, and so the population became composed of individuals with the ability. Adaptations can develop over the lifetime of individuals too, as we saw in chapter 5 for some of the abilities to thrive at high altitudes. The ability of Arctic peoples to eat a diet so high in fat that by western standards it should be unhealthful involves a third sort of explanation for how diets differ between the people of different regions. It concerns the exact nature of the fat and other components of the Arctic peoples' diet.

Let me start with some comparisons to illustrate just how much fat Arctic peoples eat. A study published in the 1980s reported that Arctic peoples had fifty percent more fat in their diet than did the British. The Arctic peoples also ate huge amounts of protein, over three times as much as the British. This predominantly animal-based diet is of course not surprising, given the Arctic environment. We do not see much fruit or edible vegetation for much of the Arctic year. Indeed, overall, the higher the latitude, the more animal food in the peoples' diet. Around ten degrees from the equator, approximately forty percent of the diet comes from animals, mostly mammals; around sixty degrees latitude, close to the Arctic Circle, we are up to a diet of approaching ninety percent animal. These figures come from Lewis Binford's wonderful compendium of information on the world's hunter-gatherer cultures, *Constructing Frames of Reference*, as did much of the information in chapter 7.

Lack of plant foods in the Arctic is not the only reason for Arctic peoples' high proportion of animal food. Mammals, especially large mammals, generally provide a far higher return of calories per time spent obtaining those calories than do plants. Consequently, even where plant foods are easy to obtain, namely in the tropics, hunter-gatherers are prepared to put a lot of effort into hunting. Where the hunted animal is large and potentially dangerous, and the hunter is male, showing off is also part of what is going on. But only part. Large mammals tend to have a higher proportion of fat than do smaller ones, and fat is highly favored.

The Arctic peoples' high-fat, high-protein diet would be severely frowned upon by most Western doctors and dietitians. We Westerners would receive strong warnings concerning heart attacks if we continued on such a supposedly unhealthful diet. Nevertheless, in a study done in the 1980s, the Arctic peoples on their high-fat, high-protein diet had, by comparison to the British people, a low incidence of heart and vascular problems and low cholesterol levels. The contrast in fact does not apply to just Brits. In general, Arctic peoples on a high-fat and -protein diet are healthier than Westerners with the same proportion of protein and especially fat in their diet.

Neanderthals too lived mostly on animals, and hence mostly on animal fat and protein. As a species they did quite well. They lasted for at least as long as the human species has so far lasted, two hundred thousand years. Presumably, they were quite healthy through that time or, again presumably, they would not have lasted as long.

I might add that the French, on their diet that's high in foie gras and cheese, are doing fine. France is the top or second-to-the-top consumer, per head, of both in the world. Yet in the World Health Organization's 2013 list of countries by life expectancy, France is thirteenth, at 82.3 years, while the USA, where so many disapprove of foie gras, is thirty-fifth, at 79.8 years. But I digress.

The low-carbohydrate Atkins diet seems to match the Arctic diet. And yet by many accounts, the Atkins diet is controversial. Various studies claim only short-term effects on weight of a low-carbohydrate diet. Weight loss seems to result from water loss, not fat loss; and worse, the Atkins diet might have no obvious effect on heart condition. If the Arctic peoples on an equivalent of the Atkins diet are healthy, why is an apparently similar diet among other peoples not unquestionably beneficial? Part of the answer is that the fat of Arctic peoples comes from wild animals, not domestic ones.

Wild-animal meat contains a higher percentage of polyunsaturated fats than does the meat of domestic animals, in the region of thirty percent as against ten percent by weight. Various fish contain especially high proportions of polyunsaturated fats, for example mackerel and salmon, and fish make up a larger proportion of the animal diet of Arctic people than that of tropical people. And as anyone who even begins to pay attention to diet and health knows, polyunsaturated fats are a lot more healthful than saturated fats. They are even good for us in moderation. Indeed, omega 3 polyunsaturated fats lower cholesterol in the blood.

Another reason why the Arctic peoples' Atkins diet is more healthful is that with fuel in such short supply, the meat is often barely cooked when eaten. It therefore still contains vitamins such as vitamin C, which are largely destroyed by cooking.

Of course, any responsible diet program advises a fair amount of exercise, along with whatever diet the program is promoting. I reported in chapter 7 that Arctic hunter-gatherers used to move farther between camps over the year than did equatorial hunter-gatherers. Equatorial hunter-gatherers used to go five to ten kilometers on average; Arctic hunter-gatherers went over forty kilometers. Most of the hunter-gatherers move camps less than once a week, even once a month. The average distance moved when looking for food might be a better indication of what in the West we would call exercise. Frank Marlowe kindly gave me some unpublished data on camp movements. Arctic peoples, the Nunamiut, moved on the order of twice as far when hunting or foraging as did tropical peoples, a dozen kilometers compared to around seven.

In sum, then, the Arctic peoples' natural diet, high in fat, is not in itself necessarily unhealthful. They ate what a good dieting program might suggest—low saturated fat, low carbohydrate, almost no sugar, vitamins from the plants in their diet, and, remember from chapter 5, plenty of vitamin D from lots of oily fish in the diet. And they exercised a lot.

In this story, the biochemistry of the Arctic peoples' diet is the crucial issue, rather than the biochemistry or physiology of their bodies. The story is still a biogeographic one, though. Peoples of different regions have different diets, which then have different consequences for their health. Arctic peoples can be healthy on their diet high in polyunsaturated fat; neighboring northern Europeans cannot be healthy on a diet high in saturated fats, however much they might wish to be.

Unhappily, obesity is now a problem among several populations of Arctic peoples—because they have moved to a Western diet high in the fat of domestic animals instead of wild ones, and they have moved to Western levels of exercise, traveling on Ski-Doos instead of by foot, for example.

The response of some people to their obesity will be to take drugs to combat, for example, high blood-cholesterol levels. I mentioned in the opening chapter that regional differences in reaction to drugs exist. "Racial diversity crucial to drug trials, treatments" was the 23 July 2014 headline in my local newspaper, *The Davis Enterprise*, in its reproduction of a *San Francisco Chronicle* piece on the fact that people from different parts of the world react differently to the same drug.

The article by Victoria Colliver was calling attention to the fact that non-Caucasians are underrepresented in drug trials. For instance, although African-Americans are twelve percent of the US population, they make up only five percent of patients in US medical trials.

Colliver reported a chilling reason for the low participation of African-Americans in drug trials—mistrust of the US medical establishment as a result of the forty-year Tuskegee experimental program. Beginning in the early 1930s, the Tuskegee experimental program enrolled six hundred rural African-Americans in Alabama. The aim was to understand how syphilis developed in the body. No problem so far. However, the subjects were told that they were receiving medical treatment for generalized weakness—when in fact they were receiving no treatment. They continued to be denied treatment, even though the medics knew that over half of them had syphilis, and knew from the mid-1940s that penicillin was available and could cure syphilis.

At best, the consequence of ignoring the patient's region of origin when prescribing drugs is that while the drug might work for native English-speaking people of European origin, it will not work for others. At worst, drugs developed for the people of one region might actually harm those of another region. Overdoses of certain drugs are easy if doctors ignore region of origin, because people of different regions have different tolerances for different drugs, in the same way that they have different tolerances for alcohol, arguably a drug itself. The *San Francisco Chronicle* piece gave an example of an Asian gentleman on a European dose of an anti-cancer drug. He suffered serious side effects. That was until his Asian oncologist realized that the dose was too high for people of Asian descent and put him on a lower dose.

Lumping people by continent as I have just done here with my mention of "an Asian gentleman" might be too crude a division. The great genetic diversity of especially Africans that I described in chapter 4 implies the possibility of equally varied reactions to drugs among people from different regions of Africa.

∽

The Japanese seem to have started consuming seaweed in the eighth century. As far as I know, scientists do not yet know when in the subsequent twelve hundred years the gene from zobellia moved into the gut bacterium of the Japanese people. We do know, though, when the genes evolved that enabled some of us as adults to digest milk. As I wrote, the answer is only five thousand years ago, maybe even less. Five thousand years ago is within historical time, within the time by which we had invented writing. In other words, we humans are still evolving biologically. Our physiology is changing as, in some environments, some individuals with particular traits survive and reproduce better than do those without the traits. Contrary to what many think, humans have not evolved by natural selection in the far past only. We are still evolving by natural selection.

CHAPTER 10

WHAT DOESN'T KILL US
HALTS US OR MOVES US

Other species influence where we can live

T his chapter is largely about the geography of the effect of pathogens and parasites on us. By contrast to the wholesale slaughter that these mostly invisible organisms visit upon us, the effect of other living things, excepting other humans, is nearly negligible.

Parasites, pathogens, and the disease, plague, and pestilence they bring have surely shaped us from our very origin. Certainly the writers of the bible knew pestilence. The Old Testament gods were continually inflicting disease upon various humans. A plague that touched every house was the last straw for the Egyptians, and they kicked Moses and his followers out of the country (Exodus 12:30-31). "Also every sickness, and every plague . . . them will the LORD bring upon thee, until thou be destroyed," says Deuteronomy 28:61, 62. And the list of pestilences goes on. Indeed, in one interpretation, pestilence was one of the four horsemen of the apocalypse, galloping alongside war, famine, and death.

Most readers of this book will be from temperate countries, and probably from cities. The diseases of high latitudes are different from those of the tropics. The diseases of cities are different from those of the countryside. In the same way that populations adapt to heat, cold, and their diet, so they adapt to the diversity of diseases that they face. Because diseases differ between regions, especially between the tropics and temperate latitudes, so the physiology of the people differs between regions. Disease affects what we are where we are.

In a previous career, I studied gorillas. Two men, Thomas Savage, a missionary, and Jeffries Wyman, a professor of anatomy at Harvard, first officially described the gorilla to science. Thomas Savage lost two wives to disease in west Africa. And yet Africans flourished in what used to be known as the white man's grave. What did the native peoples have that enabled them to survive in such disease-ridden regions? The answer, of course, is tens of thousands of years in which they evolved physiological adaptations to cope with African diseases.

Malaria is one of the more prevalent African diseases, and some of the best-known examples of adaptation to disease-causing organisms are those involved with resistance to malaria. Malaria, the disease, is caused by a single-cell parasite, a plasmodium, which enters the red blood cells. It not only feeds on the hemoglobin that carries life-giving oxygen to our tissues, but damages the red blood cells, so reducing their ability to transport oxygen.

The evolved genetic adaptation to resist malaria, in other words to resist the action of the parasite, is one that alters the nature of the wall of the red blood cells in such a way that the malaria parasite can no longer enter them. Only people who live, or have lived, or whose parents lived in malarial areas carry these sorts of genes, and hence carry an evolved resistance to the disease. Immigrating Europeans, not previously exposed to the African plasmodium, do not have the evolved genes, and they have no natural resistance to Africa's malarias.

If the Africans from malarial regions inherit the form of gene responsible for the resistant red blood cells from just one parent, they do fine in the malarial regions. However, they can suffer in non-malarial regions. They are a bit lethargic, for example. In malarial regions, the cost of lethargy is minor compared to the benefit of protection from disease. But in regions with no malaria, people with the malaria-resistant type of the gene and its accompanying lethargy can do less well than those without it. Consequently, the gene disappears from populations that have left the malarial region. Thus, African immigrants to America have lost the resistant gene over the generations and will continue to do so if the USA stays malaria-free.

But what if a child inherits the resistant gene from both parents? Then they get a double dose of its effects. And as with many medicines, too much is bad for you. With two resistant genes, the child suffers from sickle-cell anemia. The sickle-cell part of the name describes the fact that the normally disc-shaped red blood cells become seriously misshapen, often looking like a quarter moon, or a sickle blade. The anemia part indicates the lack of red blood cells and the variety of resultant symptoms, such as fever, tiredness, abdominal pain, and, in the worst cases, death.

Here we have a classic case of damned if you do and damned if you don't. The people in malarial Africa are caught between a brick and a hard place. If they carry the resistant gene trait, they might survive malaria better, and half of their children on average will survive better also. However, they will all feel tired, and one in four of their offspring might die from sickle-cell anemia. Yet if they do not carry the trait, they and their children will probably suffer from malaria, and possibly die early from it. Indeed, even if they do carry the trait, a quarter of their children on average will not inherit it at all and hence will likely get malaria.

Most people think of malaria as a tropical disease, and it pretty much is now. However, for at least the last millennium, even cold northwest Europe did not escape the affliction. Read old accounts, and the "ague" or "marsh fever" of those accounts is in fact malaria. My following brief description comes in part from Otto Knottnerus's detailed account of malaria in the countries around the North Sea from the earliest to latest records. Most of the rest is from Katrin Kuhn and her co-workers' analysis of the possible influences on malarial incidences in England and Wales in the second half of the 1800s.

For centuries, nobody knew what caused malaria. Malaria is Latin for "bad air," which is what people used to say brought malaria, presumably on account of its association with living near marshes and stagnant water. Nevertheless, the several described symptoms of ague and marsh fever, such as their regular recurrence in the patient, make the diagnosis certain. By the nineteenth century, the disease's successful treatment with quinine confirmed the interpretation.

For reasons that we do not know, malaria from the 1500s to early 1900s in Britain seemed to be extraordinarily virulent. Otto Knottnerus mentions some resistance to malaria, but does not say exactly what conferred the resistance. As recently as the early 1900s, people were dying of malaria throughout England and Wales. Kent, famous for its Romney Marshes, and the fens of East Anglia were hotbeds of the disease. Annual death rates reached one per one thousand people there in the late 1800s. By contrast, west central England and southern Wales were relatively free from the disease. Did Atlantic gales keep mosquitoes away?

With global warming and international travel, could malaria return to Britain? Katrin Kuhn and her colleagues argue not. They point out that fifty-three thousand cases of imported malaria have led to not one confirmed indigenous case. Also, Britain has continued to drain its marshes. The Anglian marshes, already considerably reduced by the mid-1900s, are now less than ten percent of their extent then. Even if the result is less protection against storm surges, at least malarial mosquitoes have largely lost their breeding grounds.

However, Susannah Townroe and Amanda Callaghan suggest that the increasing use of water barrels in Britain, or water butts as they are called there, is causing increased mosquito infestation. Of two mosquito species benefiting from the barrels, one can carry malaria, and the other both malaria and West Nile virus. Eighty percent of England's, indeed Britain's, population is urban. Urban barrels carry a greater density of mosquitoes than do rural barrels. The reason almost certainly has to do with the urban heat sink effect—concrete gets warmer than vegetation, towns get warmer than the countryside, and mosquitoes like warmth.

If malaria returned to Britain, one might imagine that East Anglians and the Kentish would be naturally better protected than others from any resurgence. However, malaria started to decline in the region from the middle of the 1800s, and, as the fens and marshes were drained and channeled, it was completely gone by the mid-1900s. With no malaria, the costs of carrying the resistant gene with its lethargic effects exceeded any benefit. Consequently, within eight generations more or less, East Anglians lost any resistant trait that they might have had. They will therefore be just as exposed as anyone else, excepting recent immigrants in Britain from malarial areas elsewhere in the world.

By the mid-1900s, resistance to malaria was bred out of the British population by the process of natural selection, by the fact that individuals without the resistant form of the gene did better than individuals with it. The mid-1900s is when I was born. Within my parents' lifetime, the genetic makeup of the British population changed as a result, in part, of evolution by natural selection. As I have said before, humankind is still naturally evolving as the environment changes.

The exact form that resistance to malaria takes varies from region to region. That means that the consequences of the deleterious effects of the resistance also differ. Thus, whereas sickle-cell anemia is characteristic of malarial regions of Africa, a form of what is termed thalassemia is more common around the Mediterranean.

The "emia" part of thalassemia comes from the Greek for blood, as in anemia. Anemia means "no blood"—because the sufferer is pale and weak. The condition protects against mosquitoes, but affects the structure of hemoglobin, and hence its ability to carry oxygen, as well as, indeed, the viability of the red blood cells. People with only one copy of the gene have small blood cells, but otherwise are okay. With two copies, tiredness and bone deformities of the face result, and babies get particularly severe anemia.

The word "thalassemia" means "anemia-by-the-sea." It refers to the fact that it was populations living in hot low-lying areas that were most likely to have the side effects of the particular resistance to malaria that produced thalassemia, because the malaria-carrying mosquitoes do well in hot, marshy areas.

Another example of a regionally confined disease producing regionally different people is sleeping sickness. The disease is prevalent in tropical Africa. So what do we find, or rather what has a group led by Martin Pollack found, but that a gene variant termed ApoL1 that protects against sleeping sickness is far more common among people of African descent than among, for instance, people of European, Chinese, or Japanese descent.

Tsetse flies transmit sleeping sickness. As with mosquitoes and the single-cell plasmodium that causes malaria, when the tsetse "bites," it injects the single-celled "trypanosome" that causes sleeping sickness. Anyone who has holidayed or worked in tsetse areas of Africa knows that tsetse fly bites are as painful as bites from horse flies. And the tsetse flies are just as difficult to kill by swatting as are horse flies. They seem to be armor-plated. Hot as it can be in African wooded savannah, the preferred environment of tsetse flies, I have preferred to have the car windows shut against the tsetse fly there rather than suffer their bite.

Trypanosomes look like small worms, two or three red blood cells in length. Unlike the variants of hemoglobin that confer resistance against malaria by making the walls of the red blood cell more impenetrable to the plasmodium, the blood plasma of Africans resistant to sleeping sickness appears to turn into a parasiticide. It kills the trypanosomes. Another well-known affliction caused by a trypanosome is Chagas disease, prevalent in Central and South America, although the trypanosome is a different one than causes sleeping sickness.

Sleeping sickness and ApoL1 are a further example of damned if you do and damned if you don't. ApoL1 confers resistance to sleeping sickness. Good. But Africans from sleeping-sickness areas who carry two copies of the ApoL1 gene are prone to various kidney diseases. Bad. For instance, people of African descent in America are up to four times more likely to suffer from kidney problems than are Americans of European descent. The African-Americans are particularly prone to one form of kidney disease, which doctors term "focal segmental glomerulosclerosis." It results from scar tissue forming on the thousands of little bulbs in the kidney, the glomeruli, responsible for filtering unnecessary chemicals from the blood. People can die from focal segmental glomerulosclerosis. Sleeping sickness can also kill. But the Africans in tsetse-fly areas are less damned if they carry the ApoL1 gene variant than if they don't—because sleeping sickness will probably kill earlier in life than will kidney disease. In America, where no sleeping sickness exists, kidney disease becomes the more pressing problem.

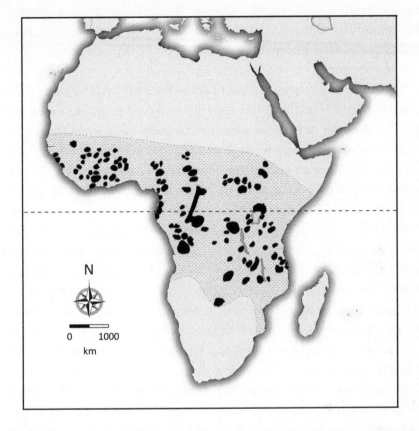

Distribution of tsetse flies (gray) and sleeping sickness (black) in Africa. *Credit: Redrawn by John Darwent from OxfordMedicine.*

A few years ago, a friend of my wife's and mine was feeling under the weather. Her doctor's clinic in California took a blood sample. An alarmingly high white blood-cell count indicated leukemia or "a parasite," her doctor said. Getting rid of a parasite is usually a lot easier than getting rid of leukemia, so the doctor and our friend were very relieved when our friend told the doctor that she had just returned from field work in Africa, Kenya to be precise. That made a parasite as the cause of our friend's lassitude more likely than leukemia.

But exactly which parasite our friend had, the California clinic could not confirm. It took a return to Kenya and a visit to a hospital whose doctors knew the local diseases to work out that what she in fact had was a disease common to the region: bilharzia. In medical terms, it is known as schistosomiasis. Bilharzia sounds like a local name for the disease, but is in fact named after Theodor Bilharz, the German doctor who first identified the parasite.

Species of a flatworm that pass part of their life cycle in water snails cause bilharzia. The flatworms leave the snail and penetrate the skin of people who wade or bathe in water in which the snails live. Once in, the parasites make their way to the liver via the host's blood vessels. There they mature into adults. They then leave the liver and migrate to the poor human's bladder or rectum, where they meet other adult bilharzia and mate with them. The eggs get back into the water in the urine and feces.

Bilharzia is common in Africa, as are so many parasites. In the first place, the snail that carries part of the schistosome's life cycle lives in most of the continent's still or slow-moving water. Additionally, Africa's general lack of sanitary facilities in rural areas means that the eggs get back into the snail-carrying water. And finally, because the disease is debilitating rather than lethal, a person and the local water's snail population can pass the parasite back and forth for years. Thankfully, bilharzia is fairly easily cured with sometimes just a single dose of the relevant medicine.

If diseases influence what we are, by influencing the evolution of physiological adaptations that combat them, they also influence where we are. "Then the angel of the LORD went forth, and smote in the camp of the Assyrians a hundred and fourscore and five thousand: and when they arose early in the morning, behold, they were all dead corpses" (Isaiah 37:36). One hundred and eighty-five thousand Assyrians dead in one night must be an exaggeration. Nevertheless, thanks to disease, the Assyrian siege of Jerusalem ended almost before it began. The story of the effectiveness of disease in invasion and war is repeated throughout history.

Africa is much closer to Europe than the Americas are. And yet Europeans had extensively colonized South America by the end of the eighteenth century, while not until well into the nineteenth century did they overrun Africa. What was the difference between the two continents? The answer is disease. Especially sleeping sickness, Ian Maudlin suggests. European incursions throughout the world have been particularly helped by the horse. Although many domestic animals suffer from sleeping sickness, horses are particularly hard hit. Without cavalry and without horses as pack animals, European invaders found it far more difficult to overrun Africa than South America. Sleeping sickness defended Africa from Europe—for a time, at least.

South America was free not only of sleeping sickness, but also of many other lethal diseases. French surveyors in the eighteenth century in what is now Ecuador suffered appallingly during their expedition to measure the circumference of the earth. Accounts have them waking up some mornings with faces so swollen from insect bites that they were nearly blind. But they did not come down with malaria or yellow fever, or any of the other diseases that killed so many "white men" in Africa. Yellow fever is so called because that is the color the skin turns as the liver is damaged.

South America certainly has its own diseases. Chagas disease is one quite well known one, which in its early stages in fact produces symptoms similar to malaria. Like sleeping sickness, it is caused by a trypanosome. And like malaria and sleeping sickness, it can eventually kill. Nevertheless, not nearly so many people suffer or die from it as suffer or die from malaria. Approximately ten thousand per year die from Chagas, which is roughly speaking the same number as die from sleeping sickness in Africa. Three times as many die from yellow fever, almost all of them in Africa. And perhaps one million die annually from malaria, again most of them in Africa.

Malaria and yellow fever arrived in South America and the Caribbean via the slave trade. The ships carried not only the slaves but also, inadvertently, the mosquitoes that transmitted malaria and yellow fever. Mosquitoes and the diseases they carry once influenced the spread of nations across the globe. William McNeill describes many instances of the effect of disease on human imperialism and expansion across the world in his fascinating *Plagues and Peoples*. More recently, his son, John McNeill, did the same with his *Mosquito Empires*. If I do not specifically mention an author as a source for statements regarding the effect of disease on us before roughly the middle of the second millennium, I will almost certainly have taken the information from William McNeill's book. After that, I will have used John McNeill's book, especially where I write about the Americas from the 1600s onward.

In addition to bringing the mosquitoes to the Americas, the Europeans also created perfect environments there for the mosquitoes. By importing livestock and tens of thousands of slaves, they provided food for the mosquitoes. And by creating swamps out of cleared forest, ponds for watering livestock, irrigation ditches, and the multitudes of containers provided by human habitation, they created the perfect breeding environment for mosquitoes—hundreds, maybe even thousands, of hectares of standing water with no fish. Moving to a similar modern change in environment, as I mentioned above, people in towns in Britain are increasingly using water butts, and correspondingly mosquitoes are doing well in British towns. As a corollary, I have already described how malaria disappeared from East Anglia with the draining of the marshy fens.

I mentioned bilharzia and its water-dwelling host, snails, a page or so previously. The ancient Egyptians were master agriculturalists, cleverly timing their irrigation cycles with the flow of the Nile. The Nile floods must have provided perfect habitat for the bilharzia-carrying snails. So it comes as no surprise that Egyptologists discovered schistosome eggs in Egyptian mummies from thirty-two hundred years ago. Imagine the care with which the mummies had to have been dissected to detect such small objects—approximately a tenth of a millimeter long. The eggs were in the kidneys, not the main part of the body in which the schistosomes lodge, implying heavy infestation. As the Egyptians had been irrigating for maybe two thousand years before that, the schistosome had had plenty of time to infest the population. A Chinese corpse from over two millennia ago also shows clear evidence of chronic bilharzia—perhaps not unpredictably, given how old Chinese rice culture is.

From disease, conquest, and slavery in the Americas, I will move to conflicts among the Europeans in the Americas. The first colonizing Europeans in South America and the Caribbean were the Spanish. By the time other European nations arrived to grab the riches of the region, the Spanish residents had become partially resistant to yellow fever and malaria. In other words, the physiology of the Spanish in the Caribbean had evolved in just a couple of centuries to be different from that of their European ancestors.

With their partial resistance to tropical diseases, the Spanish held out longer than they might have otherwise done against British attempts at takeover, thanks to the fact that in the Spanish-British wars for control of the Caribbean, up to ten times as many British troops might have died from disease as the Spanish killed directly. Consequently, the Spanish maintained the extension of their geographic range across the Atlantic, whereas the British extension initially failed.

In North America too, disease influenced who won and who lost, and so influenced our biogeography there. Disease was probably as much a factor in the rout of the British as were the rebellious American secessionists. The defeat of the British at Yorktown was the effective final event that turned America from a seditious territory into an independent state. But much as General Washington and the Comte de Rochambeau might want and be given credit for the defeat, malaria played a huge role. Some estimates indicate that although the Americans might have killed ten percent of the British forces, as much as fifty percent of them died from disease. Mosquitoes and the parasites they carried killed five times as many British soldiers as did the Americans. American forces were not so affected because they had, like the Spanish in the Caribbean, been in the country longer and had evolved resistance to the tiny killers.

A century after Yorktown, malaria and yellow fever helped foil French attempts to build the Panama Canal. The USA succeeded with its attempt where the French failed because in the intervening period, scientists finally discovered that mosquitoes, not bad air, were what transmitted malaria and yellow fever. The title of John McNeill's book, *Mosquito Empires*, perfectly encapsulates the influence of disease on the geographic range of nations—or, in other words, on why we are what we are where we are.

If disease hindered European hegemony in the Americas, it also aided it. Legend has it that the first European invaders into the Americas deliberately infected Native Americans of both continents with smallpox by offering them infected blankets as gifts. Given other well-recorded atrocities of Europeans against native populations, the story seems likely. But even if it is untrue, the native peoples would probably have died anyway on contact with Europeans.

European colonialists brought with them in their spread across the globe from the fifteenth century onward a wealth of urban diseases, such as measles, flu, and smallpox. These diseases seem to have arrived in western

Europe in the first century of the modern era—or A.D., as it used to be known—when they were as devastating to Europeans as subsequently they were to Native Americans on first contact with them.

In Europe, many died from these diseases when they first arrived in the region. But many recovered and were now immune to the original disease. That contrasted with the native peoples of North and South America, and with Caribbean and Pacific islanders. They had never encountered these Eurasian diseases of large cities and dense populations. Consequently, the Native Americans and islanders had little or no resistance. Measles by the time of European expansion was a minor nuisance to most Westerners. But to the peoples of regions where measles was not indigenous, it was lethal. On contact with Europeans, they died in their tens of thousands. "Great was the stench of death," William McNeill records one chronicler writing of the American experience.

The same happened in India and China earlier with movements of peoples within climate zones, according to William McNeill. Between climate zones, however, and especially when peoples from temperate regions tried to move into tropical regions, disease defeated them.

Supplying more accurate and precise estimates of the numbers killed by European diseases is difficult, because so few good records exist from the time of first contact. Europeans started to explore the Pacific islands later than the Americas, and they quite often did so with scientists on board. Captain Cook's voyages of the mid-1700s provide a good example. The Royal Society of Britain funded two of his voyages, along with the scientists that accompanied him. Several would later become famous, in part because of their research during the voyages. From them and others like them, we have better confirmed historical information of the devastation wrought by European diseases in the Pacific.

Some guesses had run as high as ninety percent of Native Americans in North and South America succumbing to European city diseases. However, we know that on some of the Pacific islands, half of the native population died from the introduced measles, flu, and smallpox. That proportion indicates that guesses of ninety percent mortality for Native Americans on first contact is probably an over-estimate. So does a very recent genetic study. Geneticists can get estimates of population size by comparing present and past genetic diversity within populations. Brendan O'Fallon and Lars Fehren-Schmitz calculated a fifty-percent crash in population size of native peoples in both Americas shortly after the arrival of Columbus.

Whatever the exact numbers of native peoples killed, European diseases could depopulate regions, killing many and forcing any survivors to flee what to them must have seemed a visitation from avenging gods. Certainly, accounts exist of large areas of east coast USA formerly full of thriving Native Americans, emptied of people and strewn with their remains. European diseases helped European geographic expansionism.

On a more minor geographic scale, the introduction by Europeans of tuberculosis to susceptible populations could have helped European dominance in certain areas. For instance, European fur traders almost certainly introduced the disease into French Canada, according to Caitlin Pepperell and her co-authors. It did not have an immediate effect among the scattered small populations of Native Americans. Only when the fur trade ended, and the native populations concentrated in industrialized towns and cities, did epidemics of tuberculosis break out in full.

Africa did not escape the influence of introduced disease, either. As Brenna Henn and her co-authors report, hunter-gatherer peoples in Africa, the Dobe !Kung of the Kalahari Desert and the Hadza peoples of Tanzania, suffered from epidemics of tuberculosis and measles, among other diseases. Figures for the Hadza indicate that these introduced diseases might have caused nearly half of their deaths. It seems likely that such death rates would have aided the Europeans' scramble for Africa. Carina Schlebusch's team indicates the same drastic results for the Khoi-San of southern Africa.

Back in Eurasia, could a disease that apparently killed a quarter of the Athenian army have helped prevent the defeat of Sparta in the fifth century B.C.E.? B.C.E. is the equivalent of B.C., but rather than meaning Before Christ, it means Before the Common (or Current) Era. Could Rome's hegemony during the first centuries of the first millennium of the modern era have crumbled as a result of epidemics? Possibly, because within China an almost complete breakdown at the start of the fourth century of the modern era coincided with the arrival of an epidemic (either smallpox or measles, it seems), a change of dynasties, and a nearly fifty-percent mortality rate that might have killed nearly two and a half million people.

Indeed, disease could have delayed the unification of China by over five hundred years. William McNeill points out that people settled in and developed the temperate surrounds of the northern Yellow River Valley that long before they did the same in the southern Yangtze River Valley. Even now, southern China suffers from malaria.

If epidemics and plagues affect people, and of course they do, people affect plagues. I have mostly written on Europeans introducing diseases to the countries they invaded. Europe too was invaded, and suffered the fate of people in the countries that it invaded. The maximum expansion of the Mongol Empire coincided with arrival of bubonic plague, or the Black Death, in Europe. Traffic along the Silk Roads could have been a route of introduction of this famous plague too.

Sometimes people deliberately moved the plague carriers. Western ranchers were one example. In attempts to rid grazing land of prairie dogs to make way for cattle, they would travel hundreds of miles with animals infected with the plague, according to William McNeill. Shades of deliberate introduction of smallpox into Native American peoples here? Plague is still endemic in many parts of the world, not least western America, where its reservoir is the region's several species of rodent, including prairie dogs, and their fleas.

I move now from invisible or nearly invisible other species and their effects on us to those that we hunt and that hunt us. Between school and university, I worked as a research assistant in the Queen Elizabeth National Park of Uganda. One day, I was helping a senior researcher in a little-visited area of the park when our vehicle broke down. I, as a mere assistant, was sent to get help. On foot. Fascinating as the African bush is, it is also frightening. Lions live there, and leopards, not to speak of buffalos, elephants, hippos. Every rustle in the bushes, every movement of a branch, and I froze, skin prickling.

Lions, tigers, wolves, and bears fill fairy tales, folklore, myths, and nursery rhymes. Cave paintings show them in abundance. And the bible too mentions them. The Old Testament god, not content with having people continually kill each other, "sent lions among them, which slew some of them" because "they feared not the Lord" (2 Kings 17:25-26).

Predators have killed hominins for millions of years. Two million years ago, an eagle killed the famous Taung child, a three-year-old

Australopithecus africanus. Predators kill us still. I remember as a child avidly reading John Patterson's 1907 *The Man-Eaters of Tsavo* about his final successful killing of two lions that had killed some tens of Indian and African laborers building the Kenya-Uganda "Lunatic Express" rail line. Tigers reportedly kill well over a hundred people a year in Asia, seemingly especially in the Sunderbans region of Bangladesh.

Other predators in the USA include mountain lions or cougars, and, in both the USA and Europe, bears and wolves. All happen to be minor predators of humans, despite our very public fear of them. Nevertheless, go to one of my and my wife's favorite local national parks, the Pt. Reyes National Seashore, and you will see that cougars feature largely in warning signs scattered through the Park. And yet cougars have killed only eight people in the whole USA since 1970. Bears kill an average of two or three people a year in the USA, while wolf kills of humans are essentially unknown. That contrasts with Europe, and especially Asia, where Little Red Riding Hood had every reason to fear the big bad wolf. Donna Hart and Robert Sussman quote a source that counted over a hundred people in Estonia killed by wolves in the first half of the 1800s (as listed in Lutheran Church records), almost all of them children. Predators large enough to take humans include not just mammals. I have already mentioned in chapter 8 the Komodo dragon as a predator. Crocodiles and alligators too are justly feared.

The question here, though, is not whether predators kill humans at all ever, but whether predators affect where humans lived or live in the world. The late Alan Turner argued that hominins will have found it difficult to get into Europe in large numbers until about half a million years ago. That was not because large Pleistocene carnivores previously prevented them from entering, but because not until then did Europe's large-bodied scavenging carnivores largely disappear, so leaving the carcasses for us. That is, if the hominins could avoid being killed by the predators that produced the carcasses.

An interesting idea, certainly. But how do we test it? The problem with scientifically testing ideas concerning human evolution is that much of it is a one-time event. Science finds it difficult to do anything with one-time events. We pejoratively, but with some good reason, call hypotheses based on them "just so" stories. With no other test available for what it is we are trying to explain, we can make up any story we like regarding the event. That is why when we try to understand humans, comparisons with similar situations in other animals are so important.

A huge literature exists on the topic of the extent to which predation, being eaten, controls animal populations. Not much of it, though, asks whether being eaten affects where species live. Predation can depress population density, hence grouse estates killing the birds of prey and foxes that eat the grouse. But I do not know of evidence that predators have affected the geographic range of species.

If predators did not affect our diaspora from Africa, what about prey? The abundance of large-bodied ungulates in Asia and Europe must have aided our spread out of Africa and across the world, especially as the animals will have been unused to the then-modern weaponry in the hands of bands of cooperating hunters. Huts in Siberia heavily framed with the bones, jaws, and tusks of mammoths indicate the importance of large herbivores to our survival. Could humans have moved into Siberia without those mammoths to provide us with food and, oddly enough, shelter? At the other end of the scale of prey size, did the abundance of oysters and their easy capture help the spread of the human species out of Africa and across the world?

෪

The moral of the story of my friend with bilharzia, of the stories of sickle-cell anemia and thalassemia, and of the potential for malaria and yellow fever in Britain's rain-barrel-dwelling mosquitoes is that doctors need to be familiar with the biogeography of disease. The symptoms of mild forms of sickle-cell anemia, thalassemia, and malaria can be similar to those of flu. Doctors seeing a patient are more likely to get the diagnosis correct if they know where the patient has come from. Correctly distinguishing the conditions is crucial because the treatments differ. Malaria can usually be treated with drugs, but severe cases of sickle-cell anemia require intermittent blood transfusion, while severe thalassemia can require lifelong transfusions or a bone-marrow transplant.

Not only does susceptibility to disease vary with geographic region, but so too does response to medications. Some Caucasian patients with a headache are less likely to be cured with codeine than some patients from Asia, because more Caucasians than Asians, maybe three to five times as many, have a non-working form of the enzyme that converts codeine to its active form, morphine.

Different susceptibilities, different treatments—these are why nowadays websites such as PubMed Health or the Centers for Disease Control show that a risk factor for several diseases is the person's country of origin. The "country," though, is often a whole continent. Given how much other genetic variation exists within continental populations, we probably need more precise knowledge of the geography of diseases, and of evolved resistance to them.

CHAPTER 11

MAD, BAD, AND DANGEROUS
TO KNOW

We are bad for many species, even if we help a few

I f other species, especially disease-causing organisms, have affected our geographic distribution, as I described in the last chapter, then we in turn have certainly affected the distribution of other species—thousands of them. Thus I wrote in chapter 4 that a plot of our dispersal throughout the world from Africa in the late Pleistocene would closely overlay a plot of the time and place of extinction of scores of species large enough to be worth eating. That might include Neanderthals.

Four fifths of Australian animal species of forty-four kilograms (100 lb) and over went extinct within just a few thousand years of our arrival there forty-five thousand years ago. The red kangaroo is Australia's largest surviving native species. Males, twice the size of females, weigh up to roughly eighty kilograms. Australia's giant kangaroo, extinct shortly after humans arrived in Australia, weighed nearly three times that much.

Two thirds of the terrestrial mammals of North America that weighed forty-four kilograms or more disappeared within a few hundred years of the arrival of the Clovis culture peoples roughly thirteen thousand years ago. A number had wonderful names—gomphothere, glyptodon, stegomastodon, and, my favorite, the wonderfully named *Bootherium bombifrons*. The gomphothere and stegomastodon were elephant-like species. Glyptodon was a massive armadillo. And the bootherium was a relative of the musk ox, but half again as heavy.

Humans arrived in Madagascar two thousand years ago, and within a few hundred years all lemurs weighing over ten kilograms had gone, more than twenty species. One was larger than a gorilla. The dozen or so species of Madagascar's huge elephant birds all disappeared then too. One of them might have weighed three hundred and fifty kilograms or more. That's twice the size of a male gorilla. New Zealand had moas before our arrival in the 1300s. Moas are like the emu of Australia, large and flightless. One weighed over two hundred kilograms, compared to the forty kilograms of today's emu. All eleven moa species went extinct almost immediately after our arrival in New Zealand.

Remember the dodo? It was a twenty-kilogram flightless pigeon that lived on the island of Mauritius, something like a thousand kilometers into the Indian Ocean east of Madagascar. In the mid-1600s, the first humans on the island found plenty of them. A century later, maybe even only half a century later, they were all gone. Sailors caught and killed them for fun and for food. Large, flightless birds, they were easy prey. We have historical records of the slaughter. Those writings and drawings are all that we have left of the real bird. So rapidly did we exterminate the dodo that nobody preserved one before they disappeared. No stuffed specimen of the real animal exists in any museum. If you see one in a museum, it is not the real thing: it is a model created with chicken feathers, using pictures from the 1600s as a guide.

The extinctions of large-bodied species followed so immediately upon our arrival that some use the term "blitzkrieg" to describe our impact, especially our impact in North America. The word originated with the start of the Second World War. It is German, and means "lightning war."

So, we humans arrive in a new region. Scores of species that have survived many millennia go extinct within a few centuries or less of our arrival. What can we conclude but that we drove them to extinction?

However, a lot of archeology and dating underpins the two first statements of fact. And a major lapse in logic lies behind the third sentence, the claim, the inference. Do these two sentences remind anyone of "Cheese is yellow. The moon is yellow. Therefore the moon is made of cheese"?

Is the moon really yellow? In other words have people got their facts right? Maybe not. People have argued that many of the species that some say humans killed in fact went extinct before humans arrived in the regions. Either that, or we do not know when they went extinct.

As for the inference that the moon is made of cheese, plenty of other substances are yellow. The moon could be made of them. In the case of the end-Pleistocene extinctions, the climate was rapidly changing as humans arrived for the first time in some parts of the world. Our entry into Eurasia from Africa coincided with the rapid approach of the peak of the ice age. By contrast, the climate was warming rapidly as humans entered the Americas. Indeed, it was probably precisely because it was warming that we could enter the Americas.

Can science, then, separate humans from climate as the cause of the extinctions? As with any argument, we need to start with the facts. If one side's or both sides' facts are wrong, we have as scientists nothing left to argue about, even if politicians have never allowed facts to get in the way of a partisan argument. Let us take a look at some of the facts. What about the claim that some, maybe many, of the species that humans are meant to have exterminated went extinct before humans arrived?

It turns out that the majority of these species that supposedly died out before humans arrived in at least North America were the rarest species. If something is rare, then it is difficult to find. If we cannot find it, not only do we infer that it is extinct, but we infer that it went extinct before it really did.

I will give a modern example. I mentioned a few paragraphs ago the extinction of New Zealand's moas. My uncle lives in New Zealand. On my wife's and my last two visits to see him, we went out to the small island of Tiritiri Matangi off the east coast of the North Island. We went there to find the takahe, a bird that looks like a large bright blue moorhen. For fifty years, ornithologists thought the takahe extinct. Last seen in 1898, so rare was it that not until 1948 was a small population discovered in the mountains of the South Island of New Zealand. With escaped introduced predators a threat on the mainland, the New Zealand Department of Conservation moved birds to several isolated and predator-free islands, one of them Tiritiri Matangi, where they now flourish.

Tyler Faith and Todd Surovell tested for the potential bias that rarity introduces into inferred dates of extinction. They tested for what we might term the takahe effect. They asked whether the extinct species were the rarest of all the species.

They used the number of remains of now-extinct mammal species unearthed by paleontologists and archeologists in North America to calculate the chances of the rare species being missed. They showed very clearly that the fewer the remains, the longer ago was the last date of recorded occurrence, which was the previously assumed date of extinction. Conversely, the more the remains, the closer was the last date of recorded occurrence to eleven thousand years ago.

From the graph of the relationship between the number of remains per species and the date when last recorded, they could mathematically model the likely date of extinction of those species that left few bones for us modern humans to find. They concluded that the majority of species thought to have gone extinct *before* humans arrived almost certainly in fact went extinct very close to the time humans arrived. If doctors must first do no harm, scientists must first get their facts right.

Sometimes getting the facts right is difficult, though. Australia is notoriously dry, and extraordinarily flat. Those are not the right conditions for burying dead animals sufficiently well so that they can be found and identified thousands of years later. According to Judith Field and her co-workers, the resulting paucity of fossils means that we do not know when two thirds of Australia's fauna went extinct. Field was writing on only Australian extinctions. An analysis by Graham Prescott and a group from the University of Cambridge lists estimated dates of extinction on all the world's major continents over the last seven hundred thousand years. From their list I calculated the proportion of extinctions whose dates were known to within ten thousand years. According to their collation, Australia is the least known continent. We know to within ten thousand years the date of extinction of less than half—forty-three percent—of

Australia's genera. For Eurasia, the figure is half, and for the two Americas it is two thirds each.

However, even if we knew that all of a continent's animals went extinct as humans arrived, we could not conclude that humans killed off the animals if the climate was also changing as they arrived. As I have indicated, the climate was indeed changing as humans arrived in Eurasia, the Americas, and maybe Australia. So was it climate or humans that killed the woolly mammoths, the saber-toothed cats, the ground sloth, the giant wombats, and so on?

An argument for humans, not climate, is the fact that it was the large-bodied species that were hit the hardest. We know that these are the ones that hunting (male) humans go after nowadays. In several hunting or fishing cultures, the women bring in more kilograms of flesh in total than do the men, and bring it in more reliably. However, besides the fact that the large-bodied species have more meat on them, they tend to have a higher proportion of fat than do smaller-bodied species. The hunter thus gains kudos not only for his bravery in killing a large animal, but also for bringing in one with large amounts of highly desired fat.

Occasionally we have direct evidence of the hunting. Take, for example, the study by Guadalupe Sanchez and her co-workers of a Clovis site of a little over thirteen thousand years ago in Mexico. There the archeologists unearthed from the same level both gomphothere bones and stone arrow and spear heads, strongly suggesting hunting of the animals. Gomphotheres look like elephants. Several species had come and gone in the Americas long before humans ever got there. But the latest lasted until the arrival of humans.

Such finds are rare, though, and archeologists usually have just the bones to go on. A catch attends the assumption that lots of big bones at a site means preferential hunting of large animals. Among modern hunter-gatherers, small animals are quite often eaten away from the residence site, whereas large-bodied prey is transported back to camp. Assuming that past hunting cultures did the same, it would be the large bones that are more likely to be found by archeologists. Also, large bones last longer than do small ones.

Nevertheless, Todd Surovell and Nicole Waguespack argue that the preference for large-bodied prey is real. In brief, in North America rabbits are so common and so easy to catch that even though some of them were eaten off-site, on-site remains should be hundreds of times more common than deer remains. But the opposite is the case. Indeed, the largest-bodied North American mammals, mammoths, are the most common of all. Yes, big bones preserve better. Yes, hunters are more likely to have brought back to camp a mammoth leg than a rabbit leg. But on balance, a preference for big game seems evident.

A counter-argument from the climate-only camp to the apparent preference for large-bodied species evident in their greater likelihood of extinction would be that the larger species need more land and food than do the smaller. To offer a personal analogy, I have no resident wild deer in my half-hectare garden, but at least ten wild rabbits. As the environment changes, it is the larger-bodied animals, then, that are the ones less likely to find patches of remaining suitable environment large enough to support a self-sustaining population.

In response, proponents of the idea that humans caused the extinctions point out that the species that went extinct as humans arrived had survived several previous precipitous changes in climate, from warm to ice and back again. Or if they did not survive those previous changes of climate, and some did not, we see no indication that large-bodied species were more likely to disappear than small-bodied ones when the only change in the environment was climate. Why should all the species that survived the several major shifts in climate over the previous several hundred thousand years not have survived the last two changes, if it were not for humans?

Furthermore, if climate were the cause of most of the end-Pleistocene extinctions, why did Australia's mass extinction event happen fifty thousand years ago, as the world cooled, while in the Americas, the animals survived fine through the ice age, but died out approximately eleven thousand years ago as the world warmed?

Added to any climate-only proponents' problems is the fact that when the bulk of their large-bodied species disappeared in Madagascar and New Zealand, the climate was stable by comparison to what was happening in certainly the Americas eleven thousand years ago. The remains of moas in the Māori cooking pots indicate what happened there.

In the case of Australia, not only do we not know when most of their large-bodied species died out, but opinions differ as to what the climate of Australia was doing around the time humans arrived, maybe fifty thousand years ago. Susan Rule and a group that included the Chris Johnson whom I mention again just a few paragraphs below suggest that the climate was stable, if drier than several thousand years previously. Others suggest that Australia then experienced one of its more major periods of drying. Whether it did or not, hunting had to have contributed to extinctions there, because Australia's large mammals had persisted through a previous drought that lasted twelve thousand years. Nevertheless, as Judith Field and her colleagues point out, very few sites of the right age have been excavated in Australia.

Eurasia is a more complex story than anywhere else, not because we do not know what the climate was doing, but because a pattern in the extinctions is not obvious. The climate was changing rapidly as humans arrived in the continent. It was cooling as they first entered the more southerly parts and the northeast. Remember the Yana site in Siberia, for example, occupied by thirty-two thousand years ago despite the approach of the ice age's peak cold. As humans entered the northwest, the climate was warming. If it hadn't been, the humans could not have entered the northwest, because of the massive glacier blocking their way.

A quantitative model of the relation between climate, human arrival, and extinctions by the Graham Prescott team that I have already mentioned indicates how unusual Eurasia was. They designed the model to answer the question of whether human activity or climate change was more likely to have caused the known extinctions. Into the model, they entered dates of precipitous climate change, along with firmly established dates of extinction and of the arrival of humans in Australia, Eurasia, New Zealand, and North and South America. In essence, they were asking whether extinctions were random over time, thus not coincident with climate change or the arrival of humans. For all the regions but Eurasia, the model indicated a burst of extinctions coincident with both climate change and human arrival—not surprising, as the two were indeed coincident—but more coincident with human arrival than climate change. For Eurasia, though, the model produced nearly twice as many extinctions as in fact occurred. In other words, climate and humans did not have the same effect in Eurasia as they did elsewhere in the world.

When a model produces what one might think of as a wrong answer, we have two possibilities: the model is wrong, or the world—Eurasia in this case—is acting oddly. We do not know which possibility is correct for Eurasia. However, a detailed study by Christopher Sandom and others out of Aarhus University in Denmark confirmed that surprisingly few extinctions occurred from Africa to eastern Asia and Indonesia. A map from this study shows the Americas, Australia, and western Europe as red hot with extinctions. But the Old World tropics are cold blue. To some extent, then, this study confirmed the Prescott study's findings. But in separating the Old World tropics from western Europe, the Aarhus study also indicated the importance of regional precision in analysis of causes of extinctions.

As a particular example of a Eurasian site, a recent study of well-dated deposits in a cave in east-central India indicates remarkably few extinctions. Over the last two hundred years to one hundred thousand years, only one of twenty-one genera disappeared. The unlucky one is the gelada baboon, which lives now in only Ethiopia. The others cover a range of species, carnivores of all sizes, cattle, antelope, horse, rhinoceros, hare, and so on. They are all familiar to us precisely because they did not go extinct.

An explanation for the lack of extinctions with the arrival of humankind in Eurasia is that this part of the Old World had harbored hominins for hundreds of thousands of years before the arrival of humans. Richard Klein's monumental *The Human Career* has a picture of the first unequivocal signs of hunting by humans. They are two wooden javelins dated to four hundred thousand years ago from Germany. I write "javelin," because they show no indications of hafting for any stone blade. They were carved or smoothed to a very sharp point. Klein suggests that they would have been used for stabbing, rather than throwing. Heidelberg Man or Neanderthal were the makers. If they were hunting in Eurasia then, it would surely be stretching belief to think that equally ancient hominins were not hunting in Africa at the same time, where also we do not see any burst of extinctions at any stage of human evolution there.

In exploring possible causes of something, we can put the known data into a model and see what the model tells us regarding the associations between measures of cause and measures of effect—humans and climate on the one side, extinctions on the other. Alternatively, we can make predictions of what we should see in the data were one cause operating.

Chris Johnson chose this second route to enlightenment. He tested a prediction from the hypothesis that human hunting caused the extinctions. We humans are a terrestrial species that in open country can move far and fast. We prefer to go after large-bodied species. And we do our hunting largely in the daytime. Therefore, he induced, terrestrial, large-bodied species in open habitat should have been at most danger from us. Small-bodied nocturnal species in woodland should have been relatively safe. In a survey of extinctions in the four major continents and Madagascar, Johnson indeed found that after accounting for body size and rate of reproduction (large-bodied and slowly reproducing species are more likely to go extinct, whatever the cause), the open-country terrestrial species were more likely to have gone extinct than the woodland arboreal species. It is difficult to see how a changing climate would have produced this result.

Whatever sort of species humans hunted, we would have had the greatest impact on populations of species already reduced to low numbers. Nobody says that the species that survived the previous major warmings and coolings of the climate flourished during those periods of changing climate, even if some of them in fact did. Mammoths, for example, seemed to have done so.

A key difference, though, between previous climate changes and the last one of the Pleistocene is that during the previous ones, humans were not there to finish off the remaining reduced populations. Therefore, as the climate changed for the better, their populations could expand with the expanding favorable environment. But around the end of the last ice age, the animals encountered for the first time these odd bipedal carnivorous mammals, humans, and perhaps more importantly, experienced the humans' unique ability to kill at a distance.

An example of the effects of such a killing combination could be California's Channel Island mammoths. They disappeared near eleven thousand years ago, almost as soon as Clovis hunters arrived in California. But the globe was warming then, and a warming globe means rising sea levels as ice caps melt. Island animals will often have found themselves on shrinking turf. Between the height of the ice age and when humans arrived, the Channel Islands might have lost half their area, depending on their topography. Smaller islands mean smaller populations. Smaller populations mean higher probability of extinction from the slings and arrows of outrageous environmental fortune. They also mean greater susceptibility to hunting.

The Prescott study and the Sandom study came to essentially the same conclusion regarding the influence of climate and humans. Both climate and humans have caused extinctions, they show, but both studies implicate humans in more extinctions.

The climate-or-humans debate is slowly changing from "Humans did it—No, climate did" to attempts to refine the information, and to refine understanding of which species might have been more affected by climate, which by humans, and why the difference.

For instance, a rather complex analysis by Eline Lorenzen and over fifty other authors argues that it looks as though humans had nothing to do with the demise of the woolly rhinoceros, or of the musk ox in Eurasia. Their evidence? They found too little overlap in time or space of humans with these two species. Archeological sites of humans contain very few of the bones of these two species. And in fact, the woolly rhinoceros, along with the Eurasian woolly mammoth, *increased* in numbers after humans entered the region.

We know that present-day musk ox are particularly susceptible to a warming climate. "Warm" is a relative term here. The musk ox's upper limit appears to be 10°C or so. I do not know the temperature limit for a bovid almost as hairy as the musk ox, the yak, but I do know from trips to the Bhutanese Himalayas that about three and a half thousand meters is their lower altitudinal limit. Below it, they overheat, yak herders say, and go into decline. On the way up the Himalayas, we changed from our horses and mules to the yaks at that altitude, because the yaks with their thick coats withstand the cold better than do the horses and mules. On the way down, we changed back again because the horses and mules did not mind the heat.

Perhaps the most famous extinction caused by climate is that of the Irish elk, which in fact should be known as the giant deer, given that it ranged far outside Ireland. It features in the cave paintings of Lascaux, and we know it occurred across Eurasia, all the way to Siberia. Be that as it may, whatever the cause of its extinction elsewhere, this deer disappeared from Ireland over a thousand years before we find any signs of humans in Ireland. We have no signs of it there after ten and a half thousand years ago, yet humans did not get to Ireland until nine and a half thousand years ago. The Irish elk's extinction coincides with an intensely cold period, termed the Younger Dryas. The name comes from the attractive Arctic flowers that appeared then, as detected in pollen in the soil. Female Irish elk probably could not find enough browse to survive the winter, and of course if females starve, the population dies out.

The popular story that the giant antlers of the Irish elk evolved to be so huge that the males could no longer support them is a myth. We know from the contrast in size of the males and females that this was a species in which a male mated with many females. In such species, it is the fate of females, not males, that affects the population, given that most males do not mate in any year.

If the question of whether climate change or humans caused extinctions on continents is still to be worked out, islands are a different story. Our arrival on many was so recent, the effects of our arrival so well documented, the speed of extinctions so fast by comparison to climate change, that nobody can deny that humans caused the extinctions.

Island species are particularly susceptible to human predation because, as I explained when writing about Flores, predators on islands are often quite small in size (large-bodied ones cannot find enough food), or even absent from small islands altogether. Therefore, the other animals are unafraid of large, moving objects, such as predatory humans.

At the start of this chapter, I described the extinctions of all elephant birds and large lemurs on Madagascar, of all moas on New Zealand, and of the dodo on Mauritius as humans arrived. Madagascar had no predators larger than a fox, while New Zealand and Mauritius had effectively no predators at all.

Visit the Galapagos, as my wife and I have been lucky enough to do, and you can see the effect of lack of predators in the past. To move through a colony of nesting boobies, we had to step over them. We could easily imagine the ease with which sailors could have killed hundreds of blue-footed boobies (a seabird), as the seventeenth-century sailors did with the dodo.

Island species are particularly vulnerable to not just hunting, but to nearly any change that humans cause. Islands are almost by definition small. Populations of their species are therefore small. Small populations and areas are easy for humans to destroy, however we implement our destruction. Across the Pacific, humans caused the extinction of as many as a thousand species, especially large-bodied, flightless ones that occurred on only one or a few islands.

So far, I have either not specified how the presence of humans might have caused extinctions, or I have pointed the finger at over-hunting. In fact, humans have probably caused extinctions in a variety of other ways in addition to over-hunting. I do not here confine myself to the Pleistocene extinctions. Since then, we have continued to drive species to extinction, likely with an accelerating effect.

As we humans spread through the world, we cleared and burned brush and forest. Later on, we even farmed. We know that nowadays we are increasingly driving species to extinction by destroying their environment.

Maybe it was our destruction of habitat, then, not hunting, that caused the extermination of so many species as we spread to virgin regions. Destruction of habitat would, like hunting, exterminate the largest-bodied species because in the remaining habitat suitable for them, their populations would necessarily be smaller than would be the populations of small-bodied species.

An obvious way humans change the environment is by setting fire to it. One study suggests that fire, not hunting, caused the extinction of at least one Australian species, the giant emu-like genyornis. It was two meters tall and might have weighed two hundred kilograms. A study of the carbon isotopes in its eggshells shows a huge change in diet as humans arrived, indicating that a varied environment with plenty of browse and grass changed to a far less nutritious fire-resistant scrub.

But just because humans might have sent the genyornis into oblivion by burning its environment does not mean that we exterminated all species in the same way. Susan Rule and her colleagues argue that in Australia, most of the extinctions preceded the expansion of fire there that occurred with the arrival of humans. Rather than humans causing fire that caused the extinctions, the disappearance of the large-bodied animals that we hunted to extinction led to an increase in brush, with no large animal left to browse it, and so led to increased fire.

A general way humans caused and cause extinctions is by reducing particular environments—forests, for example—to remnants of their former size. We do it with fire, and we do it with bulldozers, axes, chainsaws, drainage channels—almost any way the ingenious human mind can think of. The smaller the patch, the fewer native species will find enough habitat to survive. As importantly, the smaller the patch, the more the external environment makes it to the center of the patch, whether the external environment is a drying wind or an invasive species.

The principle that applies to these reduced habitats is the same as for the real islands of chapter 6. Small "islands" of forest contain fewer species than do large "islands" of forest, for many of the same reasons that small real islands in an ocean contain fewer species than do large real islands.

If some species survived our destruction of their habitat, and survived our introduction of new diseases, could they have succumbed to competition for resources? David Burney has pointed out that some of the very large-bodied lemurs survived in Madagascar for centuries after humans arrived. These were not primitive humans with poor weapons. If hunting wiped out the lemurs, the extermination should have occurred sooner, Burney argues. Maybe then, competition with livestock for water was the cause of the lemurs' disappearance, especially in the dry west of the country.

One Australian predator could have been outcompeted by one of the species that humans introduced to the continent something like four thousand years ago. The introduced species that I am talking of is the dingo, which probably came across as a domestic animal. Once arrived, it went feral. Australia's largest mammalian predator then was the thylacine, commonly known as the Tasmanian tiger, because it was striped, or Tasmanian wolf, because it was dog-like. It was probably extinct in Australia by roughly a thousand years ago, but it survived in the wild in Tasmania until the last one was shot in 1930. The very last died in a zoo in 1936. Recently, Melanie Fillios and two others suggested that the dingo with its larger brain outcompeted the thylacine for prey in mainland Australia.

The dingo did not get to Tasmania, which might explain why thylacines lasted longer there, despite a bounty on their head, because people thought that they killed sheep.

I described a few pages ago, how across the world, terrestrial, diurnal, open-country mammals were the ones most likely to go extinct as humans arrived in a region. The fact that our domesticated animals are also mostly ground-dwelling, diurnal species that live in more or less open habitat suggests another way humans might have caused extinctions of some native species as we settled the globe.

If the domestic species used the same environment as the native ones, maybe they came into enough contact with them to transmit novel diseases. We know that domestic disease can wipe out wildlife. In the late 1800s, millions of Africa's large ungulates—buffalo and wildebeest, for example—died from rinderpest introduced into Africa with cattle.

Ross MacPhee of the American Museum of Natural History in New York has argued strongly that we need to consider the effects of introduced disease on extinctions. Humans did not have livestock when they entered Eurasia, Australia, or the Americas, so disease is an unlikely cause of any of the sudden extinctions of large-bodied animals there. However, humans certainly took livestock to Madagascar and New Zealand, and we know that our introduction of myxomatosis to Australia devastated the Australian rabbit population, as it was meant to. Maybe, then, a disease of chickens killed the Madagascar elephant birds or the New Zealand moas. Certainly, many small Hawaiian bird species have been exterminated by avian malaria and avian pox, perhaps brought accidentally to the islands in the early 1800s along with exotic birds and the mosquito vector.

And then what about the domestic predators that we take with us? I have already mentioned the dingo. Cats have not yet been blamed for any extinctions that I know of. However, a recent estimate of several billion wild birds and mammals killed in the USA by free-ranging domestic cats gained a lot of publicity, along with criticism that the figure of several billion was too extreme.

One of the more famous devastating introductions of a predator that we have caused is that of the brown tree snake of Australia, New Guinea, and neighboring islands to the island of Guam in the northwest Pacific.

We brought it there in the mid-1900s, probably accidentally. With indigenous birds unused to any snakes, it rapidly drove to extinction several of the island's ground-nesting bird species. Similarly, in many forests in New Zealand, introduced rats and cats have silenced the glorious, melodious, ringing symphony of the native birds.

Just as humans might have killed the thylacine indirectly via our introduction of the dingo, we are almost certainly indirectly sending many species to extinction these days by warming the climate. As Camille Parmesan has been telling us for years, we are pushing species in northern latitudes north. We are pushing species up mountains everywhere as the low temperatures that they need persist only at higher and higher altitudes.

Moreover, most projections indicate that we have barely begun the global warming. As species in the Northern Hemisphere retreat north from the southern parts of their range, they are likely to run out of habitat. They will reach the limits of the protected area that they depend on, or they will find towns and agriculture in the way of movement. Conservationists are therefore talking about how to plan future reserves to take account of the shift in the geographic range of many species. We might have to think of setting aside land as protected areas now for species that will not reach them for decades. Others are talking of creating wildlife corridors—strips of woodland, for example—along which animals and plants might be able to move through land currently used by humans.

A major problem even if we provide corridors is that species might not be able to move as fast as their environment is changing. Carrie Schloss and co-workers calculated that in some regions in the Americas, over a third of mammal species might be left behind by the speed of warming. More vitally, many species will probably not be able to move at all, because we humans have destroyed their habitat north or south of where they are now.

A different sort of problem also attends the movements toward the poles. Over huge geographic areas, we do not know exactly what is happening. As usual, most of the studies of what species are doing are implemented in northern temperate regions, because that is where most biologists live. So as Camille Parmesan points out, we are missing information on Africa, tropical Asia, and South America. And apart from the famous browning of

Australia's coral reefs, we know far too little regarding what is happening at sea.

By comparison to latitudinal movement, movement up mountains to stay in a cooler range of temperatures and the associated habitat can have dire consequences, even if humans are not in the way. The reason, of course, is that eventually the species run out of mountain. An example is the gelada baboon of Ethiopia. Today it is reduced to a tiny geographic range at the top of the Simien mountains of Ethiopia at altitudes of between seventeen hundred and forty-two hundred meters. There, this large terrestrial monkey still finds the grasslands on which it depends. Global warming will not only further reduce the gelada's range, but it will extend upward the range at which crops can be grown. The limit to barley, the highest crop at present, is approximately thirty-five hundred meters. This limit, and the limits of the other crops, will rise. Even at thirty-five hundred meters, the gelada faces competition from the cattle, sheep, and goats of the ever-increasing human population of Ethiopia.

In the USA, an example of a mountain species in danger is the pika, a small rabbit-like animal that lives on and in scree slopes at high altitudes. A quarter of the populations censused in the Great Basin mountains of western America have gone extinct. No houses up there, no ski lodges, no crops, no livestock grazing, no people. It has simply gotten too warm.

Most of the extinctions that humans have caused are through sheer carelessness. We are not trying to drive species to extinction. But they disappear anyway as a result of what we do. That was not the case with smallpox. Smallpox is an example of one of two species, both viral diseases, that humans deliberately eradicated. We reduced their near-worldwide geographic distribution to zero.

Smallpox was a long-standing Old World disease until explorers and invaders introduced it into the Americas in the sixteenth century. I described its devastating effects there in chapter 10. We eliminated it in the "wild" in the late 1970s. However, vials of the virus still exist in containment laboratories, along with intense debate on whether they should be destroyed or not. If extremist Muslim clerics in various parts of the world would allow polio inoculations of their people, that disease too could be globally eliminated. The geography of culture, in this case religion, affects the geography of disease, and so affects us. Extremists in Nigeria, Afghanistan, and Pakistan are preventing polio inoculations, and so the disease persists there, although eradicated almost everywhere else in the world. Let me stress "extremists." Reasonable Muslims object to a fatwa against the polio inoculation program.

Nobody that I know of has objected to the elimination of rinderpest in the wild, another lethal Old World viral disease. It affected even-toed ungulates—cattle, sheep, buffalo, and so on. Millions died, wildlife as well as domestic animals. Well into the 1900s, plagues swept through continents. The disease acted so quickly that corpses strewed fields, plains, savannahs. However, a massive and massively expensive vaccination program finally succeeded with the 2011 declaration that rinderpest no longer existed. As with the smallpox virus, so with the rinderpest virus, it still lives in laboratories scattered through the world, some of them apparently *not* high-security containment laboratories.

Those species that we drive to extinction, we first reduce to remnant populations. Biologists used to more or less unthinkingly assume that these last survivors would persist in the center of their range, the place where conditions for them are probably the best. However, Rob Channell and Mark Lomolino showed a dozen years ago that two thirds of plant and animal species persist only at the edge of their former range. Such exclusion to the periphery is particularly obvious in Australia and North America, where over three quarters of native species persist now only at the edges of their former distribution.

So the giant panda, which used to range over much of the southern quarter of China, has retreated to the northwest periphery of its range against the foothills of the eastern Tibetan plateau, where few humans live. The California condor used to range over the whole of western, southern, and eastern USA, but now, as its name indicates, persists in the wild only in California, on the southwestern edge of its former range. And I have already described how the Tasmanian tiger used to range all over Australia, but even before Europeans reached the continent, it had retreated to Tasmania, at the far southeastern tip of its former range.

That species on the way to extinction should find their final refuge on an island, as in the case of the Tasmanian tiger, is not unusual. Less well known than the Tasmanian tiger is the giant kangaroo, whose last refuge was also Tasmania. That fact is relevant to the debate about whether climate or mankind was the main cause of extinctions. In a cooling world, should not this most southerly kangaroo population have been the first to go if climate were the main cause? That it went later correlates with the later arrival of humans in Tasmania than in southern Australia. If you have been dipping into the book here and there, I will point you to chapters 2 and 3 for details of our dispersal across the world.

Everyone knows what a mammoth is. Fewer know that the last refuge of the Eurasian mammoth was Wrangel Island, in the Arctic Ocean off the coast of northeastern Siberia. They were finally exterminated there just four thousand years ago, which was four thousand years after their disappearance from the mainland.

Two factors probably combine to produce this phenomenon of islands as last refuges. One is that islands tend to be on the perimeter of a species' geographic range, the rest of which is on the mainland. The other is that humans finally reached islands only after they had expanded across mainlands.

The fact that we often exterminate large-bodied species first, whether by hunting or by, say, destruction of habitat, means that our presence and what we do changes the nature of animal communities. That change can then affect us.

Costa Rica is famous for its tropical forests. Naturalists visit the country by the thousands to see its rich fauna and flora. However, in Costa Rica today, the tropical forest that used to cover the country exists only as tiny remnants of its once far more continuous extent. As I explained in chapter 8, predators are often the first sort of animals to disappear when natural habitat is destroyed, because they need so much land. When the predators go, their former prey increases in numbers. And that increase can lead to problems. A PhD student at my university, Elena Berg, discovered one of them to her cost.

The capuchin monkey, sometimes called the organ-grinder monkey, is hunted and eaten by birds of prey. With the decimation of Costa Rica's forests, the larger of these birds, such as the harpy eagle, have disappeared. Capuchins have consequently increased hugely in numbers. Now, capuchins do not eat just fruit and leaves. They will eat almost anything of the right size—beetles, lizards, small mice—and they especially like birds' eggs and nestlings. And here lay Elena Berg's problem. Her PhD thesis necessitated finding nestlings of Costa Rica's white-throated magpie jay and putting rings on their legs so that she could identify them later. But fast as the birds laid, along came the capuchins and ate the eggs or nestlings. It took Elena Berg far longer to ring a sufficiently large sample for statistical analysis than she had initially hoped. How seriously this new threat of an increased population of capuchins affects the jay's population, we will have to wait and see.

Without capuchin monkeys, the jays would do fine in Costa Rica as long as a few trees remained. But we can rarely guess the knock-on effects of our destruction. Little forest—few birds of prey. Few birds of prey—many monkeys. And who would have thought it, but many monkeys means few jays. And few jays means a PhD student who found data harder to collect.

This issue of knock-on effects of the changes that humans cause is another book in itself. I will give just a few more examples. Dams create lakes. Those lakes create islands as they cover the lower slopes of any hills in the dam's catchment area. Species that need large tracts of land to survive disappear. Predators are one of these. Meat-eaters need ten times as much land per individual as do foliage-eaters. A classic paper by John Terborgh and a group of researchers described how on islands created by a dam in Venezuela, the loss of predators led to a ten-fold to hundred-fold increase in numbers of rodents, howler monkeys, iguanas, and leaf-cutter ants. These are all plant-eaters. The result of the population explosion of these herbivores was a halving of the number of saplings on the smallest islands in the dammed lake, compared to forest on the surrounding mainland. There, surviving predators in the larger tracts of forest help keep the herbivore populations in check, and so help saplings to survive to maturity. Once the mature trees die, then with too few trees to replace them on the islands, we could see extra erosion of the land, and faster silting of the lake.

In effect, the forest on the Venezuelan lake's small islands was not as productive without the large predators as it was with them. A decrease in productivity of the land with loss of species can be more direct. Fewer plant species growing means less complete use of the soils' nutrients and water. Less complete use of the nutrients and water means less green stuff above ground. Some of the studies in fact indicate that loss of species can have as great an adverse effect on productivity of the land as drought does.

A main reason for the drop of productivity with loss of species, as opposed to simply loss of individual trees, is that different species use different resources in the environment. For instance, their roots go to different levels in the soil. Lose a shallow-rooting species, and the environment has lost the benefits of the nutrients in that level of soil. In arid lands, deep-rooted species are particularly important because except in the severest drought, they can continue to produce the leaves, flowers, and seeds on which humans and other animals can subsist through the dry season. The effect of extinctions depends in part, then, on the exact species removed.

The other part of the effect of loss depends on the number of species removed. The more removed, the greater the cumulative effect, and the more likely it is that a key species is lost, for example a species that disperses the seeds of a fruit tree. Less productive land means more hungry people. Rich countries can afford to compensate the move to monoculture by using fertilizer. Poor countries in which people rely on products from the natural environment cannot.

That does not mean that rich countries are immune to the knock-on effects of extinctions. According to the Centers for Disease Control and Prevention, over sixteen hundred people have died in the USA from West Nile virus in the fifteen years since its arrival in 1999 on the east coast of the country. We do not know how it arrived. It quickly reached the west coast, and is spreading through Europe also. Sixteen hundred deaths is a lot less than the tens of thousands of deaths annually in the USA due to careless driving, and nothing by comparison to deaths from disease in tropical countries. Nevertheless, we in the west are not used to death by infectious disease these days. The West Nile deaths get a lot of publicity, especially as they might be a harbinger of things to come with increased international travel, and global warming, which will contribute to the spread of disease to new geographic regions in ways that we cannot foresee.

West Nile virus is not a disease of only humans. It infects birds too. An infected bird is a carrier of the disease until it dies from the virus. The California Department of Public Health is asking us to report dead birds, so that they can track the advance of the disease. Felicia Keesing, Lisa Belden, and colleagues report findings suggesting that a cause of the increased incidence of West Nile fever and its associated deaths among humans might be a drop in the diversity of bird species. They suggest the possibility that some of the surviving bird species might be extra-effective carriers of the disease. How is that possible? For a start, more of the surviving species might be carrier species. And then, exacerbating that problem, the population size of these surviving species could increase, because the other species are no longer there to compete with them. A greater proportion of species as carriers, each of them in larger populations, means more routes of transmission to humans.

The topic of the geography of disease and its knock-on effects leads us to the increasing incidence of allergies in the USA. At the entrances of supermarkets in my home town are machines that dispense disinfectant cloths that we can use to wipe the handles of the shopping carts. Maybe we, the rich in the West, should not be so hysterical regarding cleanliness. I say this because of a study in eastern Finland, involving three levels of biogeographic diversity. One was the diversity of bacteria on the skin of adolescents. Another was the biodiversity of the general region of the adolescents' homes. And the third was the biodiversity of the gardens of the adolescents' homes.

In the region of Finland studied, the cleaner the adolescent, as judged by fall in the diversity of a particular form of bacteria on their skin (the regional home of the bacteria), the greater the chance that the adolescents were allergic to something, as judged by measured antigens in the blood. The mechanism seemed to work via an anti-inflammatory chemical produced by the body. The fewer sorts of bacteria on the skin, the less of the protective chemical.

The authors did not ask about washing habits. They judged cleanliness by only the bacterial diversity of the skin. But they found that rural adolescents were better protected from allergies because of their bacterial ecosystem. Who knows whether the rural adolescents washed less than urban adolescents. However, rural people do live surrounded by a greater variety of environments than do urbanites, and that greater variety of environments correlates with a greater variety of bacteria, and fewer allergies.

The way, then, for urban people to protect their children from allergies is, for a start, to get rid of those sanitizing wipes. Then, following the lessons learned from the white-throated magpie, the Venezuelan dam islands, the loss of deep-rooting plants, and the spread of West Nile virus in the USA coincident with loss of bird species, urban families need to encourage a greater variety of plant species outside their homes with the resultant potential for a hitherto unforeseen increase in the variety of bacteria in the vicinity of the homes, and a consequent decreased incidence of allergies.

In the previous chapter, I wrote regarding how indigenous diseases kept invaders out of regions, as well as of the devastating effects on residents of new disease organisms introduced by invaders. If the indigenous peoples suffer from the introduced diseases, the invaders have helped those species by spreading them around the globe. All organisms that we spread benefit from our actions, in the sense that we have extended their geographic range. Whether that extension benefits us or not is another question. Some of the organisms that we spread, we now wish that we had not. The spread of others, we are happy with.

In the last chapter, I wrote on the dispersal across the world of a variety of disease organisms—smallpox, measles, malaria, yellow fever, and tuberculosis, to name a few. These diseases have benefited from our spreading of their geographic range, even as we humans have not. The diseases have caused too many human deaths to count, and led in part to the extermination of several Native American cultures. In this chapter on our effect on other species, I have so far mentioned the extermination of several Hawaiian bird species as a result of our introduction to the islands of avian malaria and avian pox. Let me add to the list our global spread of HIV retrovirus from West Africa, and our global dispersal of a flu virus annually. In the West, flu is a mild disease, especially if we are sensible enough to get inoculated. Elsewhere, several thousand can die. History provides some example of how virulent flu can be. The 1919 global flu pandemic killed millions of people. Right now, we are successfully spreading the Ebola virus across West Africa. Formerly a disease organism of just the depths of African forests, it has, thanks to us, spread its geographic range into African cities. It has also spread into high-containment facilities in European and U.S. cities, if only for a few days each time so far.

Turning from diseases to larger organisms whose geographic range humans have considerably expanded, they include among many others the Caribbean cane toad and the dingo in Australia, the horse in Australia and the Americas, and twenty or so European songbirds now wild in New Zealand, and several in North America too. Two of the most common introduced birds in North America are the starling and the house sparrow. We brought the house sparrow and the starling to the USA in the 1800s, and now their geographic range covers most of the continent.

With reference to rabbits in Australia, I mentioned earlier in this chapter our introduction of myxomatosis there. Myxomatosis is a virus disease with horrible effects on rabbits—tumors, swollen blinded eyes, lassitude, fever, and only eventually, maybe after two weeks, death. As I said, we deliberately introduced the disease into Australia to reduce the rabbit population, which itself we introduced in the 1700s, probably as an easy source of food. In the mid-1800s, the animals went feral on the continent. We all know about rabbits' breeding habits, and within a couple of decades of their release, they were a plague. With myxomatosis, the rabbit population crashed, but many of the animals were naturally resistant, and numbers soon built back, if never to previous levels. So now the rabbit and myxomatosis have a geographic range halfway across the world from where they started, thanks to humans.

Other devastating extensions of geographic range include the brown tree snake to Guam, which I have already mentioned, and rats and cats all over the world. In many forests in New Zealand, introduced rats and cats have silenced, as I also previously described, the beautiful songs of the native birds. Several plant species that we have introduced beyond their native range do so well that they become a nuisance. Kudzu is a famous example. Native to eastern and southeastern Asia, it is now smothering native plants over large swaths of Australia and the USA. Water hyacinth is another famous example. An Amazonian species originally, it clogs waterways across much of the tropical world. Nevertheless, on islands, the introduction of plants has for the most part simply increased the number of plant species on the islands. The native plants have continued to flourish. Why this unusually benign outcome should be the case is a mystery that botanists and ecologists have yet to understand.

If we have extended the geographic range of nuisance species such as kudzu and water hyacinth, we have considerably extended the geographic range of many plant species beneficial to us. I am talking of our crop species. We have extended the geographic range of so many, some of them worldwide, that a full list would be impossible. I will simply mention the two that I most like to see on my plate. Potatoes came from South America, but now grow on all continents. China is the lead producer, at three times as much as the USA grows annually. Rice came originally from China, which is still the world's top producer. As with potatoes, rice now grows on all continents, including where I live in the Central Valley of northern California.

Our spreading of species around the world might even have saved some from extinction. Here in my current home town of Davis, California, the ginkgo trees provide one of the most beautiful fall/autumn colors. Their leaves then are bright gold. Tens of millions of years ago, various species of ginkgo grew throughout most of the Northern Hemisphere. By roughly two and a half million years ago, though, the genus had essentially disappeared. Although people talk of climate change as the culprit, nobody really knows the cause, given the huge changes of climate that have taken place over the last few tens of millions of years. However, one species survived in a tiny area of central south China and in central eastern China. Scientists used to think that monks had established the eastern population with seeds from the southern population. However, the eastern population is so genetically diverse that it could not have come from just a few seeds. It must be a relict population from before the species' near extinction. From its introduction into Europe in the early 1700s, and into the USA in the late 1700s, the ginkgo has flourished and is now a widespread species again. A word of warning for any landscapers: you should plant male ginkgoes, because the females produce many messy, smelly seeds.

On a grander biogeographic scale, global warming is already allowing species to move north, as I described a few pages previously. There, I pointed out the problems of the shift of range. But not all species will suffer. Some will even benefit. Going north is not necessarily going to lead to extinction. Look at the scores of species in the Arctic tundra where just twenty thousand years ago, all we would have seen would have been thousands of square kilometers of near-lifeless ice sheets. Look at the most visited part of the USA's Yosemite National Park, its valley. Where it is not covered with hotel, camping site, shop, and parking lot, it is lushly green, filled with woodland, meadows, and all the attendant species. Fifteen thousand years ago, a glacier filled it. More quantitatively, Holland, over the thirty years to the start of the current millennium, gained nearly twice as many lichen species as it already had because so many species had moved in from the south.

Eventually scores of species, even hundreds of them, are going to extend their range thanks to global warming, maybe as many as are going to lose their range. Biogeographically, a range extension is a benefit to the species involved. It has been a benefit to us, and it is now a benefit to us, including a biogeographic benefit. Without global warming, the first Americans would not have entered the Americas. Going in the other direction, the geographic range of human activity will very soon extend to encompass the whole of the Arctic Ocean. Polar bears might suffer, but we humans, along with many other species, are going to have open to us an immense previously inaccessible ocean. It might be politically incorrect to suggest that global warming can be a benefit, but the English language is full of aphorisms connecting change to opportunity.

Finally, we humans are, each of us, a breathing, walking home to thousands of forms of bacteria. We might each of us have a different complement of bacteria. Each of us is a different geographic region. In fact, even each part of our body is a different biogeographic region for microbes. The bacteria on our scalp are different from the ones on our cheeks, which are different from those in our mouth, which are different from those on our ears, which differ from the ones on our chest, arms, groin, legs, feet.

We knew a long time ago about the three species of louse that lousy people had on their body. Each prefers a different part of the body. But the study of what is termed our "microbiome" is a new burgeoning field of enquiry. One of the world's premier science journals, *Science*, carried in its 8 June 2012 issue a whole section on our microbiome. Ecologists such as Elizabeth Costello have entered the field. She is using general ecological theory to understand the human microbiome as an ecosystem. Biogeographers have yet to enter the field. Conversely, microbiologists interested in how and why the microbiome differs from one part of the body to another would do well to immerse themselves in the field of biogeography, because those are the fundamental questions of the field. A report in 2014 by Stephanie Schnorr and co-workers that the Hadza of eastern Africa have a more diverse gut microbiome than do Italians might be a start.

I began this chapter on our effects on other species with accounts of all the large-bodied mammals that went extinct as humans arrived in their part of their world. From the counts of the ones we drove to extinction, I omitted our closest relatives, the other hominins—all of them large-bodied mammals, of course. Erectus weighed on the order of sixty kilograms. And what do we see, but that Erectus went extinct in southeast Asia on the island of Java at around thirty thousand years ago, which could have been when humans first got to the island. Neanderthals (remember, I do not classify Neanderthals as humans) blinked out fifty to forty thousand years ago in Europe, about the same time as humans got there. Is it just a coincidence that Erectus and Neanderthal were large-bodied mammals, and that they went extinct as humans arrived in their last refuges?

It could be. As must be clear now from this chapter, the climate was plummeting to the depth of the last ice age as humans entered Eurasia. All the indications are that Neanderthals were retreating south ahead of the advancing cold, finally ending in southern Spain—a refuge (or refugium in biogeographical terms) where so many Brits go now to escape the British winters. Some genetic data indicate a very low genetic diversity in Neanderthal populations even in Siberia before the retreat to southern Spain. That low degree of genetic diversity indicates a population size small enough over a period of perhaps a hundred thousand years to indicate that the Neanderthals might have been on their way out even without that final push from invading humans. Not only that, but so small were the remaining bands of Neanderthals, they were apparently mating with half-siblings. Such is the rather extraordinary level of detail that we can now get from ancient DNA in recovered bones.

Another finding matches the supposition of internal decline of the Neanderthals in the face of increasing cold. It is that in the Caucasus, the region between the Black and Caspian seas, Neanderthals might have disappeared by thirty-seven thousand years ago at the latest, which is some thousands of years before humans apparently got there. Rachel Wood and colleagues' conclusion that Neanderthals disappeared from the mountains of southern Spain something like fifty thousand years ago, before humans arrived there, is more evidence of Neanderthal retreat in the face of increasing cold. The same could have happened with Erectus in southeast Asia, where the ice age produced a belt of aridity, perhaps leaving Erectus on only Java in a population that was too small to survive.

Further support for the idea that cold, not humans, drove Neanderthals to extinction is the fact that, as I write, nobody has found an archeological site that shows unequivocal evidence of the simultaneous presence of both Neanderthals and modern humans. As I have said before, though, current findings of the earliest presence of a species are highly unlikely to in fact be the earliest. Absence of evidence is not evidence of absence of contact. In the present context, we can see this from the fact that dates for human presence in western Europe get earlier and earlier. Nor indeed are we likely to find evidence of the very latest surviving Neanderthal. In that case, we are likely to miss the first contacts.

Supporting the idea of contact between late Neanderthals and early humans, some archeologists argue that the nature of the stone-tool culture of the latest Neanderthals indicates that that culture was influenced by contact with early humans. I will come later in this chapter to genetic evidence of contact. Furthermore, Neanderthals, like many of the other now-extinct large-bodied mammals that I have just discussed, survived two previous ice ages. Why should they not have survived this last one—except for the arrival of humans?

The question is especially pertinent in light of a highly detailed analysis by William Banks and colleagues of the nature of the environment that Neanderthals and humans lived in until roughly forty thousand years ago. The study argued that Neanderthals and humans favored the same sort of habitat, namely mixed woodland and grassland. Forty thousand years ago, that environment stretched across much of western Europe. Neanderthals should have been there if humans were neither constraining where Neanderthals could live nor killing them. Yet by forty thousand years ago, only one of the two species was present across most of that environment favorable for both, and only that one emerged at the end of the ice age. What else can one conclude, as the title of the Banks paper states, other than that arriving humans competitively excluded Neanderthals from remaining in their favored habitat, and so caused their extinction?

The "else" that one can conclude is coincidence, as John Stewart suggested in his detailed account of the demise of the Neanderthals. One case of simultaneous disappearance might be coincidence. Scores of cases, as in the disappearance of scores of other large mammals as humans arrive, seems like cause, though. Humans—as a final "Et tu, Brute" stab on top of the other metaphoric stabs of increasing cold and disappearing habitat resulting from the approach of the peak of the ice age—seems to me a not unlikely scenario in all cases, Neanderthals included.

Let me add a final twist to the story. We know that in human history, people ate other people. Despite the classic Flanders and Swan song, *The Reluctant Cannibal*, with its refrain of "Eating people is wrong," cannibalism is widespread across human cultures, if relatively rare. Cut marks on Neanderthal bones indicate that they too ate each other.

If humans were the final blow, how did we deliver that blow? Studies of animals have long indicated two forms of competition, direct and indirect. In direct competition, individuals or populations fight one another for resources. With indirect competition, one simply uses resources more efficiently than the other. It gets to them first, or exploits them faster.

Given what humans have done to other peoples in historic and modern times, deliberate genocide of Neanderthals and Erectus is not impossible. However, with little evidence of simultaneous presence, we currently have to assume indirect competition.

Archeologists and others had thought that modern humans used a wider variety of plant foods than did Neanderthals, so providing us with more food than the Neanderthals used. More food is always an advantage, but perhaps especially so at a time of increasing cold. However, a 2014 report by Amanda Henry and colleagues of the variety of microscopic plant remains on teeth and tools showed that Neanderthals ate just as varied a plant diet as did modern humans. Not only that, they cooked the plants too.

Nevertheless, it seems unlikely that with our more advanced tools, humans could not do better than Neanderthals. Evidence that we might have done better is the finding that early human sites in Iberia (Spain plus

Portugal) and southern France contain twice the amount of rabbit remains as Neanderthal sites. The significance of rabbits as prey is their great abundance. Also, because they live in warrens, a hunter can easily catch a lot of them at one time with the right equipment. In this case, that would be nets if the rabbits could be frightened from their warrens.

Humans did not have only better tools than did the Neanderthals. We were also more efficient walkers and, according to Dan Lieberman, more efficient runners too. We might even have had hunting dogs back then, which can markedly increase not only hunters' ability to find and capture prey, but also to carry it back to camp, as research summarized by Pat Shipman recently showed. With better tools, including maybe dogs, and better locomotion, humans will have gotten to game and other food before the Neanderthals did.

Add an abundant food source, rabbits, that Neanderthals did not use, and humans would have bred faster than did Neanderthals, meaning even more of us to exploit the land, in effect denuding it for the Neanderthals. Maybe humans even unintentionally frightened the game enough so that Neanderthals with their more primitive weapons could not get sufficiently close to kill it. Whichever scenario we choose, humans would have replaced Neanderthals without ever meeting them—without, so to speak, a shot being fired.

But meet them we humans did, it seems. And we appear to have mated with them. As teams working with Svante Pääbo's laboratory in the Max Planck Institute for Evolutionary Anthropology in Leipzig, Germany, have shown, Neanderthal genes are in Europeans and western Asians of today. Not only that, but Denisovan genes are in east Asians, Native Americans, New Guineans, and Melanesians.

I first introduced Denisovans in chapter 2. Genetically distinct from any other Eurasian hominin, Denisovans are known from less than a handful of bones. The discoverers have not yet decided whether they want to claim that it is a new species, and so have not given it any scientific name.

The genetic contributions of Neanderthals and Denisovans to the modern human population are quite small. Perhaps two percent of modern European and western Asian genes come from Neanderthals. Denisovan genes might make up five percent of the Pacific peoples' genes, but less than half a percent of Asian and Native American genes.

Not only might humans have bred with Denisovans and Neanderthals, but geneticists suspect yet another, so far unnamed, hominin that interbred with Denisovans. And in the cradle of mankind, Africa, it looks as though modern humans bred with a second mystery hominin approximately thirty-five thousand years ago. That other hominin split from the line leading to humans several hundred thousand years ago, around the same time as the Denisovans did. All we know concerning it is its genes and what they might tell us of its physiology and anatomy.

One major benefit from our Neanderthal and Denisovan genetic inheritance, and presumably the reason why their genes persist in us today, is a generalized immunity to disease. Humans moved into areas already occupied by these other species, into populations of these other species that had already adapted to the diseases in the region, and, through mating with them, we acquired their resistance to the diseases.

These findings come from the amazing Svante Pääbo lab that I mentioned just above. This research group is sequencing and identifying the DNA of earlier and earlier fossils. The Neanderthal and Denisovan bones that provided the DNA that I have just been writing on are at least fifty thousand years old. The Pääbo lab is obtaining more and more complete readings of the whole genome of these early hominins. Indeed, they reckon that they have a complete Neanderthal genome. The result is that they can, as I have already indicated, identify genes for resistance to particular diseases—and, for example, genes for blue eyes.

Although I have fairly emphatically indicated that humans mated with Neanderthals and Denisovans, and although the Pääbo laboratory findings seemed incontrovertible, I hope that, by now, readers have realized that in science, we can rarely be one hundred percent sure of interpretation. Each side might be sure, but the world is complex, and someone somewhere can usually see something wrong or incomplete with a finding.

In the case of the Denisovan and Neanderthal genes that we carry in our cells, an alternative way we could have gotten them other than interbreeding is simply inheritance from a common ancestor. After all, we share genes with even sea urchins. One argument against the idea that the shared genes arise simply because of inheritance from a common ancestor is that none has yet been found in any African population.

The argument regarding where the shared genes came from is ongoing as I write, among people far more versed in the science of genetics than I am. To close, I will say that unlike climate-change deniers, who refuse to accept the phenomenon because climatologists argue regarding the details, the geneticists and anthropologists studying Neanderthals accept the presence of Neanderthal genes in modern humans and are trying to clarify how they got there.

∽

The world's population is increasing everywhere, but it is increasing fastest in the tropical continents, and fastest of all in Africa. Combine that growth with the biodiversity of the tropics, and the inescapable consequence is going to be another pulse of extinctions, a burst of disappearances of other species' geographic ranges. We have exterminated and are continuing to exterminate many more species than we have saved or are saving. But we have saved and are saving at least some species.

What about our own species, *Homo sapiens*? What have *Homo sapiens* populations done to each other? What are we doing? What will we do?

CHAPTER 12

CONQUEST AND COOPERATION

Humans are bad for each other,
even if we occasionally help one another

n the last two chapters, I presented evidence indicating that other species have influenced what we are where we are. In turn, we humans have influenced not only where other species were, but whether they even continued to exist. I ended the last chapter asking what human populations have done to each other, are currently doing to each other, and will continue to do to each other. Now, I will review what seems to be the incontrovertible fact that we humans have done to each other as we have done to other species. We have affected where other cultures are, and often whether they continued to exist. Much of the effect has been deleterious. But occasionally, a human culture has benefited another, extending its range and helping it survive.

Starting with Exodus and the flight of the Israelites from Egypt, and the Israelites' god's promise in Exodus 23:31 to "deliver the inhabitants of the land into your hand; and thou shalt drive them out before thee," through the next twelve books of the bible to 2 Chronicles we get a nonstop account of war, rape, looting, and genocide as Middle Eastern peoples compete for land. Only the book of Ruth provides a break from the slaughter and mayhem.

My school history books repeated this theme of lethal competition between peoples. They concentrated of course on the battles that the English won, especially against the French. Agincourt and Crécy loomed large in my history lessons. I write "battles that the English won" because that is what we English were taught. In fact, at both engagements a large contingent of Welsh made up the "English" army. The books could not ignore the Battle of Hastings, 1066 and all that, which the French won. But they presented it to us as the battle that originated the English monarchy. The mnemonic to remember the succession of English kings and queens does not begin with the Anglo-Saxon kings, but with the invading Frenchman, William I. "Willy, Willy, Harry, Steve," starts the mnemonic, signifying kings William I, William II, Henry I, and Stephen. English schools have erased the memory of our defeated Anglo-Saxon ancestors, or at least they had when I attended them as a pupil.

Humans seem to have been unpleasant to each other since our very beginnings. The Kennewick Man from nine thousand years ago, discovered in Washington State, has a stone arrow head or spear head in his hip bone. The iceman, Ötzi, died a violent death in the Italian Alps over five thousand years ago, killed by either or both of a blow to the head and an arrow into the shoulder that penetrated an artery. Archeological digs of villages and towns in ancient Britain find many of them strongly fortified. It is not much of an exaggeration to say that European history until 1945 was one long tribal competition for land—one long battle to maintain and expand each people's geographic range.

The bible's accounts of man's competition with man have tens of thousands at a time slaughtered. The numbers are probably exaggerations, but the frequency of the described slaughters indicates a high rate of attempted genocide, and the attempted reduction of another population's geographic range to zero. Jared Diamond in his *The Third Chimpanzee*, by which he means us modern humans, lists seventeen genocides throughout human history that killed over 10,000 people, and six that killed more than a million. The results of the occupation of East Timor by Indonesia from the 1970s to the '90s—supported, incidentally, by the USA, Britain, and

Australia—is an example of the former. Turkey's annihilation of Armenians during the First World War, the Khmer Rouge slaughter of Cambodians in the 1970s, and of course, Nazi Germany's slaughter of Jews and other minorities, are examples of the latter.

The Third Chimpanzee was published twenty years ago. In that time, over five million have died of butchering and starvation in the eastern Democratic Republic of the Congo, as various factions in the region have battled to extend their group's range in the eastern DRC.

The slaughter by the British of the aboriginal Tasmanians is notorious. Jared Diamond describes this too. The British hunted the Tasmanians down with dogs, as if the native aborigines were vermin. Lines of settlers swept across the land, killing all before them. A population of a few thousand across the whole of Tasmania was reduced in just thirty years by disease and slaughter to perhaps less than two hundred. In the mid-1830s, the British shipped these last few survivors to Flinders Island, off the northeast corner of Tasmania. Fourteen years later, fewer than fifty remained. Once more they were translocated, this time to a settlement in southern Tasmania designed formerly for convicts.

Among the very last native Tasmanians was Truganini, who died in 1876. So poorly known were the Tasmanians and their cultures that her name has at least six spellings, with Trugernanner a common one. In 1905, the last speaker of any Tasmanian language died, but nobody knows her Tasmanian name. She was given the name Fanny Cochrane, and eventually married William Smith, an Englishman. According to the Wikipedia website on her, the only audio recordings of a Tasmanian language are hers. You can hear a small portion on YouTube. Just type in Fanny Cochrane Smith. A people who had ranged over sixty-two thousand square kilometers was reduced in a little over a century—through slaughter, starvation, and disease brought on by another people, the British invaders—to just a few photographs and a wax cylinder recording.

I have written of the Tasmanians almost as if they were one culture. They were not. In a detailed analysis of the few records available, Claire Bowern recognizes five distinct language families in Tasmania. As an indication of how distinct they are, she identified only twenty-four words that all the languages shared out of the nearly thirty-five hundred that she counted.

Not only was that proportion of shared words extremely small, but many of them were for items introduced by the British, such as cattle. So the British did not exterminate just "the Tasmanians." They exterminated the Northern Tasmanians, the Bruny Islanders, the Oyster Bay peoples, the Northeastern Tasmanians, and the Western Tasmanians.

Australia did not escape British depredations either. There too, the native peoples were hunted down as if they were vermin. "We have seized upon the country, and shot down the inhabitants," wrote one settler. Some complained of the brutality. "Mr. Bligh, with a party of the police . . . drove the poor creatures . . . into the river—and commenced butchering them forthwith. . . . Thus terminated the foulest deed—and shall I say, the foulest day—Maryborough ever witnessed." These quotes are from chapter 7 of David Day's detailed history of the European annexation of Australia.

American colonialists did the same to the Yana peoples of northern California in the mid to late 1800s. Once again, Jared Diamond describes their fate. The last few score of them were deliberately hunted and killed, until only four remained, hiding in what is now the Lassen National Forest. My wife and I love to hike in this conservation area, but cannot imagine trying to survive there. Three of the outcasts died after surveyors ransacked their camp. The lone survivor, Mr. Ishi, emerged from the forest three years later in 1911, to live out the remaining five years of his life in Berkeley.

A picture of Ishi is in Jared Diamond's *The Third Chimpanzee*. Ishi features also in Daniel Nettle and Suzanne Romaine's *Vanishing Voices*, as does a smiling Red Thunder Cloud, the last speaker of Catawba Sioux, as well as a picture of a pensive Tefvic Esenç, the last person well-versed in Ubykh, a language of the Caucasus, now roughly comprised of Georgia, Armenia, and Azerbaijan. Not all the featured "last speakers" were male. Marie Smith Jones is there, the last speaker of Eyak, a language from southern Alaska. Closer to my British origins, they have a photograph of Ned Maddrell, the last person to have spoken Manx since his childhood. Red Thunder Cloud died in 1996, Tefvic Esenç in 1992, Ned Maddrell in 1974, and Marie Jones in 2008.

Nettle and Romaine feature a few more "last speakers," including Roscinda Nolasquez, the last Cupeño, a Native American people of southern California, who died in 1987. The language did not quite die with her, however. It remains as a book of a hundred and ninety-eight pages of Cupeño oral history and songs co-edited by Nolasquez and Jane Hill. Nolasquez was ninety-five years old when she died.

In Europe, perhaps the last fluent native speaker of only Cornish, Dolly Pentreath, died in her eighties two centuries ago. She became famous enough in her lifetime to have her portrait painted, and Nettle and Romaine have a photograph of her gravestone, emplaced nearly a century after her death.

Roscinda Nolasquez did not die in her birthplace. The US government used a Supreme Court decision in 1903 to throw her and all her people off their land and force them to move to the 52-square-kilometer Pala Indian Reservation in San Diego County.

The forced move was just one in a long history on the continent. Native Americans formerly covered essentially the whole of the United States. As European and other colonialists arrived in the east, so the native peoples were pushed west. Now their reservations cover less than three percent of the land. These reservations are not, to put it mildly, in the most productive lands of the USA. The largest are all in the west, and therefore in some of the most arid parts of the United States. The largest of all, the Navajo Nation, is in the Four Corners region, well known for its magnificent scenery, but definitely not for a green scenery. It is almost all canyon and desert.

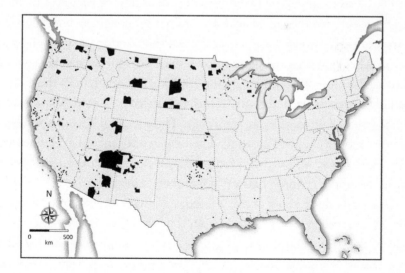

Native American reservations, many not in the most fertile regions of the USA. *Credit: Redrawn by John Darwent from WikiCommons.*

Exclusion of subordinated peoples to the periphery of their range, and hence often to marginal habitat, is fundamentally and chillingly the same as what we have done to many non-human species, a process that I detailed in the last chapter. And as for species, so for cultures, the process has happened all over the world.

In Asia, the mountains of Taiwan off the south coast of China are effectively a reservation, ringed by lowlanders speaking a form of Chinese. In the highlands, the native Taiwanese speak on the order of ten languages of Austronesian origin—the exact number depends on how one classifies the languages.

Ned Maddrell's Manx was the former main language of the Isle of Man, a small island halfway between northern England and Northern Ireland. Maddrell attributed his speaking of Manx to living in a small village on the island. The main towns, by contrast, were swamped by English. Similarly Scottish Gaelic, or Erse, survived better in the outer islands of Scotland and northern Scotland, Irish Gaelic in the west of Ireland, and Welsh in the west of Wales. In essence, the expansion of the English pushed the native languages and speakers to the edge of their former range.

The Cornish language of Cornwall in Britain and the Breton of Brittany in France suffered the same fate. Extraordinarily enough, records allow the retreat of these two languages over the last thousand years to be mapped. Cornish began withdrawing west down the Cornwall peninsula maybe as early as the eleventh century. By the 1500s, it was gone from the eastern half of the peninsula, and by the 1800s it persisted only at the very western tip, in one twentieth of its former range. The retreat of Breton also began more or less a thousand years ago. Then it extended inland to Rennes and Nantes. However, its westward compression essentially stopped by the start of the twentieth century, halfway down the peninsula.

Deliberate destruction of cultures does not have to entail the taking of life, as I described a page or so back in the case of the British extermination of the Tasmanians. One can, for example, simply ban the speaking or teaching of a particular language. And hegemonic powers have done so. English-speaking powers have been strong proponents of such bans in North America, Australia, and Britain itself.

Henry VIII ordered the Irish to speak English instead of their native Gaelic. The English government in the mid-1500s did not directly ban the speaking of Welsh, but instead it banned anyone speaking Welsh from holding land, or holding office, or having freedom to congregate. Under these circumstances, what could the Welsh do but speak English? The Australian government's Human Rights and Equal Opportunity Commission reports that up to the 1970s, Australian government practice and policy banned and discouraged the speaking of aboriginal languages.

It was only in 1967 that Britain passed an act allowing Welsh to be used in courts and other government venues, and not until the 1990s that the British government gave Welsh full equality with English as an official language. Before that, in the 1800s, a school pupil caught speaking Welsh might have had to wear a wooden slab around their neck, and pass it to any other child that they heard speaking Welsh. The last child with the slab at the end of the week received a beating. A governmental report in the mid-1800s into the state of education in Wales described the practice as cruel, but at the same time described the Welsh language as a "manifold barrier to the moral progress . . . of the people."

The report, however, is full of contradictions. Alongside the denunciation of the language of the people, the Welsh themselves are "far superior to the same class of Englishman in being able to read the bible . . . skilled in discussing with eloquence and subtility [sic] abstruse points of polemic theology"; they have a "natural ability and capacity for instruction"; "they desire it [to be taught English] to the full extent of their power" in order to improve their lot in life. Indeed, well-paying jobs in the cities required English. In banning Welsh, then, England was living up to the later Tory accusation of being a nanny state, being strict for its subjects' own good.

The eradication of a language and, with it, a culture, is often a side effect of other influences. Immigrants voluntarily adopt their host country's language, for example, or residents adopt the language of the powerful invaders. A lingua franca spread through Papua New Guinea because people able to speak to more others were more likely to get jobs—shades of English in Wales a hundred years earlier.

These examples are all from Daniel Nettle and Suzanne Romaine's *Vanishing Voices*, where anyone interested in, or wanting to be horrified by, the eradication of languages can find many other examples.

Red Thunder Cloud, the last speaker of Catawba Sioux, whom I previously mentioned, might have been the last speaker of Catawba, but it turns out that he had not many Native American genes in him, if any. Ives Goddard, editor of volume 17 of the *Handbook of North American Indians*, describes him as a colorful character, and gives his genealogy. Red Thunder Cloud was born Cromwell Ashbie Hawkins West, to Cromwell West and Roberta Hawkins. Ms. Hawkins's father was William Hawkins, one of the first African-American lawyers in Baltimore. Her mother was Ada McMechen, daughter of George McMechen and Mildred Harris. And so on.

Red Thunder Cloud, of British origin, in appropriating the appearance and language of the Catawba Sioux, emphasizes that Europeans had taken over North America. One estimate has less than half a million Native Americans remaining in the USA in 1850, but thirty million non-natives. The effect of such swamping is epitomized in Vladimir Arseniev's *Dersu the Trapper*, where he writes about the erosion of traditional Siberian culture by immigrating Chinese. "Among them [the Siberians], I found one old woman who still remembered her native language. . . . At first, only with difficulty, she recalled eleven words . . . now she had completely lost her sense of nationality, even her own mother-tongue."

Extinction of languages is not new. K. David Harrison reckons that perhaps twelve thousand languages might have been spoken by pre-agricultural people, compared to not far from seven thousand after the development of agriculture. In chapter 7, I mentioned Daniel Nettle's finding that length of the growing season correlated with diversity of languages—the richer the land, the more cultures could coexist there. One continent stood out in the Nettle analysis as depauperate in languages by comparison to the length of its growing season: South America. Disease brought by the Spanish conquerors, along with deliberate genocide, seem likely causes of the linguistic poverty.

I have mentioned how island cultures are particularly susceptible to extinction. The British extermination of the Tasmanians is a stark example.

In the Nettle study, one island stood out as depauperate in languages by comparison to the length of its growing season—Cuba. The Spaniards were not obviously more genocidal on Cuba than on mainland South America, but with nowhere for the local peoples to go, the Spanish exterminations of the indigenous Cubans were devastating.

Humankind is by no means unique in its effect on other populations' geographic ranges. Species compete within and between themselves, and probably have done so ever since species existed. A species moves into a new region, often these days introduced there by humans, and the native species disappears. Witness the disappearance of the red squirrel from most of England after the gray squirrel arrived from America in the nineteenth century.

In previous chapters, I have discussed how the environment influences our anatomy and our physiology. It also influences our behavior. Thomas Malthus argued over two hundred years ago that humans almost inevitably find themselves in a perpetual state of competition for food, and therefore for land, because our numbers always build to fill the space and consume the resources available.

Poor weather can exacerbate the competition by reducing the amount of food available. For instance, the French Revolution of 1789, one of Europe's major social upheavals, followed five years of poor harvests, capped by the appalling winter of 1788/89. Rivers froze, and the resulting massive spring thaw flooded thousands of hectares. Certainly, the bad weather was not the only cause of the Revolution. "Let them eat cake" epitomizes the government's ineptitude and the various other contributing influences.

Untangling the immediate causes of war, and its effects on the geography of peoples, is difficult. Nevertheless, some recent studies seem to confirm the links between climate, competition, and war, and therefore between climate, competition, and change in the geographic range of cultures.

David Zhang and his colleagues looked at records from 1500 to 1800 for Europe, the period of the so-called Little Ice Age. The cold reached its nadir from roughly 1550 to 1650, as did the General Crisis of the time, a period of extreme social unrest throughout the continent. Europe's major rivers and canals froze solid in the winters of this period. Here is the sequence of seemingly connected events, substantiated by Zhang and his co-authors with rigorous statistical testing: cold → a near-halving of crop yield → tripled grain prices → lowered wages → famine toward the end of the period resulting in three times the number of plagues because of the people's weakened state, and a drop in the average person's height of one centimeter → fifty percent increase in number of wars → doubling of war deaths → declining population with millions dead → and a geographic effect on where peoples were, namely a tripling of migration rates.

Zhibin Zhang and colleagues investigated the effects of climate on unrest in China from the beginning of the first century of the Common Era (formerly A.D.) to the end of the nineteenth century. They too found that cold weather might have correlated with unrest and, quite often, dynastic change. The form of civic turmoil was quite specific. Nomadic invasions from the north increased in cold weather. The authors surmised that an abbreviated growing season in the north led to food shortages, including for livestock. The result was invasion. To add insult to injury, cool periods seem to be associated with drought, which favors locust plagues, which of course exacerbate food shortages.

One can imagine that competition and range change would be more influenced by weather in the distant past than in modern times. Even now, though, conflict and the fight for land can be tied to the weather. Climatologists know well these days the El Niño-La Niña weather cycles tied to periodic warming and cooling across the tropical Pacific Ocean. These conditions produce unusual drought and wet, so affecting food supplies. Characteristically, the tropics are more affected than are higher latitudes.

Correspondingly, Solomon Hsiang and his co-authors found that in the half century from 1950 to 2004, conflicts were twice as frequent in the tropics in El Niño years compared to La Niña years. The study defined a "conflict" as more than twenty-five dead in a battle. Whether the climate was wetter or drier, warmer or colder in these El Niño-La Niña periods, they did not say, because the effect that the El Niño-La Niña cycle has on climate varies with region. The cycle has far less effect on higher latitudes, especially northern latitudes, than it does on equatorial latitudes. Substantiating the argument that climate affects conflict, the study found no effect of the cycles on conflicts at higher latitudes.

The Hsiang team tested all sorts of other possible influences on conflict in addition to El Niño-La Niña cycles, such as sex ratios of populations, degree of urbanization, nature of political system, the especially conflicted state of various regions in Africa, and so on. One can imagine, for example, that because higher-latitude countries are generally richer than tropical countries, and often have more stable political systems, the higher-latitude populations would suffer less from the same intensity of El Niño bad weather that affects tropical countries.

None of these other factors correlated as strongly with the frequency of conflict as did climate. The study concluded that across the globe in the half century from 1950, perhaps one fifth of conflicts could have been initiated in part by poor weather.

Solomon Hsiang and a new group of co-authors then analyzed the results of all previous rigorous studies of the possible influence of climate on conflict. This time they analyzed the correlation between climate and not just wars, but also individual acts of violence, such as battery and murder, as well as overthrow of governments. Their analysis went as far back as did the published quantitative studies that they could find to collate. The oldest they found was an analysis of violence among Native Americans in the Rocky Mountains of the USA twelve thousand years ago. Most of the studies, though, were of conflicts in the last century.

From a total of forty-five published studies, they found a link between climate change and violence in thirty-five of them. They indexed climate change in a variety of ways, and in their estimate of violence they included both war and domestic violence. The studies of conflict between groups of peoples and of institutional overthrow, including governmental overthrow, came from all over the world. Fifteen of the studies were of violence between individuals, most in the USA, but India, Australia, and Tanzania featured too. The overall finding was that higher-than-normal temperatures, and drought, rather than the opposite, seemed to cause violence, both general and personal.

Finally, in perhaps the most detailed study yet done of the influence of climate on crime by individuals, Matthew Ranson showed in a county-by-county analysis across the USA for the thirty years from 1980 to 2009 that murders and assaults increased as the maximum temperature increased.

Global warming will therefore produce massive increases in violence in the USA, he predicts. Who would have thought that the topic of biogeography would include not just crime, but the ability to predict needed increases in means of crime prevention?

For sure, not all conflicts result from adverse effects of climate. Humans are very good at dreaming up excuses to fight—hence, the results of a study on conflict in East Africa that implicated climate, but showed that climate was a minor factor among many.

Whatever the climate-linked cause of violence, the resulting fighting can then prevent any agriculture. The consequent famine can drive people out of the area. As I write now in 2014, hundreds of thousands of starving Somalis have moved to refugee camps in the neighboring countries of Ethiopia and Kenya.

If bad weather can help induce conflict and migration, so can good weather. By ensuring abundant fodder for horses, a few decades of extraordinarily good weather could have fueled Genghis Khan's invasion of Eurasia in the early 1200s. Mara Hvistendahl reported this idea of Amy Hessl and Neil Pederson's, which arose from their work dating climate change by the use of tree rings. Narrow rings indicate poor growth, which implies poor weather; broad rings indicate good growth in good weather. Indeed, Ulf Büntgen and co-workers suggest that in general, cultural expansions, such as those of the Celts and Romans two thousand or so years ago, coincided with centuries of good weather.

However good the weather, if it weren't that the steppe was so easy for a mounted army to travel across, Genghis Khan would surely not have been so successful. More quantitatively, a recent study of factors that might explain the rise of large-scale societies in Africa and Eurasia showed that a major influence, besides advanced war technology, was ease of movement, as indicated by the presence of flat steppe environment that allowed cavalry and chariots.

Before I move on to the next topic, let me emphasize that neither the original authors nor I propose that poor weather alone causes murder or wars. It is one influence among many, one element in the proverbial perfect storm. Social, political, and economic factors all play a part. That is almost too obvious to need saying. But decades of writing and reading have taught me how easy it is to be misunderstood. In sum, geography (climate and topography) can in certain times and places influence where humans are through influencing motivations for invasions and war.

We cannot blame all extinctions of peoples on competition for land, or even on side effects of competition. Rare things, animate or inanimate, are more likely to disappear than are common things. And most things, including cultures, are rare. That, it turns out, is a fact of life.

Take words written by Shakespeare. More than twenty-five thousand different words (of a total of around thirty thousand) turn up in his writing fewer than ten times. Less than a thousand occur over a hundred times. Only three ("and," "I," and "to") appear over twenty thousand times. So, over eighty percent of Shakespeare's words are rare, in the sense that he used them fewer than ten times each. Of classes of items in my kitchen cupboards (saucepans, crockery, etc.), it took eight of them to match the number of the most common class of item, cutlery. In December 2011, I counted each week of the month the number of birds coming to feed at our bird tables. The most common species consistently numbered more individuals than the next five most common species together. In case a US birder is reading this, the most common was the junco; the next five were fox sparrow, brown towhee, western towhee, scrub jay, and acorn woodpecker.

Something can be common because it is widespread, or common because there is lots of it wherever it happens to exist. An example of the former might be sand; of the latter, gold in a vault. If we look at our closest animal relatives, the non-human primates, we find that across the breadth of geographic range sizes, a hundred and sixty-one species are in the bottom ten percent of the distribution of sizes. That is, approximately half of species are rare by this measure of the extent of the world that they live in. By contrast, only one species, the vervet monkey (or green monkey, or grivet, as it is also called) is in the top ten percent of geographic range sizes. Only one species is thus "common"—in other words, less than one percent of all species.

Taking population densities as our measure of commonness and rarity, the contrast is not as extreme, but nevertheless of the same order of magnitude. We see over fifty times as many rare species—they are in the bottom ten percent of the range of population sizes—as common species—in the top ten percent of population sizes.

Turning to human languages, we see the same contrast. Most are rare; few are common, whether we count the area of the world that each covers, or the number of speakers of each. Use the same bottom ten percent and top ten percent of area of geographic range sizes that I used for the comparison of non-human primates, and two thirds of languages are rare, compared to just one twentieth of them that are common.

Count numbers of speakers, and nearly three quarters of the world's indigenous languages are spoken by fewer than thirty thousand people each. Hundreds of stadiums throughout the world could easily hold thirty thousand people. By contrast, only one twentieth, just five percent, of the world's languages have over a million speakers each.

We get the same sort of ratios for geographic range sizes, population sizes, and densities of hunter-gatherer societies. Take population sizes recorded by the first anthropologists to count them, and the distinction between bottom and top ten percent of the range of land area covered by the societies. Two thirds of the societies lay in the bottom ten percent of areas, and only one hundredth of the societies in the top ten percent. The former had fewer than fifteen hundred people each, the latter a minimum of over thirteen thousand.

Rarity, then, is common. That fact is a fascinating phenomenon in itself. But why else should we care about the observation that most things, including cultures, are rare? One answer is that we tend to value rare things. That is why De Beers kept most of its diamonds in its vaults, instead of flooding the market with them.

More importantly, rare living things, whether species or cultures, are more likely to go extinct than are common ones. A species or culture surviving in only a small area of forest is more likely to find its home clear-cut than a species or culture that lives in a large area of forest. A species or culture surviving on only a small island is more likely to find its home devastated by natural disaster. That happened to the Tamboran people, culture, and language on the island of Sumbawa in Indonesia with the 1815 explosion of Mt. Tambora. All gone in an instant. I described the global effects of this explosion in chapter 2.

The native Tasmanians that I wrote on earlier in this chapter could well have been doomed, even if the British had not arrived. They existed as small populations on a small island, already losing useful artifacts because of lack of skilled artisans, as I described in chapter 8. Cultures that live in a small range are more likely to experience poor conditions over the whole of the range than are cultures that live in a large range. With loss of technology and consequent potential loss of ability to cope with change, it might not have taken too many consecutive years of poor weather to have reduced the Tasmanian populations below a level from which they could not recover.

Already the populations of tiny low-lying oceanic islands, such as the Marshall Islands and the Maldives, are suffering from the rise in sea levels induced by climate warming. Rising sea levels of course threaten all coastal populations. However, the smaller the island, the less likely it is that the interior is sufficiently high to escape flooding, or sufficiently large for a viable population to survive.

Across the world, humans now speak maybe sixty-nine hundred languages. Nearly five hundred of those are spoken by less than one hundred people each. Maybe a hundred and thirty languages are now spoken by fewer than ten people. One of my university's reception halls could seat all of the speakers of those hundred and thirty languages. As we might expect from the fact that cultures in the tropics cover smaller territories than they do outside the tropics, three quarters of these endangered languages are in the tropics.

Overall, ninety-four percent of the world's languages are spoken by only six percent of the world's people. Worse, estimates are that over four hundred and seventy of the rare languages are spoken now by only a few elderly people, perhaps even only one elderly person. As I have already described, Daniel Nettle and Suzanne Romaine's book *Vanishing Voices* contains several photographs of the last remaining speaker of a language—Eyak, Manx, Ubykh. If language is culture, the geographic range of each of those cultures has been reduced to wherever the last speaker roams. All of the pictures show old people. Their final range, then, was probably not very large, just the distance between their home, a shop or two, and their doctor.

In North America currently, two hundred and nine languages are spoken. But only forty-six of those are spoken by children. How can the other hundred and sixty-three languages, three quarters of the total, not disappear within a generation?

Conquering powers do not always completely exterminate or severely limit a people or their culture. Sometimes they force migration, as during the slave trade, and therefore expand the geographic ranges of people and at least parts of their culture. Because of the slave trade, African people, culture, and genes abound in the Americas, the Caribbean, and elsewhere.

The biogeographic consequences are similar with the "Highland Clearances," the expulsion by rich landowners of Scots small-farmers to make way for sheep farming in the 1700s and 1800s. Canada and Australia then received large influxes of Scots genes. Around this period was when Australia received yet more British genes, along with Irish genes, with the arrival of a total of over a hundred thousand convicts.

Thankfully, we do not always compete or force emigration. We cooperate too, even if sometimes we cooperate to compete better. Who needs reminding that, for example, Italy allied with Germany in the Second World War, and the USA with Britain. However, here I want to talk about peaceful cooperation and peaceful expansion.

I am British. But I now live far from Britain. I live in California, eight thousand kilometers away from my parents' country. And I am being paid to work here. I have been lucky enough to work also in various African countries, as well as in Japan. I obviously could not have shifted my range so greatly were it not for the welcome of those other countries. I am just one person. Worldwide, millions of people have moved. Great shifts of range have happened peacefully, or relatively peacefully, anti-immigration political parties notwithstanding.

Millions of members of the former British Empire from outside Britain now live in Britain. Even more millions from other parts of the world now live in the USA. Chinatown in San Francisco is thousands of kilometers from any town of China. While some competition has definitely occurred among residents and immigrants, and nearly always some right-wing part of the host countries tries to resist present and future immigration, the cultural diversity of the cities of Britain and the USA exists for the most part because of mutual cooperation between residents and immigrants.

The *Times Atlas of World History* and the *Times Complete History of the World* contain hundreds of maps of the movements of peoples across the face of our planet. The majority of such books and maps show aggressive invasions and wars, and the contraction and expansion of peoples' geographic ranges that resulted. Cultures come and go, affected by competition, and by the global fact that rare things, inanimate and animate, species and cultures, are more likely to disappear than are common ones. Affected as we humans are by the same biological and environmental influences as are so many other species, our history is our biogeography. Historians of the movement of peoples, of the effect of climate and the environment on the geography of peoples or nations, are, in part, biogeographers.

If many languages and cultures are disappearing, a few, happily, are retreating from the brink of extinction. Some have come back with the help of people of other cultures. Two Britons, Desmond and Grace Derbyshire, in conjunction with FUNAI, the Brazilian National Indian Foundation, saved the Hixkaryana language and peoples. They did so by supplying medical help, by studying the culture, and by encouraging pride in it.

∽

I will close this chapter with mention of the work of two organizations whose aim is to save cultures and languages from extinction. One is Survival International, to which my proceeds from this book will go. Survival International is dedicated to helping "tribal peoples defend their lives, protect their lands and determine their own futures." The other is the Summer Institute of Linguistics, or SIL. SIL is the organization behind Ethnologue, a website listing all the world's languages that I and others have used as a superb source of information on the geography of language, and hence of cultures. The Summer Institute of Linguistics is particularly interested in encouraging survival of languages. It achieves its aims with a large measure of community development, clinics, schools, and advocacy, as well as with research, teaching, and the easy and free availability of its Ethnologue website.

These two organizations cannot possibly save all rare and disappearing languages and cultures. But at least they have slowed the process of extinction. Cultures whose geographic range would be nonexistent without them manage to hang on still because of them, and because of other like-minded institutions, along with the work of local peoples.

CHAPTER 13

EPILOGUE

Are we going to last the distance?

I n *Humankind*, I have endeavored to show how our global distribution and the type of organism we humans are has been shaped by the same biogeographical forces that shape other species. We have not been created in our own image. We are not a special creation.

As is the case for other animal species, the human species had a single geographic origin from which we expanded. As among other animal species, humans differ anatomically and physiologically in different regions as a result of individuals of certain genetic types surviving better and leaving more offspring than other types in those environments. As is the case for other animal species, climate affects where humans can live, and geographic barriers affect where humans can go. Even today, geographic barriers keep populations separate. As we see less diversity of animals (and plants too) at high latitudes than at lower latitudes, and less diversity on islands, especially small islands, than on mainlands, so too do we see less cultural diversity of humans at high latitudes and on islands. As other animal species have different diets in different parts of the world, and different adaptations to those diets, so do humans. Like other animal species, we humans compete and cooperate among ourselves and with other species such that we drive some species and cultures to extinction, but help expand the range of other species and cultures.

One fact that our knowledge of biogeography and evolution reveals is that eventually all species go extinct. At present, though, judging from the size of our population and the extent of our global distribution, we are thriving. When I was born, in 1950, the global human population numbered two and a half billion. A little over sixty years later, sometime between October 2011 and March 2012, we reached seven billion. According to some projections, we could well number ten billion in another sixty years. By then, we might even have extended our geographic range to include colonies on our moon and on Mars.

But for how much longer will we flourish? Humans as a species are the sole surviving branch of a once-spreading tree of hominin ancestors that at various times had three, four, even five branches living simultaneously across the Old World. More than fifteen of our close relatives, hominins, have tread the earth before us. All are now extinct. We know how long fourteen of them existed. What do they tell us concerning the length of existence of a hominin species? How much longer have we human hominins to go?

One of our ancestral relatives, *Homo erectus*, lasted one and a half million years. Others lasted from a little over a million years to just three hundred thousand years. Our average ancestor existed for six hundred thousand years. To be just an average human-like species, just an average hominin, humans need to persist for another four hundred thousand years, twice as long as we have existed so far.

I am not optimistic that we will last that four hundred thousand years into the future. Uniquely, our species uses each year now more resources than the earth produces in the year, meaning that we are depleting our capital. Glaciers and ice caps melt at a rate not before recorded. Yet we are too short-sighted, too willfully ignorant, too greedy to do anything about it. I include myself. I live a rich Western lifestyle, using way more resources than I put back into the earth. Almost everyone in the West does so. But will any one of us adopt the more sustainable lifestyle of the majority in the Third World? Almost certainly not.

The struggle for existence over the millennia, the process of evolution by natural selection, has left in each of us a desire for resources that was necessary to survival in the past. It kept us alive, reproducing, and migrating in a world in which resources were often difficult to acquire. Now in a world in which our technology allows exploitation on a massive scale, that same greed makes us fat, and it makes us lethal—to ourselves as well as to the world.

But our exploiting brains are also cooperative brains, as my doctorate advisor, Robert Hinde, continually stresses. We compete by cooperating as much as we compete by fighting. Twice as many mountain gorillas now live in the Virunga Volcanoes region of east-central Africa as when I started studying them in the 1970s. We have returned wolves to Yellowstone National Park. Whereas, as a child in England, I rarely if ever saw a bird of prey, these days kestrels, kites, buzzards are almost continually in sight, even in heavily populated and agricultural southern England. If we can maintain and extend the geographic range of other species, maybe we can have hope for our own species.

CITATIONS

I have used many more works than I cite here. Where the story is well known, I cite no authors, or cite only a summary or two. Where the story is less well known, I cite some of the latest studies, especially if the findings are not yet into textbooks or popular books.

I have listed the chapter topics in order of their appearance within each chapter. Within each topic, sources are listed with the latest first—because the latest should contain a summary of what has gone before, or at least reference to it.

CHAPTER 1. PROLOGUE

A.H. Harcourt. 2012. *Human Biogeography*. Ch. 1.
Humans biogeographically similar to other species—(Foley, 1987)
"Race" as a spurious concept—(Edgar & Hunley, 2009)
Chimpanzee subspecies more genetically varied than all humans—(Stone et al., 2002)
Geography of milk digestion—(Durham, 1991)
Regional differences in response to drugs—(Bustamante et al., 2011), (Need & Goldstein, 2009), (Bloch, 2004), (Taylor et al., 2004), (Exner et al., 2001), (Dimsdale, 2000)
Criticism of "race"-based science—(Kahn, 2007)
Lack of diversity of subjects in studies of genetics of disease—(Bloch, 2004), (Bustamante et al., 2011), (Need & Goldstein, 2009)
The Flores hobbit finding—(Brown et al., 2004)
The metric and imperial measuring systems—(Marciano, 2014)

CHAPTER 2. WE ARE ALL AFRICAN

A.H. Harcourt. 2012. *Human Biogeography*. Ch. 2.

Skeletal differences between early and late modern humans—(Klein, 2009)

Modern human origin 200,000 years ago—Fossil evidence (McDougall et al., 2005); Genetics (Ingman et al., 2000)

Hominin evolutionary tree and terminology—(Wood, 2010), (Klein, 2009)

Hominin brain sizes—(McHenry & Coffing, 2000){Ingman, 2000 #1663;Ingman, 2000 #1663}

Homo erectus in Dmanisi, Georgia 1.8 mya—(de Lumley & Lordkipanidze, 2006)

Dmanisi skull and number of *Homo* species—(Lordkipanidze et al., 2013)

Origins in sub-Saharan Africa—(Schlebusch et al., 2012), (Henn, Bustamante, et al., 2011), (Henn, Gignoux, et al., 2011), (Ramachandran et al., 2005)

Newly discovered hominin (Denisova) in Eurasia—(Krause et al., 2010)

Out of Africa—(Reyes-Centeno et al., 2014), (Mellars et al., 2013), (Eriksson et al., 2012), (Stewart & Stringer, 2012), (Petraglia et al., 2010), (Oppenheimer, 2009), (Pope & Terrell, 2008), (Oppenheimer, S., 2003)

Early presence in Saudi Arabia—(Armitage et al., 2011), (Petraglia et al., 2010)

Climate and Neanderthals in Middle East—(Hallin et al.)

Causes of exodus from Africa (climate, population size)—(Klein & Steele, 2013), (Blome et al., 2012; Eriksson et al., 2012), (Henn et al., 2012), (Stewart & Stringer, 2012), (Matthews & Butler, 2011), (Swain et al., 2007), (Ambrose, 2003), (Ingman et al., 2000), (Ambrose, 1998)

Climate and hominin evolution—(Stewart & Stringer, 2012), (Donges et al., 2011), (Scholz et al., 2007)

Earliest humans in Europe—(Benazzi et al., 2011), (Soares et al., 2010)

Climate and humans in Greenland—(D'Andrea & Huang, 2011), (Dugmore et al., 2007)

Mt. Toba and the exodus from Africa—(Witze & Kanipe, 2015), (Lane et al., 2013), (Louys, 2012), (Storey et al., 2012), (Ambrose, 2003), (Oppenheimer, C., 2003), (Ambrose, 1998)

Dispersals back into and within Africa—(Hodgson et al., 2014), (Campbell & Tishkoff, 2010), (Patin et al., 2009), (Campbell & Tishkoff, 2008), (Henn et al., 2008), (Aldenderfer, 2003), (Jones, 1995), (Thomson, 1887)

Dispersal to Asia, New Guinea, and Australia—(Reyes-Centeno et al., 2014), (Pagel et al., 2013), (Pugach et al., 2013), (van Holst Pellekaan, 2013), (Demeter et al., 2012), (Eriksson et al., 2012), (Stewart & Stringer, 2012), (Kayser, 2010){Oppenheimer, 2003 #1117;Stewart, 2012 #1611}, (Marwick, 2009), (Oppenheimer, 2009), (Pope & Terrell, 2008), (Hudjashov et al., 2007), (Aldenderfer & Yinong, 2004), (O'Connell & Allen, 2004)

40,000 year gap between first and subsequent arrivals in Australia—(Pugach et al., 2013), (Pearson, 2004), (Savolainen et al., 2004)

Early presence in India dispute—(Mellars et al., 2013), (Petraglia et al., 2010)

Humans, Neanderthals in India 70 kya—(Petraglia et al., 2010), (Petraglia et al., 2007)

Earliest in Eurasia—(Oppenheimer, S., 2003), (Goebel, 1999)

Into Tibet—(Qi et al., 2013), (Aldenderfer & Yinong, 2004), (Aldenderfer, 2003)

Dispersal to Japan—(Adachi et al., 2011)

Siberian route to East Asia—(Fu et al., 2014), (Reyes-Centeno et al., 2014)

Into western Europe—(Deguilloux et al., 2012), (Eriksson et al., 2012), (Stewart & Stringer, 2012), (Vigne et al., 2012), (Higham & Compton, 2011), (Benazzi et al.,

2011), (Soares et al., 2010), (Cavalli-Sforza, 2000), (Cavalli-Sforza et al., 1994), (Barbujani & Sokal, 1990)

Into Britain—(Hughes et al., 2014), (Oppenheimer, 2006){Oppenheimer, 2006 #1466;Benazzi, 2011 #1577}

Earliest Arctic site in Eurasia, Yana Rhinoceros Horn—(Pitulko et al., 2004)

Northeastern Siberian environment—(Hoffecker et al., 2014), (Meiri et al., 2014), (Zazula & Froese, 2003)

Earliest accepted American sites and dates—Texas (Waters et al., 2011); Monte Verde (Dillehay et al., 2008), (George et al., 2005), (Dillehay, 1999)

Earlier so far largely unaccepted sites and dates in the Americas—(Fariña & Tambussol, 2014), (Lahaye et al., 2013), (Guidon, 1986)

Peopling of the Americas—(Rasmussen et al., 2014), (Rademaker et al., 2013), (Eriksson et al., 2012){Raghavan, 2013 #1583;Raghavan, 2014 #1583}, (Reich et al., 2012){Reich, 2012 #1603;Reich, 2012 #1603}, (Pitblado, 2011), (Schroeder et al., 2007), (Barton et al., 2004), (Schurr, 2004), (Aldenderfer, 2003), (Oppenheimer, S., 2003), (Roosevelt et al., 2002)

Clovis culture and Clovis first debate—(Pitblado, 2011), (Haynes, 2002), (Roosevelt et al., 2002), (Fiedel, 2000)

Peopling of the Andes—(Rademaker et al., 2013), (Aldenderfer, 2003)

Back into Siberia—(Reich et al., 2012){Reich, 2012 #1603;Reich, 2012 #1603}

Agriculture and population expansion and movement—(Deguilloux et al., 2012), (Gignoux et al., 2010), (Vigne, 2008)

Dispersal of culture or of people?—(Hughes et al., 2014), (Chen et al., 2012), (Deguilloux et al., 2012), (Skoglund et al., 2012), (Forster & Renfrew, 2011), (Gignoux et al., 2010), (Cavalli-Sforza, 2000), (Kirch, 2000){Cavalli-Sforza, 1994 #859;Kirch, 2000 #1668}, (Cavalli-Sforza et al., 1988)

Fate of residents when agriculture and agriculturalists arrive—(Deguilloux et al., 2012), (Skoglund et al., 2012), (Gignoux et al., 2010), (Bramanti et al., 2009), (Richards et al., 2000)

Residents adopt language of the male invaders—(Forster & Renfrew, 2011), (Kemp et al., 2010), (Helgason et al., 2001), (Barbujani & Sokal, 1990)

Subsequent global movements—(Hellenthal et al., 2014), (Pickrell et al., 2014), (Cavalli-Sforza, 2000), (Cavalli-Sforza et al., 1994)

European genes in Southern African Khoi-San peoples—(Hellenthal et al., 2014), (Pickrell et al., 2014)

Global movements of peoples invalidates concept of "race"—(Hunley et al., 2009)

Expansion into Pacific—(Wollstein et al., 2010), (Oppenheimer, S., 2003), (Kirch, 2000)

Lapita culture—(Kirch, 2000), (Kirch, 1997)

Arrival in New Zealand—(Knapp et al., 2012), (Murray-McIntosh et al., 1998)

Small number of founding women—New Zealand (Penny et al., 2002); Madagascar (Cox et al., 2012)

Arrival in Madagascar—(Cox et al., 2012){Cox, 2012 #1653;Cox, 2012 #1653}, (Gommery et al., 2011), (Razafindrazaka et al., 2010), (Hurles et al., 2005), (Burney et al., 2004), (Oppenheimer, S., 2003)

CHAPTER 3. FROM HERE TO THERE AND BACK AGAIN

A.H. Harcourt. 2012. *Human Biogeography*. Ch. 2.

Size of African exodus—(Liu et al., 2006), (Harpending et al., 1998)

Modeling the coastal route—(Field & Lahr, 2005)

Shellfish foraging—(Archer et al., 2014){Archer*, 2014 #1753;Archer, 2014 #1753}, (Taylor et al., 2011), (Steele & Klein, 2008), (Field & Lahr, 2005), (Bird & Bird, 2000), (Walter et al., 2000), (Meehan, 1977)

Out of Africa and to rest of world—(Bar-Yosef & Belfer-Cohen, 2013), (Eriksson et al., 2012), (Benazzi et al., 2011), (Petraglia et al., 2010), (Oppenheimer, 2009), (Pope & Terrell, 2008), (Field & Lahr, 2005), (Oppenheimer, 2003)

Sahul (New Guinea, Australia)—(Bar-Yosef & Belfer-Cohen, 2013), (van Holst Pellekaan, 2013), (Kayser, 2010), (Hudjashov et al., 2007)

India—(Majumder, 2010)

Europe—(Deguilloux et al., 2012), (Soares et al., 2010), (Zwyns et al., 2012), (Pope & Terrell, 2008), (Oppenheimer, 2006), (Oppenheimer, 2003), (Goebel, 1999)

Asia—(Stoneking & Delfin, 2010)

Japan—(Adachi et al., 2011), (Erlandson, 2002)

Americas—(Chatters et al., 2014), (Raghavan et al., 2014), (Dixon, 2013), (de Saint Pierre et al., 2012), (Eriksson et al., 2012), (Pitblado, 2011), (O'Rourke & Raff, 2010), (Fagundes et al., 2008), (Tamm et al., 2007), (Oppenheimer, 2003), (Jablonski, 2002), (Mandryk et al., 2001), (Anderson & Gillam, 2000), (Fiedel, 2000), (Fladmark, 1979)

Origins of Siberian ancestors of original Americans (Mal'ta site)—(Raghavan et al., 2014){Raghavan, 2013 #1583;Raghavan, 2014 #1583}

From Americas to Siberia—(Reich et al., 2012){Reich, 2012 #1603;Reich, 2012 #1603}, (Karafet et al., 1997)

Coastal route first—(Pitblado, 2011), (Fagundes et al., 2008), (Mandryk et al., 2001), (Fiedel, 2000), (Fladmark, 1979)

Speed of Clovis advance—(Hamilton & Buchanan, 2007), (Waters & Stafford, 2007)

Siberian (and European) origins of all native Americans—(Raghavan et al., 2014) {Raghavan, 2013 #1583;Raghavan, 2014 #1583}, (Schroeder et al., 2007)

Amazon forest not retreating during last glacial—(Colinvaux et al., 1996)

Solutrean culture in America?—Yes (Stanford & Bradley, 2012); No (Raghavan et al., 2014), (Straus et al., 2005)

Diet of Paraguayan Aché—(Hill et al., 1984)

Amazon forest persisting through Pleistocene—(Costa et al., 2000), (Colinvaux et al., 1996)

Across the Pacific—(Tumonggor et al., 2013), (Soares et al., 2011), (Cox et al., 2010), (Wollstein et al., 2010), (Gray et al., 2009), (Kayser et al., 2008), (Penny et al., 2002), (Kirch, 2000)

Different findings depending on genes studied—(Tumonggor et al., 2013), (Kayser et al., 2008)

Sex differences in migration—(Tumonggor et al., 2013), (Cox et al., 2010), (Stoneking & Delfin, 2010), (Hage & Marck, 2003)

Genghis Khan's Y-chromosome genes—(Zerjal et al., 2003)

New Guineans are not Asians—(van Holst Pellekaan, 2013), (Cox et al., 2010), (Stoneking et al., 1990)

Australian desert forcing coastal route to southeast—(Oppenheimer, 2009)

Ancient travel between Polynesia and the Americas—(Thomson et al., 2014),

(Roullier et al., 2013), (Jones et al., 2011), (Storey et al., 2007), (Clarke et al., 2006), (Ballard et al., 2005), (Erickson et al., 2005)

To Madagascar—(Cox et al., 2012), (Serva et al., 2012), (Hurles et al., 2005)

Migrations within Africa—(Campbell & Tishkoff, 2010)

Eurasian genes in Africans and movement back into Africa—(Hellenthal et al., 2014), (Pickrell et al., 2014), (Abi-Rached et al., 2011)

Speed of diaspora—Sahul (Macaulay et al., 2005), (Beaton, 1991); Americas (Meltzer, 2004), (Surovell, 2003), (Surovell, 2000), (Fiedel, 2000), (Beaton, 1991), (Martin, 1973); European pastrolaists (Vigne, 2008); Marginal value theorem (Stephens & Krebs, 1986); Australian cane toads (Brown et al., 2013), (Lindström et al., 2013)

Individuals at front of expanding wave are different—(Brown et al., 2013), (Lindström et al., 2013), (Moreau et al., 2011)

CHAPTER 4. HOW DO WE KNOW WHAT WE THINK WE KNOW?

A.H. Harcourt. 2012. *Human Biogeography*. Ch. 2.

New finds and new interpretations—(Matthew, 1939), (Matthew, 1915)

Using language to discern movements—(Pagel et al., 2013), (Greenhill et al., 2010), (Greenberg, 1987)

Findings from archeology, genetics, linguistics match because people move with their culture—(Cavalli-Sforza, 2000), (Cavalli-Sforza et al., 1988)

Rejection of early Australian dates—(O'Connell & Allen, 2004)

Neigboring San communities with different origins—(Hellenthal et al., 2014), (Pickrell et al., 2014)

Earliest sophisticated culture, including art, in South Africa—(Henshilwood et al., 2011), (Wadley et al., 2011)

Earliest accepted American sites and dates—Texas (Waters et al., 2011); Monte Verde (Dillehay et al., 2008), (George et al., 2005), (Dillehay, 1999)

Earlier so far largely unaccepted sites and dates in the Americas—(Fariña & Tambusso, 2014), (Lahaye et al., 2013), (Guidon, 1986)

New theories accepted by the young—(Hull et al., 1978)

Human origins in Africa—Genetic evidence, (Henn et al., 2012), (Tishkoff et al., 2009), (Ingman et al., 2000), (Cann et al., 1987); Archeological evidence (McDougall et al., 2005)

Genetic evidence for origins in Africa, and spread across the world—(Henn et al., 2012), (Li et al., 2008), (Weaver & Roseman, 2008), (Liu et al., 2006), (Tishkoff & Verrelli, 2003), (Cavalli-Sforza et al., 1994), (Cann et al., 1987)

Humans globally less genetically diverse than chimpanzees regionally—(Stone et al., 2002)

Other species tell us about our diaspora—(Grindon & Davison, 2013), (Roullier et al., 2013), (Tanabe et al., 2010), (Moodley et al., 2009), (Searle et al., 2009), (Wilmshurst et al., 2008), (Larson, Albarella, et al., 2007), (Larson, Cucchi, et al., 2007), (Linz et al., 2007), (Clarke et al., 2006), (Matisoo-Smith & Robins, 2004), (Falush et al., 2003), (Austin, 1999)

CHAPTER 5. VARIETY IS THE SPICE OF LIFE

A.H. Harcourt. 2012. *Human Biogeography*. Ch. 5.

Skin color—not adaptive—(Darwin, 1871, I, 276).

Fur color—adaptive—Primates (Kamilar & Bradley, 2011); Cattle (Finch & Western, 1977)

Skin color in humans—adaptive—(Jablonski & Chaplin, 2013), (Khan, 2010), (Jablonski & Chaplin, 2002), (Jablonski & Chaplin, 2000), (Norton et al., 2007)

Date of evolution of pale skin of Europeans—(Wilde et al., 2014)

Skin color—genetics and a fish—(Lamason et al., 2005)

Oldest monument, Nabta Playa, Egypt—(Malville et al., 1998)

Size, shape, and temperature: Bergmann and Allen effects—Animals—(Freckleton et al., 2003), (Meiri & Dayan, 2003), (McNab, 2002), (Ashton et al., 2000)

Size, shape, and temperature: Bergmann and Allen effects—Primates—(Harcourt & Schreier, 2009)

Size, shape, and temperature: Bergmann and Allen effects—Humans—(Gilligan et al., 2013), (Little, 2010), (Tilkens et al., 2007), (Molnar, 2006), (Bogin & Rios, 2003), (Ruff, 2002), (Binford, 2001), (Katzmarzyk & Leonard, 1998), (Holliday & Falsetti, 1995), (Roberts, 1978)

Shape and athleticism—(McDougall, 2009), (Entine, 2001)

Size, shape, and temperature: Bergmann and Allen effects—Small size of pygmy peoples—(Jarvis et al., 2012), (Perry & Dominy, 2009), (Migliano et al., 2007)

Size, shape, and nutrition—(Stinson & Frisancho, 1978)

Size, shape, and temperature: Bergmann and Allen effects—Neanderthal—(Walker et al., 2011), (Weaver & Steudel-Numbers, 2005), (Finlayson, 2004), (Holliday & Falsetti, 1995)

Warm southern Spain, so Neanderthal there not cold-adapted—(Domínguez-Villar et al., 2013), (Walker et al., 2011), (Finlayson, 2004), (Carrión et al., 2003)

Neanderthal body proportions because of active lifestyle?—(Hora & Sladek, 2014), (Gilligan et al., 2013), (Shaw & Stock, 2013), (Higgins & Ruff, 2011), (Walker et al., 2011), (Finlayson, 2004)

Neanderthal, human sex differences, and hunting—(Bird & Bird, 2008), (Kuhn & Stiner, 2006), (Chilton, 2004)

Metabolism and temperature—General, animals—(McNab, 2002); Heat (Taylor, 2006), (Moran, 2000); Salt loss (and hypertension) (Campbell & Tishkoff, 2008), (Young, 2007), (Young et al., 2005); Pygmy peoples (Young et al., 2005); Cold (Little, 2010), (Froehle, 2008), (Moran, 2000), (Frisancho, 1993); Genetics (Hancock & Di Rienzo, 2008), (Young et al., 2005); Polar explorers (Huntford, 1999), (Cherry-Garrard, 1922)

Behavior and temperature in Tasmania—(Gilligan, 2007)

Fat islanders and efficient metabolism—Animals—(McNab, 2002); Humans (Genné-Bacon, 2014), (Molnar, 2006), (Bindon & Baker, 1997)

Lifetime development of high-altitude physiology and anatomy—(Frisancho, 2013), (Weitz et al., 2013)

Tibetan, Andean, and Ethiopian high-altitude vigor—Anatomy, physiology—Humans—(Huerta-Sánchez et al., 2014), (Beall, 2013), (Frisancho, 2013), (Weitz et al., 2013), (Alkorta-Aranburu et al., 2012), (Scheinfeldt et al., 2012), (Moore et al., 2011), (Beall et al., 2010), (Bigham et al., 2010), (Simonson et al., 2010), (Yi et al., 2010), (Julian et al., 2009), (Bennett et al., 2008), (Beall, 2007), (Henderson et al., 2005), (Moore, 2001), (Moore, 1998), (Frisancho, 1993)

High-altitude vigor—Animals—(McNab, 2002), (Schmidt-Nielsen, 1997)

Tibetan, Andean, and Ethiopian genetics of high-altitude vigor—(Beall et al., 2010), (Simonson et al., 2010), (Yi et al., 2010)

Rapid evolution of high-altitude abilities—(Yi et al., 2010)

Earliest Tibetan, Andean sites—(Rademaker et al., 2013), (Aldenderfer & Yinong, 2004)

Animals adapt to high altitudes—(Li et al., 2014), (Keller et al., 2013), (Qiu et al., 2012), (McNab, 2002)

Differences between the sexes—Shape and metabolism—(Ruff, 2002), (Stinson, 2000); Donner Party (Grayson, 1993); Dutch hunger (Banning, 1946); Shipwrecks (Elinder & Erixson, 2012), (Frey et al., 2010); Mayflower (Johnson, 1994-2013)

CHAPTER 6. GENE MAPS AND ROADS LESS TRAVELED

A.H. Harcourt. 2012. *Human Biogeography*. Ch. 3, 4.

General—Humans—(Cavalli-Sforza, 2000)

Origins—Genetic accident and regional differences—(Weaver et al., 2007), (Mielke et al., 2006), (Oppenheimer, 2006), (Cavalli-Sforza, 2000), (Cavalli-Sforza et al., 1994), (Diamond, 1987)

Origins—Founder effects and disease—Tay-Sachs—(Risch et al., 2003); porphyria (Diamond, 1987)

Origins—Language—(Lewis, 2009)

Sahara wet in the past—(Osborne et al., 2008)

Ice caps, ice-free corridor, and entry into America—(Dixon, 2013), (Mandryk et al., 2001), (Burns, 1996), (Fladmark, 1979)

Early presence of bison in the ice-free corridor—(Dixon, 2013)

Water as barrier—(Barbujani & Sokal, 1990), (Meggers, 1977); Moses (Drews & Han, 2010)

Geographic barriers in Europe cause regional cultural and genetic differences—(Novembre et al., 2008), (Barbujani & Sokal, 1990)

Mountains as barriers—(Nettle, 1996), (Barbujani & Sokal, 1990)

Mountains as barriers, judged by late arrival at high altitude—(Qi et al., 2013), (Rademaker et al., 2013)

Mountains "higher" in the tropics—(Cashdan, 2001a), (Janzen, 1967)

Distance, economics, lack of transport, as barriers—(Robb, 2007), (Molnar, 2006), (Harrison, 1995), (Cashdan, 2001a), (Perry, 1969)

Parasites and pathogens as barriers to movement—(Dunn et al., 2010), (Fincher & Thornhill, 2008b), (Fincher & Thornhill, 2008a)

Xenophobia and culture as barriers—(Currie & Mace, 2012), (Fincher & Thornhill, 2012), (Hünemeier & Gómez-Valdés, 2012), (Fincher & Thornhill, 2008b), (Majumder, 2010), (Cashdan, 2001b), (Rabbie, 1992), (Milton, 1991), (Dow et al., 1987)

CHAPTER 7. IS MAN MERELY A MONKEY?

A.H. Harcourt. 2012. *Human Biogeography*. Ch. 5.

Definitions of species, cultures, languages—(Nettle, 1999), (Simpson, 1961)

Diversity of cultures—main sources of raw information—Hunter-gatherers—(Binford, 2001); Languages (Gordon, 2005), (Goddard, 1996)

Often high cultural diversity where high biological diversity—(Burnside et al., 2012), (Gorenflo et al., 2012), (Loh & Harmon, 2005), (Maffi, 2005), (Stepp et al., 2005), (Sutherland, 2003), (Moore et al., 2002), (Nettle & Romaine, 2000), (Nettle, 1999)

Geographic range size smaller near the equator—Primates—(Harcourt, 2006), (Harcourt, 2000b); Mammals (Ruggiero, 1994); Humans (Currie et al., 2013), (Mace & Pagel, 1995)

Territoriality of humans—(Reséndez, 2007)

Overlap of geographic ranges of non-human primates, and relevance to humans—(Harcourt & Wood, 2012)

Tropical biodiversity and its explanations—(Brown, 2014), (Lomolino et al., 2010, Ch. 15), (Harcourt, 2006), (Hawkins et al., 2003), (Allen et al., 2002), (Janzen, 1967)

Territoriality of humans—(Eerkens et al., 2014), (Reséndez, 2007)

Tropical diversity of human cultures and its environmental causes—(Gavin et al., 2013), (Currie & Mace, 2012){Currie, 2012 #1735;Currie, 2012 #1735}, (Collard & Foley, 2002), (Cashdan, 2001), (Nettle & Romaine, 2000), (Nettle, 1999), (Nettle, 1998), (Mace & Pagel, 1995)

Geographic barriers promote diversity—(Novembre et al., 2008), (Nettle, 1996), (Barbujani & Sokal, 1990)

Triangular patterns in biogeography—(Currie et al., 2013), (Harcourt & Schreier, 2009), (Brown, 1995), (Brown & Maurer, 1989)

Hotspots of biodiversity—(Harcourt, 2000a)

Hotspot of eastern Asian hunter-gatherer societies—(Stoneking & Delfin, 2010)

High diversity of some non-tropical cultures—(Codding & Jones, 2013), (Gordon, 2005), (Stepp et al., 2005), (Birdsell, 1953)

Australian territory size—(Birdsell, 1953)

Cities as hotspots of cultural diversity—(Currie & Mace, 2012), (Ottaviano & Peri, 2005)

Tropical diversity of parasites, disease organisms and their vectors, and the biogeographical consequences—(Dunn et al., 2010), (Guernier & Guegan, 2009), (Fincher & Thornhill, 2008b), (Fincher & Thornhill, 2008a), (Nunn et al., 2005), (Guernier et al., 2004), (Poulin & Morand, 2004)

Disease prevents expansion of tropical cultures—(Cashdan, 2001), (Diamond, 1997, Ch. 10), (MacArthur, 1972)

More genetic adaptations to disease than to climate—(Fumagalli et al., 2011)

CHAPTER 8. ISLANDS ARE SPECIAL

A.H. Harcourt. 2012. *Human Biogeography.* Ch. 6.

Hobbit debate—(Henneberg et al., 2014), (Kubo et al., 2013), (Brown, 2012), (Falk, 2011), (Aiello, 2010), (Meijer et al., 2010), (Berger et al., 2008), (Martin et al., 2006)

Island animal body size, other anatomy, and physiology—(Lomolino et al., 2013), (Montgomery, 2013), (Okie & Brown, 2009), (Bromham & Cardillo, 2007), (Palombo, 2007), (Lomolino, 2005), (Burness et al., 2001)

Large lizards on Flores and Australia—(Molnar, 2004), (Wroe, 2002), (Diamond, 1991)

Impoverished small islands—Cultures (Tasmania effect)—(Derex et al., 2013), (Gavin & Sibanda, 2012), (Kline & Boyd, 2010), (Powell et al., 2009), (Read, 2008), (Mellars, 2006), (Henrich, 2004), (Mellars, 1996), (Terrell, 1986), (Torrence, 1983); Diseases (Strassman & Dunbar, 1999), (Black, 1966); Species (Lomolino et al., 2010, Ch. 13), (Baz & Monserrat, 1999), (Harcourt, 1999)

Java/Flores number of languages—(Lewis, 2009)

Biodiverse meeting zones—(Linder & de Klerk, 2012), (Kingdon, 1989)

Impoverished distant islands, maybe—Cultures—(Gavin & Sibanda, 2012), (Cashdan, 2001); Species (MacArthur & Wilson, 1967)

Lemur densities—(Emmons, 1999), (Harcourt et al., 2005)

CHAPTER 9. WE ARE WHAT WE EAT

A.H. Harcourt. 2012. *Human Biogeography*. Ch. 7.

Nonhuman primates, diet and distribution—(Rowe & Myers, 2011), (Harcourt et al., 2002){Rowe, 2011 #1612;Edwards, 2007 #1797}

Asian peoples and alcohol—(Segal & Duffy, 1992)

Alcohol poisoning adaptive?—(Peng et al., 2010)

Cultural influences on alcohol's effects—(Lentz, 1999)

Milk—(Ranciaro et al., 2014), (Salque et al., 2013), (Curry, 2013), (Leonardi et al., 2012), (Gerbault et al., 2011), (Ingram et al., 2009), (Enattah et al., 2008), (Evershed et al., 2008), (Campbell & Tishkoff, 2008), (Vigne, 2008), (Burger et al., 2007), (Edwards et al., 2007), (Tishkoff et al., 2007), (Bersaglieri et al., 2004), (Beja-Pereira et al., 2003), (Holden & Mace, 1997), (Durham, 1991)

Starch and tubers—(Hancock et al., 2010), (Wrangham & Carmody, 2010), (Marlowe & Berbesque, 2009), (Perry et al., 2007), (Padmaja & Steinkraus, 1995), (Hawkes et al., 1989), (McGeachin & Akin, 1982)

Japanese and seaweed—(Hehemann et al., 2010)

Arctic peoples and fat—(Marlowe, 2005), (Cordain et al., 2002), (Binford, 2001), (Richards et al., 2000), (Baker, 1988)

Benefits of hunting large animals—(Bird et al., 2009), (Bird & Bird, 1997)

Regional responses to drugs—(Centers for Disease Control and Prevention, 2013), (Campbell & Tishkoff, 2008)

CHAPTER 10. WHAT DOESN'T KILL US HALTS US OR MOVES US

A.H. Harcourt. 2012. *Human Biogeography*. Ch. 9.

Malaria, yellow fever—(Townroe & Callaghan, 2014), (McNeill, 2010), (Campbell & Tishkoff, 2008), (Mielke et al., 2006, Ch. 6), (Knottnerus, 2002)

Sleeping sickness—(Pollak et al., 2010)

Disease rates—Chagas, malaria, sleeping sickness—(Lozano et al., 2012)

Disease prevents colonization—(McNeill, 2010), (Maudlin, 2006), (Curtin, 1998), (McNeill, 1976)

Colonization brings disease—(Henn et al., 2012), (Schlebusch et al., 2012), (O'Fallon & Fehren-Schmitz, 2011), (Pepperell et al., 2011), (Diamond, 1997, Ch. 5)

Predators—(Berger & McGraw, 2007), (Turner, 1992)

Drugs—(Meyer, 1999)

CHAPTER 11. MAD, BAD, AND DANGEROUS TO KNOW

A.H. Harcourt. 2012. *Human Biogeography*. Ch. 10.

Human-caused extinction of animal species—Hunting—(Sanchez et al., 2014), (Sandom et al., 2014), (Grund et al., 2012), (Rule et al., 2012), (Lorenzen et al., 2011), (Nikolskiy et al., 2011), (Prideaux et al., 2010), (Faith & Surovell, 2009), (Surovell & Waguespack, 2009), (Turney et al., 2008), (Miller et al., 2005),

(Surovell et al., 2005), (Johnson, 2002), (Holdaway & Jacomb, 2000), (Burney, 1997), (MacPhee & Marx, 1997), (Caughley & Gunn, 1996, Ch. 2), (Steadman, 1995); Competition (Fillios et al., 2012), (Koch & Barnosky, 2006)

Australia relatively unknown—(Field et al., 2013), (Prescott et al., 2012)

Spear heads and gomphothere bones—(Sanchez et al., 2014)

Detailed computer models, humans vs. climate—(Sandom et al., 2014), (Prescott et al., 2012)

Eurasia unusual—(Prescott et al., 2012), (Koch & Barnosky, 2006), (Stuart, 1991)

Few extinctions in Old World tropics, only one in India—(Roberts et al., 2014), (Sandom et al., 2014)

First signs of hunting—(Klein, 2009)

Environment x extinction implicates hunting—(Johnson, 2002)

Other causes of Pleistocene extinctions, particularly climate change—(Field et al., 2013), (Field & Wroe, 2012), (Fillios et al., 2012), (Pitulko & Nikolskiy, 2012), (Rick et al., 2012), (Lorenzen et al., 2011), (Nogués-Bravo et al., 2008), (MacPhee & Marx, 1997)

Both humans and climate—(Prescott et al., 2012){Prescott, 2012 #384;Prescott, 2012 #384}, (Lorenzen et al., 2011), (Nikolskiy et al., 2011), (Koch & Barnosky, 2006)

Island species susceptible—(Duncan et al., 2013), (Steadman, 2006)

Other causes—Fire—(Gillespie, 2008), (Miller et al., 2005)

Fire resulted from extinctions—(Rule et al., 2012)

Humans or climate? The argument continues—(Yule et al., 2014)

Australia fire and extinction—(Miller et al., 2005)

Fragments of habitat contain few species—(Harcourt & Doherty, 2005), (Marsh, 2003), (Laurance & Bierregaard, 1997)

Introduced species—General—(Davis, 2009); Competition (Fillios et al., 2012), (Koch & Barnosky, 2006), (Burney, 1997); Disease (MacPhee & Marx, 1997), (van Riper et al., 1986), (Warner, 1964); Predators (Loss et al., 2013), (Fritts & Rodda, 1998)

Range shift with climate warming—(Schloss et al., 2012), (Lomolino et al., 2010, Ch. 16), (Parmesan, 2006)

Parks and wildlife corridors—(Cabeza & Moilanen, 2001)

Climate and gelada baboon, pika—(Grayson, 2005), (Dunbar, 1998)

Eradication of smallpox, rinderpest—(Greenwood, 2014)

Survival on the edge—(Channell & Lomolino, 2000)

Knock-on effects—(Hanski et al., 2012), (Keesing et al., 2010), (Terborgh, 1999)

Introduced species—Beneficial effects—(Crane, 2013), (Zhao et al., 2010), (Sax & Gaines, 2008), (White, 2004), (Heinsohn, 2001)

Human microbiome—(Schnorr et al., 2014), (Costello et al., 2012), (Costello et al., 2009)

Global warming and the Arctic—(United States Environmental Protection Agency, 2014)

Neanderthal x climate—(Wood et al., 2013), (Pinhasi et al., 2012), (Stewart, 2005)

Neanderthal small population size and inbreeding—(Prüfer et al., 2014)

Neanderthal competition with humans—(Fa et al., 2013), (Shipman, 2012), (Lieberman et al., 2009), (Banks et al., 2008), (Stewart, 2005)

Neanderthal (and human) cannibalism—(Marlar et al., 2000), (Defleur et al., 1999)

Neanderthal plant diet—(Hardy et al., 2012), (Henry et al., 2011)
Humans mate with Neanderthal—(Lohse & Frantz, 2014), (Green et al., 2010)
Humans mate with Denisova—(Reich et al., 2010)
Humans mate with hominin species in Africa—(Hammer & Woerner, 2011)
Neanderthal genes in humans confer disease resistance—(Abi-Rached et al., 2011)
Humans not mate with Neanderthal (or Denisova)—(Lowery et al., 2013), (Eriksson & Manica, 2012)

CHAPTER 12. CONQUEST AND COOPERATION

A.H. Harcourt. 2012. *Human Biogeography*. Ch. 8.
General—(Malthus, 1798), (Overy, 2010)
Genocide—(Diamond, 1992, Ch. 16), Survival International http://www.survivalinternational.org/
Exclusion of species to marginal areas—(Channell & Lomolino, 2000)
Weather and violence—(Ranson, 2014), (Hsiang et al., 2013){Hsiang, 2013 #1846;Hsiang, 2013 #1846}, (Hvistendahl, 2012), (O'Loughlin et al., 2012), (Büntgen et al., 2011), (Hsiang et al., 2011), (Zhang et al., 2011), (Zhang et al., 2010)
Topography and war—(Turchin et al., 2013)
Extinction of languages, cultures—(Gorenflo et al., 2012), (Lewis, 2009), (Harrison, 2007), (Day, 2001), (Nettle & Romaine, 2000), (Harmon, 1996), (Brigandi, 1987), (Hill & Nolasquez, 1973), (Arseniev, 1928, 1939), (Government of the United Kingdom, 1847)
Exclusion of peoples to marginal areas—(Nettle & Romaine, 2000), (Bateson, 1983)
Rare things disappear, whether species or cultures—(Harmon & Loh, 2010), (Lewis, 2009), (Harrison, 2007), (Harcourt, 2006), (Harcourt et al., 2002), (Binford, 2001), (Harmon, 1996)
Saving of languages and cultures—Summer Institute of Linguistics http://www.sil.org/, Survival International http://www.survivalinternational.org

CHAPTER 13. EPILOGUE

A.H. Harcourt. 2012. *Human Biogeography*. Ch. 10.
Duration of life per hominin species—(Wood, 2010), (Foley, 2002)
Hope for humans?—(Hinde, 2011), (Hinde, 2002)

SOURCES

Abi-Rached, L. et al. 2011. "The shaping of modern human immune systems by multiregional admixture with archaic humans." *Science*, 334: 89-94.

Adachi, N. et al. 2011. "Mitochondrial DNA analysis of Hokkaido Jomon skeletons: Remnants of archaic maternal lineages at the southwestern edge of former Beringia." *American Journal of Physical Anthropology*, 146: 346-360.

Aiello, L.C. 2010. "Five years of *Homo floresiensis*." *American Journal of Physical Anthropology*, 142: 167-179.

Aldenderfer, M. and Yinong, Z. 2004. "The prehistory of the Tibetan plateau to the seventh century A.D.: perspectives and research from China and the West since 1950." *Journal of World Prehistory*, 18: 1-55.

Aldenderfer, M.S. 2003. "Moving up in the world." *American Scientist*, 91: 542-549.

Alkorta-Aranburu, G. et al. 2012. "The genetic architecture of adaptations to high altitude in Ethiopia." *PLOS Genetics*, 8: e1003110.

Allen, A.P. et al. 2002. "Global biodiversity, biochemical kinetics, and the energetic-equivalence rule." *Science* 297: 1545-1548.

Ambrose, S.H. 1998. "Late Pleistocene human population bottlenecks, volcanic winter, and the differentiation of modern humans." *Journal of Human Evolution*, 34: 623-651.

Ambrose, S.H. 2003. "Did the super-eruption of Toba cause a human population bottleneck? Reply to Gathorne-Hardy and Harcourt-Smith." *Journal of Human Evolution*, 45: 231-237.

Anderson, D.G. and Gillam, J.C. 2000. "Paleoindian colonization of the Americas: implications from an examination of physiography, demography, and artifact distribution." *American Antiquity*, 65: 43-66.

Archer, W. et al. 2014. "Early Pleistocene aquatic resource use in the Turkana Basin." *Journal of Human Evolution*, 77: 74-87.

Armitage, S.J. et al. 2011. "The southern route 'Out of Africa': Evidence for an early expansion of modern humans into Arabia." *Science*, 331: 453-456.

Arseniev, V.K. 1928, 1939. *Dersu the Trapper*. London: Secker & Warburg.

Ashton, K.G. et al. 2000. "Is Bergmann's rule valid for mammals?" *American Naturalist*, 156: 390-415.

Austin, C.C. 1999. "Lizards took express train to Polynesia." *Nature*, 397: 113-114.

Baker, P.T. 1988. Nutritional stress. In *Human Biology. An Introduction to Human Evolution*,

Variation, Growth, and Adaptability, edited by G.A. Harrison, et al., 479-507. Oxford: Oxford University Press.

Ballard, C. et al. (Eds.). (2005). *The Sweet Potato in Oceania: a Reappraisal*. Pittsburgh, Sydney: University of Pittsburgh, University of Sydney.

Banks, W.E. et al. 2008. "Neanderthal extinction by competitive exclusion." *PLOS ONE*, 3: e3972.

Banning, C. 1946. "The Netherlands during German occupation. Food shortage and public health, first half of 1945." *Annals of the American Academy of Political and Social Science*, 245: 93-110.

Bar-Yosef, O. and Belfer-Cohen, A. 2013. "Following Pleistocene road signs of human dispersals across Eurasia." *Quaternary International* 285: 30-43.

Barbujani, G. and Sokal, R.R. 1990. "Zones of sharp genetic change in Europe are also linguistic boundaries." *Proceedings of the National Academy of Sciences, USA*, 87: 1816-1819.

Barton, C.M. et al. (Eds.). (2004). *The Settlement of the American Continents. A Multidisciplinary Approach to Human Biogeography*. Tucson: University of Arizona Press.

Bateson, C. 1983. *The Convict Ships 1787-1868*. Sydney: Library of Australian History.

Baz, A. and Monserrat, V.J. 1999. "Distribution of domestic *Psocoptera* in Madrid apartments." *Medical and Veterinary Entomology*, 13: 259-264.

Beall, C.M. 2007. "Two routes to functional adaptation: Tibetan and Andean high-altitude natives." *Proceedings of the National Academy of Sciences, USA*, 104: 8655-8660.

Beall, C.M. 2013. "Human adaptability studies at high altitude: Research designs and major concepts during fifty years of discovery." *American Journal of Human Biology*, 25: 141-147.

Beall, C.M. et al. 2010. "Natural selection on EPAS1 (HIF2alpha) associated with low hemoglobin concentration in Tibetan highlanders." *Proceedings of the National Academy of Sciences USA*, 107: 11459-11464.

Beaton, J.M. 1991. Colonizing continents: some problems from Australia and the Americas. In *The First Americans: Search and Research*, edited by T.D. Dillehay and D.J. Meltzer, 209-230. Boca Raton, USA: CRC Press.

Beja-Pereira, A. et al. 2003. "Gene-culture coevolution between cattle milk protein genes and human lactase genes." *Nature Genetics*, 35: 311—313.

Benazzi, S. et al. 2011. "Early dispersal of modern humans in Europe and implications for Neanderthal behaviour." *Nature*, 479: 525-528.

Bennett, A. et al. 2008. "Evidence that parent-of-origin affects birth-weight reductions at high altitude." *American Journal of Human Biology*, 20: 592-597.

Berger, L.R. et al. 2008. "Small-bodied humans from Palau, Micronesia." *PLOS ONE*, 3: e1780.

Berger, L.R. and McGraw, W.S. 2007. "Further evidence for eagle predation of, and feeding damage on, the Taung child." *South African Journal of Science*, 103: 496-498.

Bersaglieri, T. et al. 2004. "Genetic signatures of strong recent positive selection at the lactase gene." *American Journal of Human Genetics*, 74: 1111-1120.

Bigham, A.W. et al. 2010. "Identifying signatures of natural selection in Tibetan and Andean populations using dense genome scan data." *PLOS Genetics*, 6: 1-14.

Bindon, J.R. and Baker, P.T. 1997. "Bergmann's rule and the thrifty genotype." *American Journal of Physical Anthropology*, 104: 201-210.

Binford, L.R. 2001. *Constructing Frames of Reference: An Analytical Method for Archaeological Theory Building Using Hunter-Gatherer and Environmental Data Sets*. Berkeley: University of California Press.

Bird, D.W. and Bird, R.B. 2000. "The ethnoarchaeology of juvenile foragers: Shellfishing strategies among Meriam children." *Journal of Anthropological Archaeology*, 19: 461-476.

Bird, R.B. and Bird, D.W. 2008. "Why women hunt. Risk and contemporary foraging in a Western Desert aboriginal community." *Current Anthropology*, 49: 655-693.

Bird, R.B. et al. 2009. "What explains differences in men's and women's production?" *Human Nature*, 20: 105-129.

Bird, R.L.B. and Bird, D.W. 1997. "Delayed reciprocity and tolerated theft: the behavioral ecology of food-sharing strategies." *Current Anthropology*, 38: 49-78.

Birdsell, J.B. 1953. "Some environmental and cultural factors influencing the structuring of Australian Aboriginal populations." *American Naturalist*, 87: 171-207.

Black, F.L. 1966. "Measles endemicity in insular populations: critical community size and its evolutionary implication." *Journal of Theoretical Biology*, 11: 207-211.

Bloch, M.G. 2004. "Race-based therapeutics." *New England Journal of Medicine*, 351: 2035-2037.

Blome, M.W. et al. 2012. "The environmental context for the origins of modern human diversity: A synthesis of regional variability in African climate 150,000-30,000 years ago." *Journal of Human Evolution*, 62: 563-592.

Bogin, B. and Rios, L. 2003. "Rapid morphological change in living humans: implications for modern human origins." *Comparative Biochemistry and Physiology A—Molecular and Integrative Physiology*, 136: 71-84.

Bramanti, B. et al. 2009. "Genetic discontinuity between local hunter-gatherers and central Europe's first farmers." *Science*, 326: 137-140.

Brigandi, P. 1987. "Roscinda Velasquez remembered." *Journal of California and Great Basin Anthropology*, 9: 2-3.

Bromham, L. and Cardillo, M. 2007. "Primates follow the 'island rule': implications for interpreting *Homo floresiensis*." *Biology Letters*, 3: 398-400.

Brown, G.P. et al. 2013. "The early toad gets the worm: cane toads at an invasion front benefit from higher prey availability." *Journal of Animal Ecology*, 82: 854-862.

Brown, J.H. 1995. *Macroecology*. Chicago: University of Chicago Press.

Brown, J.H. 2014. "Why are there so many species in the tropics?" *Journal of Biogeography*, 41: 8-22.

Brown, J.H. and Maurer, B.A. 1989. "Macroecology: the division of food and space among species on continents." *Science*, 243: 1145-1150.

Brown, P. 2012. "LB1 and LB6 *Homo floresiensis* are not modern human (*Homo sapiens*) cretins." *Journal of Human Evolution*, 62: 201-224.

Brown, P. et al. 2004. "A new small-bodied hominin from the late Pleistocene of Flores, Indonesia." *Nature*, 431: 1055-1061.

Büntgen, U. et al. 2011. "2500 years of European climate variability and human susceptibility." *Science*, 331: 578-582.

Burger, J. et al. 2007. "Absence of the lactase-persistence-associated allele in early Neolithic Europeans." *Proceedings of the National Academy of Sciences USA*, 104: 3736-3741.

Burness, G.P. et al. 2001. "Dinosaurs, dragons, and dwarfs: the evolution of maximal body size." *Proceedings of the National Academy of Sciences, USA*, 98: 14518-14523.

Burney, D.A. 1997. Theories and facts regarding Holocene environmental change before and after human colonization. In *Natural Change and Human Impact in Madagascar*, edited by S.M. Goodman and B.D. Patterson, 75-89. Washington, D.C.: Smithsonian Institution Press.

Burney, D.A. et al. 2004. "A chronology for late prehistoric Madagascar." *Journal of Human Evolution*, 47: 25-63.

Burns, J.A. 1996. "Vertebrate paleontology and the alleged ice-free corridor: the meat of the matter." *Quaternary International*, 32: 107-112.

Burnside, W.R. et al. 2012. "Human macroecology: linking pattern and process in big-picture human ecology." *Biological Reviews*, 87: 194-208.

Bustamante, C.D. et al. 2011. "Genomics for the world." *Nature*, 475: 163-165.

Cabeza, M. and Moilanen, A. 2001. "Design of reserve networks and the persistence of biodiversity." *Trends in Ecology and Evolution*, 16: 242-248.

Campbell, M.C. and Tishkoff, S.A. 2008. "African genetic diversity: Implications for human demographic history, modern human origins, and complex disease mapping." *Annual Review of Genomics and Human Genetics*, 9: 403-433.

Campbell, M.C. and Tishkoff, S.A. 2010. "The evolution of human genetic and phenotypic variation in Africa." *Current Biology*, 20: R166-R173.

Cann, R.L. et al. 1987. "Mitochondrial DNA and human evolution." *Nature*, 325: 31-36.

Carrión, J.S. et al. 2003. "Glacial refugia of temperate, Mediterranean and Ibero-North African flora in south-eastern Spain: new evidence from cave pollen at two Neanderthal man sites." *Global Ecology and Biogeography*, 12: 119-129.

Cashdan, E. 2001a. "Ethnic diversity and its environmental determinants: effects of climate, pathogens, and habitat diversity." *American Anthropologist*, 103: 968-991.

Cashdan, E. 2001b. "Ethnocentrism and xenophobia: A cross-cultural study." *Current Anthropology*, 42: 760-765.

Caughley, G. and Gunn, A. 1996. *Conservation Biology in Theory and Practice*. Cambridge, Massachusetts: Blackwell Science.

Cavalli-Sforza, L.L. 2000. *Genes, Peoples, and Languages*. Berkeley: University of California Press.

Cavalli-Sforza, L.L. et al. 1994. *The History and Geography of Human Genes*. Princeton: Princeton University Press.

Cavalli-Sforza, L.L. et al. 1988. "Reconstruction of human evolution: Bringing together genetic, archaeological, and linguistic data." *Proceedings of the National Academy of Sciences, USA*, 85: 6002-6006.

Centers for Disease Control and Prevention. (2013). U.S. Public Health Service Syphilis Study at Tuskegee. http://www.cdc.gov/tuskegee/timeline.htm

Channell, R. and Lomolino, M.V. 2000. "Dynamic biogeography and conservation of endangered species." *Nature*, 403: 84-86.

Chatters, J.C. et al. 2014. "Late Pleistocene Human Skeleton and mtDNA Link Paleoamericans and modern Native Americans." *Science*, 344: 750-754.

Chen, J. et al. 2012. "Worldwide analysis of genetic and linguistic relationships of human populations." *Human Biology*, 84: 553-570.

Cherry-Garrard, A.G.B. 1922. *The Worst Journey in the World: Antarctic 1910–1913*. New York: Doran.

Chilton, E.S. 2004. Gender, age and subsistence diversity in Paleoindian societies. In *The Settlement of the American Continents. A Multidisciplinary Approach to Human Biogeography*, edited by C.M. Barton, et al., 162-172. Tucson: University of Arizona Press.

Clarke, A.C. et al. 2006. "Reconstructing the origins and dispersal of the Polynesian bottle gourd (*Lagenaria siceraria*)." *Molecular Biology and Evolution*, 23: 893-900.

Codding, B.F. and Jones, T.L. 2013. "Environmental productivity predicts migration, demographic, and linguistic patterns in prehistoric California." *Proceedings of the National Academy of Sciences USA*, 110: 14569-14573.

Colinvaux, P.A. et al. 1996. "A long pollen record from lowland Amazonia: forest and cooling in glacial times." *Science*, 274: 85-88.

Collard, I.F. and Foley, R.A. 2002. "Latitudinal patterns and environmental determinants of recent human cultural diversity: do humans follow biogeographical rules?" *Evolutionary Ecology Research*, 4: 371-383.

Cordain, L. et al. 2002. "The paradoxical nature of hunter-gatherer diets: meat-based, yet non-atherogenic." *European Journal of Clinical Nutrition*, 56: S42-S52.

Costa, L.P. et al. 2000. "Biogeography of South American forest mammals: endemism and diversity in the Atlantic Forest." *Biotropica*, 32: 872-881.

Costello, E.K. et al. 2009. "Bacterial community variation in human body habitats across space and time." *Science*, 326: 1694-1697.

Costello, E.K. et al. 2012. "The application of ecological theory toward an understanding of the human microbiome." *Science*, 336: 1255-1262.

Cox, M.P. et al. 2010. "Autosomal and X-linked single nucleotide polymorphisms reveal a steep Asian-Melanesian ancestry cline in eastern Indonesia and a sex bias in admixture rates." *Proceedings of the Royal Society B*, 277: 1589-1596.

Cox, M.P. et al. 2012. "A small cohort of Island Southeast Asian women founded Madagascar." *Proceedings of the Royal Society B*, 279: 2761-2768.

Crane, P. 2013. *Ginkgo. The Tree that Time Forgot.* New Haven, USA: Yale University Press.

Currie, T.E. and Mace, R. 2012a. "Analyses do not support the parasite-stress theory of human sociality." *Behavioral and Brain Sciences*, 35: 83-85.

Currie, T.E. and Mace, R. 2012b. "The evolution of ethnolinguistic diversity." *Advances in Complex Systems*, 15: 1150006-1150026.

Currie, T.E. et al. 2013. "Cultural phylogeography of the Bantu Languages of sub-Saharan Africa." *Proceedings of the Royal Society B*, 280: 20130695.

Curry, A. 2013. "The milk revolution." *Nature*, 500: 20-22.

Curtin, P.D. 1998. *Disease and Empire. The Health of European Troops in the Conquest of Africa.* Cambridge: Cambridge University Press.

D'Andrea, W.J. and Huang, Y. 2011. "Abrupt Holocene climate change as an important factor for human migration in West Greenland." *Proceedings of the National Academy of Sciences USA*, 108: 9765-9769.

Darwin, C. 1871. *The Descent of Man, and Selection in Relation to Sex.* London: John Murray.

Davis, M.A. 2009. *Invasion Biology.* Oxford: Oxford University Press.

Day, D. 2001. *Claiming a Continent. A New History of Australia.* Harper Collins Publishers PTY Limited.

de Lumley, M.-A. and Lordkipanidze, D. 2006. "L'homme de Dmanissi (*Homo georgicus*), il y a 1 810 000 ans." *Comptes Rendus Palevol*, 5: 273-281.

de Saint Pierre, M. et al. 2012. "An alternative model for the early peopling of Southern South America revealed by analyses of three mitochondrial DNA haplogroups." *PLOS ONE*, 7: e43486

Defleur, A. et al. 1999. "Neanderthal cannibalism at Moula-Guercy, Ardèche, France." *Science*, 286: 128-131.

Deguilloux, M.-F. et al. 2012. "European neolithization and ancient DNA: an assessment." *Evolutionary Anthropology*, 21: 24-37.

Demeter, F. et al. 2012. "Anatomically modern human in Southeast Asia (Laos) by 46 ka." *Proceedings of the National Academy of Sciences, USA*, 109: 14375-14380.

Derex, M. et al. 2013. "Experimental evidence for the influence of group size on cultural complexity." *Nature*, 503: 389-391.

Diamond, J. 1987. "Observing the founder effect in human evolution." *Nature*, 329: 105-106.

Diamond, J.M. 1991. "A case of missing marsupials." *Nature*, 353: 17.

Diamond, J.M. 1992. *The Third Chimpanzee. The Evolution and Future of the Human Animal.* New York: Harper Collins Publishers Inc.

Diamond, J.M. 1997. *Guns, Germs, and Steel: The Fates of Human Societies.* New York: W.W. Norton.

Dillehay, T.D. 1999. "The Late Pleistocene cultures of South America." *Evolutionary Anthropology*, 7: 206-216.

Dillehay, T.D. et al. 2008. "Monte Verde: seaweed, food, medicine, and the peopling of South America." *Science*, 320: 784-786.

Dimsdale, J.E. 2000. "Stalked by the past: The influence of ethnicity on health." *Psychosomatic Medicine*, 62: 161-170.

Dixon, E.J. 2013. "Late Pleistocene colonization of North America from Northeast Asia: New insights from large-scale paleogeographic reconstructions." *Quaternary International*, 285: 57-67.

Domínguez-Villar, D. et al. 2013. "Early maximum extent of paleoglaciers from Mediterranean mountains during the last glaciation." *Scientific Reports*, 3: 2034.

Donges, J.F. et al. 2011. "Nonlinear detection of paleoclimate-variability transitions possibly related to human evolution." *Proceedings of the National Academy of Sciences, USA*, 108: 20422-20427.

Dow, M.M. et al. 1987. "Partial correlation of distance matrices in studies of population structure." *American Journal of Physical Anthropology*, 72: 343-352.

Drews, C. and Han, W. 2010. "Dynamics of wind setdown at Suez and the Eastern Nile Delta." *PLOS ONE*, 5: e12481.

Dugmore, A.J. et al. 2007. "Norse Greenland settlement: Reflections on the climate change, trade, and the contrasting fates of human settlements in the North Atlantic Islands." *Arctic Anthropology*, 44 (1): 12-36.

Dunbar, R.I.M. 1998. "Impact of global warming on the distribution and survival of the gelada baboon: a modelling approach." *Global Change Biology*, 4: 293-304.

Duncan, R.P. et al. 2013. "Magnitude and variation of prehistoric bird extinctions in the Pacific." *Proceedings of the National Academy of Sciences, USA*, 110: 6436-6441.

Dunn, R.R. et al. 2010. "Global drivers of human pathogen richness and prevalence." *Proceedings of the Royal Society B*, 277: 2587-2595.

Durham, W.H. 1991 *Coevolution: Genes, Culture, and Human Diversity*. Stanford: Stanford University Press.

Edgar, H.J.H. and Hunley, K.L. 2009. "Race reconciled: how biological anthropologists view human variation." *American Journal of Physical Anthropology*, 139: 1-107.

Edwards, C.J. et al. 2007. "Mitochondrial DNA analysis shows a Near Eastern Neolithic origin for domestic cattle and no indication of domestication of European aurochs." *Proceedings of the Royal Society B*, 274: 1377-1385.

Eerkens, J.W. et al. 2014. "Tracing the mobility of individuals using stable isotope signatures in biological tissues: 'locals' and 'non-locals' in an ancient case of violent death from Central California." *Journal of Archaeological Science*, 41: 474-481.

Elinder, M. and Erixson, O. 2012. "Gender, social norms, and survival in maritime disasters." *Proceedings of the National Academy of Sciences USA*, 109: 13220-13224.

Emmons, L.H. 1999. Of mice and monkeys: primates as predictors of mammal community richness. In *Primate communities*, edited by J.G. Fleagle, et al., 171-188. Cambridge, UK: Cambridge University Press.

Enattah, N.S. et al. 2008. "Independent introduction of two lactase-persistence alleles into human populations reflects different history of adaptation to milk culture." *American Journal of Human Genetics*, 82: 57-72.

Entine, J. 2001. *Taboo: Why Black Athletes Dominate Sports And Why We're Afraid To Talk About It*. New York: Public Affairs.

Erickson, D.L. et al. 2005. "An Asian origin for a 10,000-year-old domesticated plant in the Americas." *Proceedings of the National Academy of Sciences USA*, 102: 18315-18320.

Eriksson, A. et al. 2012. "Late Pleistocene climate change and the global expansion of anatomically modern humans." *Proceedings of the National Academy of Sciences USA*, 109: 16089-16094.

Eriksson, A. and Manica, A. 2012. "Effect of ancient population structure on the degree of polymorphism shared between modern human populations and ancient hominins." *Proceedings of the National Academy of Sciences USA*, 109: 13956-13960.

Erlandson, J.M. (2002). *Anatomically modern humans, maritime voyaging, and the Pleistocene colonization of the Americas*. Paper presented at the The First Americans. The Pleistocene Colonization of the New World, San Francisco.

Evershed, R.P. et al. 2008. "Earliest date for milk use in the Near East and southeastern Europe linked to cattle herding." *Nature* 455: 528-531.

Exner, D.V. et al. 2001. "Lesser response to angiotensin-converting-enzyme inhibitor therapy in blacks as compared with white patients with left ventricular dysfunction." *New England Journal of Medicine*, 344: 1351-1357.

Fa, J.E. et al. 2013. "Rabbits and hominin survival in Iberia." *Journal of Human Evolution*, 64: 233-241.

Fagundes, N.J.R. et al. 2008. "Mitochondrial population genomics supports a single pre-Clovis origin with a coastal route for the peopling of the Americas." *American Journal of Human Genetics*, 82: 583-592.

Faith, J.T. and Surovell, T.A. 2009. "Synchronous extinction of North America's Pleistocene mammals." *Proceedings of the National Academy of Sciences, USA*, 106: 20641-20645.

Falk, D. 2011. *The Fossil Chronicles.* Berkeley: University of California Press.

Falush, D. et al. 2003. "Traces of human migrations in *Helicobacter pylori* populations." *Science,* 299: 1582-1585.

Fariña, R.A. and Tambusso, P.S. 2014. "Arroyo del Vizcaíno, Uruguay: a fossil-rich 30-ka-old megafaunal locality with cut-marked bones." *Proceedings of the Royal Society B,* 281: 20132211.

Fiedel, S.J. 2000. "The peopling of the New World: present evidence, new theories, and future directions." *Journal of Archaeological Research,* 8: 39-103.

Field, J. and Wroe, S. 2012. "Aridity, faunal adaptations and Australian Late Pleistocene extinctions." *World Archaeology,* 44: 56-74.

Field, J. et al. 2013. "Looking for the archaeological signature in Australian Megafaunal extinctions." *Quaternary International,* 285: 76-88.

Field, J.S. and Lahr, M.M. 2005. "Assessment of the southern dispersal: GIS-based analyses of potential routes at oxygen isotopic stage 4." *Journal of World Prehistory,* 19: 1-45.

Fillios, M. et al. 2012. "The impact of the dingo on the thylacine in Holocene Australia." *World Archaeology,* 44: 118-134.

Finch, V.A. and Western, D. 1977. "Cattle colors in pastoral herds: natural selection or social preference." *Ecology,* 58: 1384-1392.

Fincher, C.L. and Thornhill, R. 2008a. "Assortative sociality, limited dispersal, infectious disease and the genesis of the global pattern of religion diversity." *Proceedings of the Royal Society, London.B.,* 275: 2587-2594.

Fincher, C.L. and Thornhill, R. 2008b. "A parasite-driven wedge: infectious diseases may explain language and other biodiversity." *Oikos,* 117: 1289-1297.

Fincher, C.L. and Thornhill, R. 2012. "Parasite-stress promotes in-group assortative sociality: The cases of strong family ties and heightened religiosity." *Behavioral and Brain Sciences,* 35: 61-79.

Finlayson, C. 2004. *Neanderthals and Modern Humans. An Ecological and Evolutionary Perspective.* Cambridge: Cambridge University Press.

Fladmark, K.R. 1979. "Routes: alternative migration corridors for early man in North America." *American Antiquity,* 44: 55-69.

Foley, R. 1987. *Another Unique Species. Patterns in Human Evolutionary Ecology.* Harlow: Longman Scientific & Technical.

Foley, R.A. 2002. "Adaptive radiations and dispersals in hominin evolutionary ecology." *Evolutionary Anthropology,* Suppl. 1: 32-37.

Forster, P. and Renfrew, C. 2011. "Mother tongue and Y chromosomes." *Science,* 333: 1390-1391.

Freckleton, R.P. et al. 2003. "Notes and comments. Bergmann's rule and body size in mammals." *American Naturalist,* 161: 821-825.

Frey, B.S. et al. 2010. "Interaction of natural survival instincts and internalized social norms exploring the *Titanic* and *Lusitania* disasters." *Proceedings of the National Academy of Sciences, USA,* 107: 4862-4865.

Frisancho, A.R. 1993. *Human Adaptation and Accomodation.* Ann Arbor: University of Michigan Press.

Frisancho, A.R. 2013. "Developmental functional adaptation to high altitude: review." *American Journal of Human Biology,* 25: 151-168.

Fritts, T.H. and Rodda, G.H. 1998. "The role of introduced species in the degradation of island ecosystems: a case history of Guam." *Annual Review of Ecology and Systematics,* 29: 113-140.

Froehle, A.W. 2008. "Climate variables as predictors of basal metabolic rate: new equations." *American Journal of Human Biology,* 20: 510-529.

Fu, Q. et al. 2014. "Genome sequence of a 45,000-year-old modern human from western Siberia." *Nature,* 514: 445-449.

Fumagalli, M. et al. 2011. "Signatures of environmental genetic adaptation pinpoint pathogens as the main selective pressure through human evolution." *PLOS Genetics,* 7: DOI: 10.1371/journal.pgen.1002355.

Gavin, M.C. et al. 2013. "Toward a mechanistic understanding of linguistic diversity." *Bioscience*, 63: 524-535.

Gavin, M.C. and Sibanda, N. 2012. "The island biogeography of languages." *Global Ecology and Biogeography*, 21: 958-967.

Genné-Bacon, E.A. 2014. "Thinking evolutionarily about obesity." *Yale Journal of Biology and Medicine*, 87: 99-112.

George, D. et al. 2005. "Resolving an anomalous radiocarbon determination on mastodon bone from Monte Verde, Chile." *American Antiquity*, 70: 766-772.

Gerbault, P. et al. 2011. "Evolution of lactase persistence: an example of human niche construction." *Philosophical Transactions of the Royal Society, London, Series B.*, 366: 863-877.

Gignoux, C.A. et al. 2010. "Rapid, global demographic expansions after the origins of agriculture." *Proceedings of the National Academy of Sciences, USA*, 108: 6044-6049.

Gillespie, R. 2008. "Updating Martin's global extinction model." *Quaternary Science Reviews*, 27: 2522-2529.

Gilligan, I. 2007. "Resisting the cold in ice age Tasmania: thermal environment and settlement strategies." *Antiquity*, 81: 555-568.

Gilligan, I. et al. 2013. "Femoral neck-shaft angle in humans: variation relating to climate, clothing, lifestyle, sex, age and size." *Journal of Anatomy*, 223: 133-151.

Goddard, I. (Ed.). (1996). *Handbook of the North American Indians. Vol 17. Languages.* Washington, D.C.: Smithsonian Institution.

Goebel, T. 1999. "Pleistocene human colonization of Siberia and peopling of the Americas: An ecological approach." *Evolutionary Anthropology*, 8: 208-227.

Gommery, D. et al. 2011. "Oldest evidence of human activities in Madagascar on subfossil hippopotamus bones from Anjohibe (Mahajanga Province)." *Comptes Rendus. Palevol*, 10: 271-278.

Gordon, R.G. (Ed.). (2005). *Ethnologue: Languages of the World, 15th ed. http://www.ethnologue. com/*. Dallas, TX: SIL International.

Gorenflo, L.J. et al. 2012. "Co-occurrence of linguistic and biological diversity in biodiversity hotspots and high biodiversity wilderness areas." *Proceedings of the National Academy of Sciences, USA*, 109: 8032-8037.

Government of the United Kingdom. (1847). *Reports of the Commissioners of Enquiry into the State of Education in Wales.* London: UK Government.

Gray, R.D. et al. 2009. "Language phylogenies reveal expansion pulses and pauses in Pacific settlement." *Science*, 323: 479-483.

Grayson, D.K. 1993. "Differential mortality and the Donner Party disaster." *Evolutionary Anthropology*, 2: 151-159.

Grayson, D.K. 2005. "A brief history of the Great Basin pikas." *Journal of Biogeography*, 32: 2103-2111.

Green, R.E. et al. 2010. "A draft sequence of the Neandertal genome." *Science*, 328: 710-722.

Greenberg, J.H. 1987. *Language in the Americas.* Stanford, California: Stanford University Press.

Greenhill, S.J. et al. 2010. "The shape and tempo of language evolution." *Proceedings of the Royal Society, London.B.*, 277: 2443-2450.

Greenwood, B. 2014. "The contribution of vaccination to global health: past, present and future." *Philosophical Transactions of the Royal Society, London, B*, 369: 20130433.

Grindon, A.J. and Davison, A. 2013. "Irish *Cepaea nemoralis* land snails have a cryptic Franco-Iberian origin that is most easily explained by the movements of Mesolithic humans." *PLOS ONE*, 8: e65792.

Grund, B.S. et al. 2012. "Range sizes and shifts of North American Pleistocene mammals are not consistent with a climatic explanation for extinction." *World Archaeology*, 44: 43-55.

Guernier, V. and Guegan, J.-F. 2009. "May Rapoport's rule apply to human associated pathogens?" *Ecohealth*, 6: 509-521.

Guernier, V. et al. 2004. "Ecology drives the worldwide distribution of human diseases." *PLOS Biology*, 2: 0740-0746.

Guidon, N.G.D. 1986. "Carbon-14 dates point to man in the Americas 32,000 years ago." *Nature*, 321: 769-771.

Hage, P. and Marck, J. 2003. "Matrilineality and the Melanesian origin of Polynesian Y chromosomes." *Current Anthropology*, 44: 121-127.

Hallin, K.A. et al. "Paleoclimate during Neandertal and anatomically modern human occupation at Amud and Qafzeh, Israel: the stable isotope data." *Journal of Human Evolution*, 62: 59-73.

Hamilton, M.J. and Buchanan, B. 2007. "Spatial gradients in Clovis-age radiocarbon dates across North America suggest rapid colonization from the north." *Proceedings of the National Academy of Sciences USA*, 104: 15625-15630.

Hammer, M.F. and Woerner, A.E. 2011. "Genetic evidence for archaic admixture in Africa." *Proceedings of the National Academy of Sciences USA*, 108: 15123-15128.

Hancock, A.M. and Di Rienzo, A. 2008. "Detecting the genetic signature of natural selection in human populations: models, methods, and data." *Annual Review of Anthropology*, 37: 197-217.

Hancock, A.M. et al. 2010. "Human adaptations to diet, subsistence, and ecoregion are due to subtle shifts in allele frequency." *Proceedings of the National Academy of Sciences, USA*, 107: 8924-8930.

Hanski, I. et al. 2012. "Environmental biodiversity, human microbiota, and allergy are interrelated." *Proceedings of the National Academy of Sciences USA*, 109: 8334-8339.

Harcourt, A.H. 1999. "Biogeographic relationships of primates on south-east Asian islands." *Global Ecology and Biogeography*, 8: 55-61.

Harcourt, A.H. 2000a. "Coincidence and mismatch in hotspots of primate biodiversity: a worldwide survey." *Biological Conservation*, 93: 163-175.

Harcourt, A.H. 2000b. "Latitude and latitudinal extent: a global analysis of the Rapoport effect in a tropical mammalian taxon: primates." *Journal of Biogeography*, 27: 1169-1182.

Harcourt, A.H. 2006. "Rarity in the tropics: biogeography and macroecology of the primates." *Journal of Biogeography*, 33: 2077-2087.

Harcourt, A.H. 2012. *Human Biogeography*. Berkeley: University of California Press.

Harcourt, A.H. et al. 2002. "Rarity, specialization and extinction in primates." *Journal of Biogeography*, 29: 445-456.

Harcourt, A.H. et al. 2005. "The distribution-abundance (i.e. density) relationship: its form and causes in a tropical mammal order, Primates." *Journal of Biogeography*, 32: 565-579.

Harcourt, A.H. and Doherty, D.A. 2005. "Species-area relationships of primates in tropical forest fragments: a global analysis." *Journal of Applied Ecology*, 42: 630-637.

Harcourt, A.H. and Schreier, B.M. 2009. "Diversity, body mass, and latitudinal gradients in primates." *International Journal of Primatology*, 30: 283-300.

Harcourt, A.H. and Wood, M.A. 2012. "Rivers as barriers to primate distributions in Africa." *International Journal of Primatology*, 33: 168-183.

Hardy, K. et al. 2012. "Neanderthal medics? Evidence for food, cooking, and medicinal plants entrapped in dental calculus." *Naturwissenschaften*, 99: 617-626.

Harmon, D. 1996. "Losing species, losing languages: connections between biological and linguistic diversity." *Southwest Journal of Linguistics*, 15: 89-108.

Harmon, D. and Loh, J. 2010. "The Index of Linguistic Diversity: a new quantitative measure of trends in the status of the world's languages." *Language Documentation and Conservation*, 4: 97-151.

Harpending, H.C. et al. 1998. "Genetic traces of ancient demography." *Proceedings of the National Academy of Sciences, USA*, 95: 1961-1967.

Harrison, G.A. 1995. *The Human Biology of the English Village*. Oxford: Oxford University Press.

Harrison, K.D. 2007. *When Languages Die*. Oxford: Oxford University Press.

Hawkes, K. et al. 1989. Hardworking Hadza grandmothers. In *Comparative Socioecology. The Behavioral Ecology of Humans and Other Mammals*, edited by V. Standen and R.A. Foley, 341-390. Oxford: Blackwell Scientific Publishers.

Hawkins, B.A. et al. 2003. "Energy, water, and broadscale geographic patterns of species richness." *Ecology* 84: 3105-3117.

Haynes, G. 2002. *The Early Settlement of America. The Clovis Era*. Cambridge: University of Cambridge.

Hehemann, J.H. et al. 2010. "Transfer of carbohydrate-active enzymes from marine bacteria to Japanese gut microbiota." *Nature*, 464: 908-914.

Heinsohn, T.E. 2001. Human influences on vertebrate zoogeography: animal translocation and biological invasions across and to the east of Wallace's Line. In *Faunal and Floral Migrations and Evolution in SE Asia-Australasia*, edited by I.M. Metcalfe, et al., 153-170. Lisse: A.A. Balkema Publ.

Helgason, A. et al. 2001. "mtDNA and the islands of the North Atlantic: Estimating the proportions of Norse and Gaelic ancestry." *American Journal of Human Genetics*, 68: 723-737.

Hellenthal, G. et al. 2014. "A genetic atlas of human admixture history." *Science*, 343: 747-751.

Henderson, J. et al. 2005. "The EPAS1 gene influences the aerobic-anaerobic contribution in elite endurance athletes." *Human Genetics*, 118: 416-423.

Henn, B.M. et al. 2011. "Reply to Hublin and Klein: Locating a geographic point of dispersion in Africa for contemporary humans." *Proceedings of the National Academy of Sciences, USA*, 108: E278.

Henn, B.M. et al. 2012. "The great human expansion." *Proceedings of the National Academy of Sciences, USA*, 109: 17758-17764.

Henn, B.M. et al. 2011. "Hunter-gatherer genomic diversity suggests a southern African origin for modern humans." *Proceedings of the National Academy of Sciences, USA*, 108: 5154-5162.

Henn, B.M. et al. 2008. "Y-chromosomal evidence of a pastoralist migration through Tanzania to southern Africa." *Proceedings of the National Academy of Sciences, USA*, 105: 10693–10698.

Henneberg, M. et al. 2014. "Evolved developmental homeostasis disturbed in LB1 from Flores, Indonesia, denotes Down syndrome and not diagnostic traits of the invalid species *Homo floresiensis*." *Proceedings of the National Academy of Sciences*, 111: 11967-11972.

Henrich, J. 2004. "Demography and cultural evolution: how adaptive cultural processes can produce maladaptive losses: the Tasmanian case." *American Antiquity*, 69: 197-214.

Henry, A.G. et al. 2011. "Microfossils in calculus demonstrate consumption of plants and cooked foods in Neanderthal diets (Shanidar III, Iraq; Spy I and II, Belgium)." *Proceedings of the National Academy of Sciences, USA*, 108: 486-491.

Henshilwood, C.S. et al. 2011. "A 100,000-year-old ochre-processing workshop at Blombos Cave, South Africa." *Science*, 334: 219-222.

Higgins, R.W. and Ruff, C.B. 2011. "The effects of distal limb segment shortening on locomotor efficiency in sloped terrain: implications for Neandertal locomotor behavior." *American Journal of Physical Anthropology*, 146: 336-345.

Higham, T. and Compton, T. 2011. "The earliest evidence for anatomically modern humans in northwestern Europe." *Nature*, 479: 521-524.

Hill, J. and Nolasquez, R. (Eds.). (1973). *Mulu'wetam: The First People. Cupeño Oral History and Language*: Malki Museum Press.

Hill, K. et al. 1984. "Seasonal variance in the diet of Aché hunter-gatherers in eastern Paraguay." *Human Ecology* 12: 145-180.

Hinde, R.A. 2002. *Why Good is Good. The Sources of Morality*. London: Routledge.

Hinde, R.A. 2011. *Changing How We Live: Society from the Bottom Up*. Nottingham: Spokesman.

Hodgson, J.A. et al. 2014. "Early back-to-Africa migration into the Horn of Africa." *PLOS Genetics*, 10: e1004393.

Hoffecker, J.F. et al. 2014. "Out of Beringia?" *Science*, 343: 979-980.

Holdaway, R.N. and Jacomb, C. 2000. "Rapid extinction of the moas (Aves: Dinornithiformes): model, test, and implications." *Science*, 287: 2250-2254.

Holden, C. and Mace, R. 1997. "Phylogenetic analysis of the evolution of lactose digestion in adults." *Human Biology*, 69: 605-628.

Holliday, T.W. and Falsetti, A.B. 1995. "Lower-limb length of European early-modern humans in relation to mobility and climate." *Journal of Human Evolution*, 29: 141-153.

Hora, M. and Sladek, V. 2014. "Influence of lower limb configuration on walking cost in Late Pleistocene humans." *Journal of Human Evolution*, 67: 19-32.

Hsiang, S.M. et al. 2013. "Quantifying the influence of climate on human conflict." *Science*, 341: 1235367.

Hsiang, S.M. et al. 2011. "Civil conflicts are associated with the global climate." *Nature*, 476: 438-441.

Hudjashov, G. et al. 2007. "Revealing the prehistoric settlement of Australia by Y chromosome and mtDNA analysis." *Proceedings of the National Academy of Sciences, USA*, 104: 8726-8730.

Huerta-Sánchez, E. et al. 2014. "Altitude adaptation in Tibetans caused by introgression of Denisovan-like DNA." *Nature*, 512: 194-197.

Hughes, S. et al. 2014. "Anglo-Saxon origins investigated by isotopic analysis of burials from Berinsfield, Oxfordshire, UK." *Journal of Archaeological Science*, 42: 81-92.

Hull, D.L. et al. 1978. "Planck's principle: Do younger scientists accept new scientific ideas with greater alacrity than older scientists?" *Science*, 202: 717-723.

Hünemeier, T. and Gómez-Valdés, J. 2012. "Cultural diversification promotes rapid phenotypic evolution in Xavánte Indians." *Proceedings of the National Academy of Sciences USA*, 109: 73-77.

Hunley, K.L. et al. 2009. "The global pattern of gene identity variation reveals a history of long-range migrations, bottlenecks, and local mate exchange: Implications for biological race." *American Journal of Physical Anthropology*, 139: 35-46.

Huntford, R. 1999. *The Last Place on Earth. Scott and Amundsen's Race to the South Pole*. New York: Modern Library.

Hurles, M.E. et al. 2005. "The dual origin of the Malagasy in Island Southeast Asia and East Africa: evidence from maternal and paternal lineages." *American Journal of Human Genetics*, 76: 894-901.

Hvistendahl, M. 2012. "Roots of empire." *Science*, 337: 1596-1599.

Ingman, M. et al. 2000. "Mitochondrial genome variation and the origin of modern humans." *Nature*, 408: 708-713.

Ingram, C.J.E. et al. 2009. "Lactose digestion and evolutionary genetics of lactase persistence." *Human Genetics*, 124: 579-591.

Jablonski, N.G. (Ed.). (2002). *The First Americans. The Pleistocene Colonization of the New World*. San Francisco: California Academy of Sciences.

Jablonski, N.G. and Chaplin, G. 2000. "The evolution of human skin coloration." *Journal of Human Evolution*, 39: 57-106.

Jablonski, N.G. and Chaplin, G. 2002. "Skin deep." *Scientific American*, 287: 74-81.

Jablonski, N.G. and Chaplin, G. 2013. "Epidermal pigmentation in the human lineage is an adaptation to ultraviolet radiation." *Journal of Human Evolution*, 65: 671-675.

Janzen, D.H. 1967. "Why mountain passes are higher in the tropics." *American Naturalist*, 101: 233-249.

Jarvis, J.P. et al. 2012. "Patterns of ancestry, signatures of natural selection, and genetic association with stature in Western African Pygmies." *PLOS Genetics*, 8: 299-313.

Johnson, C. (1994-2013). Mayflower History.com. 2014, http://mayflowerhistory.com/

Johnson, C.N. 2002. "Determinants of loss of mammal species during the Late Quaternary 'megafauna' extinctions: life history and ecology, but not body size." *Proceedings of the Royal Society B.*, 269: 2221-2227.

Jones, R. 1995. "Tasmanian archaeology—establishing the sequences." *Annual Review of Anthropology*, 24: 423-446.

Jones, T.L. et al. (Eds.). (2011). *Polynesians in America. Pre-Columbian contacts with the New World*. Altamira Press.

Julian, C.G. et al. 2009. "Augmented uterine artery blood flow and oxygen delivery protect Andeans from altitude-associated reductions in fetal growth." *American Journal of Physiology. Regulatory, Integrative and Comparative Physiology*, 296: R1564-R1575.

Kahn, J. 2007. "Race in a bottle." *Scientific American*, 297: 40-45.

Kamilar, J.M. and Bradley, B.J. 2011. "Interspecific variation in primate coat colour supports Gloger's rule." *Journal of Biogeography*, 38: 2270-2277.

Karafet, T. et al. 1997. "Y chromosome markers and trans-Bering Strait dispersals." *American Journal of Physical Anthropology*, 102: 301-314.

Katzmarzyk, P.T. and Leonard, W.R. 1998. "Climatic influences on human body size and proportions: ecological adaptations and secular trends." *American Journal of Physical Anthropology*, 106: 483-503.

Kayser, M. 2010. "The human genetic history of Oceania: Near and remote views of dispersal." *Current Biology*, 20: R194-R201.

Kayser, M. et al. 2008. "Genome-wide analysis indicates more Asian than Melanesian ancestry of Polynesians." *American Journal of Human Genetics*, 82: 194-198.

Keesing, F. et al. 2010. "Impacts of biodiversity on the emergence and transmission of infectious diseases." *Nature* 468: 647-652.

Keller, I. et al. 2013. "Widespread phenotypic and genetic divergence along altitudinal gradients in animals." *Journal of Evolutionary Biology*, 26: 2527-2543.

Kemp, B.M. et al. 2010. "Evaluating the farming/language dispersal hypothesis with genetic variation exhibited by populations in the Southwest and Mesoamerica." *Proceedings of the National Academy of Sciences USA*, 107: 6759-6764.

Khan, B.S.R. 2010. "Diet, disease and pigment variation in humans." *Medical Hypotheses*, 75: 363-367.

Kingdon, J. 1989. *Island Africa. The Evolution of Africa's Rare Animals and Plants*. Princeton: Princeton University Press.

Kirch, P.V. 1997. *The Lapita Peoples: Ancestors of the Oceanic World*. Oxford, UK: Blackwell Publisheers.

Kirch, P.V. 2000. *On the Road of the Winds. An Archaeological History of the Pacific Islands Before European Contact*. Berkeley: University of California.

Klein, R.G. 2009. *The Human Career: Human Biological and Cultural Origins. 3rd. ed*. Chicago: University of Chicago.

Klein, R.G. and Steele, T.E. 2013. "Archaeological shellfish size and later human evolution in Africa." *Proceedings of the National Academy of Sciences, USA*, 110: 10910-10915.

Kline, M.A. and Boyd, R. 2010. "Population size predicts technological complexity in Oceania." *Proceedings of the Royal Society, London.B.*, 277: 2559-2564.

Knapp, M. et al. 2012. "Complete mitochondrial DNA genome sequences from the first New Zealanders." *Proceedings of the National Academy of Sciences*, 109: 18350-18354.

Knottnerus, O.S. 2002. Malaria around the North Sea: A survey. In *Climatic Development and History of the North Atlantic Realm: Hanse Conference Report*, edited by G. Wefer, et al., 339-353. Berlin: Springer-Verlag.

Koch, P.L. and Barnosky, A.D. 2006. "Late Quaternary extinctions: state of the debate." *Annual Review of Ecology and Systematics*, 37: 215-250.

Krause, J. et al. 2010. "The complete mitochondrial DNA genome of an unknown hominin from southern Siberia." *Nature*, 464: 894-897.

Kubo, D. et al. 2013. "Brain size of *Homo floresiensis* and its evolutionary implications." *Proceedings of the Royal Society B*, 280: 10.1098/rspb.2013.0338.

Kuhn, S.L. and Stiner, M.C. 2006. "What's a mother to do? The division of labor among Neandertals and Modern Humans in Eurasia." *Current Anthropology*, 47: 953-963.

Lahaye, C. et al. 2013. "Human occupation in South America by 20,000 b.c.: the Toca da Tira Peia site, Piaui, Brazil." *Journal of Archaeological Science*, 40: 2840-2847.

Lamason, R.L. et al. 2005. "SLC24A5, a putative cation exchanger, affects pigmentation in zebrafish and humans." *Science*, 310: 1782-1786.

Lane, C.S. et al. 2013. "Ash from the Toba supereruption in Lake Malawi shows no volcanic winter in East Africa at 75 ka." *Proceedings of the National Academy of Sciences, USA*, 110: 8025-8029.

Larson, G. et al. 2007. "Ancient DNA, pig domestication, and the spread of the Neolithic into Europe." *Proceedings of the National Academy of Sciences USA*, 104: 15276-15281.

Larson, G. et al. 2007. "Phylogeny and ancient DNA of Sus provides insights into neolithic expansion in Island Southeast Asia and Oceania." *Proceedings of the National Academy of Sciences USA*, 104: 4834-4839.

Laurance, W.F. and Bierregaard, R.O. (Eds.). (1997). *Tropical Forest Remnants. Ecology, Management, and Conservation of Fragmented Communities.* Chicago: University of Chicago Press.

Lentz, C. 1999. Alcohol consumption between community ritual and political economy: case studies from Ecuador and Ghana. In *Changing Food Habits. Case Studies from Africa, South America and Europe*, edited by C. Lentz, 155-180. Amsterdam: Harwood Academic Publishers.

Leonardi, M. et al. 2012. "The evolution of lactase persistence in Europe. A synthesis of archaeological and genetic evidence." *International Dairy Journal*, 22: 88-97.

Lewis, M.P. (Ed.). (2009). *Ethnologue: Languages of the World, 16th ed.* http://www.ethnologue.com/. Dallas, TX: SIL International.

Li, J.Z. et al. 2008. "Worldwide human relationships inferred from genome-wide patterns of variation." *Science*, 319: 1100-1104.

Li, Y. et al. 2014. "Population variation revealed high-altitude adaptation of Tibetan mastiffs." *Molecular Biology and Evolution*, 31: 1200-1205.

Lieberman, D.E. et al. 2009. Brains, brawn, and the evolution of human endurance running capabilities. In *The First Humans—Origin and Early Evolution of the Genus Homo*, edited by F.E. Grine, et al., 77-92. New York: Springer Science + Business Media B.V.

Linder, H.P. and de Klerk, H.M. 2012. "The partitioning of Africa: statistically defined biogeographical regions in sub-Saharan Africa." *Journal of Biogeography*, 39: 1189-1205.

Lindström, T. et al. 2013. "Rapid shifts in dispersal behavior on an expanding range edge." *Proceedings of the National Academy of Sciences, USA*, 110: 13452-13456.

Linz, B. et al. 2007. "An African origin for the intimate association between humans and *Helicobacter pylori*." *Nature*, 445: 915-918.

Little, M.A. 2010. Geography, migration, and environmental plasticity as contributors to human variation. In *Human Variation. From the Laboratory to the Field.*, edited by C.G. Mascie-Taylor, et al., 157-181. New York: CRC Press.

Liu, H. et al. 2006. "A geographically explicit genetic model of worldwide human-settlement history." *American Journal of Human Genetics*, 79: 230-237.

Loh, J. and Harmon, D. 2005. "A global index of biocultural diversity." *Ecological Indicators*, 5: 231-241.

Lohse, K. and Frantz, L.A.F. 2014. "Neandertal admixture in Eurasia confirmed by maximum-likelihood analysis of three genomes." *Genetics*, 196: 1241-1251.

Lomolino, M.V. 2005. "Body size evolution in insular vertebrates: generality of the island rule." *Journal of Biogeography*, 32: 1683-1699.

Lomolino, M.V. et al. 2010. *Biogeography, 4th. ed.* Sunderland, Mass.: Sinauer Associates.

Lomolino, M.V. et al. 2013. "Of mice and mammoths: generality and antiquity of the island rule." *Journal of Biogeography*, 40: 1427-1439.

Lordkipanidze, D. et al. 2013. "A complete skull from Dmanisi, Georgia, and the evolutionary biology of early *Homo*." *Science*, 342: 326-331.

Lorenzen, E.D. et al. 2011. "Species-specific responses of Late Quaternary megafauna to climate and humans." *Nature* 479: 359-364.

Loss, S.R. et al. 2013. "The impact of free-ranging domestic cats on wildlife of the United States." *Nature Communications*, 4: Article 1396, 1391-1397.

Louys, J. 2012. "Mammal community structure of Sundanese fossil assemblages from the Late Pleistocene, and a discussion on the ecological effects of the Toba eruption." *Quaternary International*, 258: 80-87.

Lowery, R.K. et al. 2013. "Neanderthal and Denisova genetic affinities with contemporary humans: introgression versus common ancestral polymorphism." *Gene*, 530: 83-94.

Lozano, R. et al. 2012. "Global and regional mortality from 235 causes of death for 20 age groups in 1990 and 2010: a systematic analysis for the Global Burden of Disease Study 2010." *Lancet*, 380: 2095-2128.

MacArthur, R.H. 1972. *Geographical Ecology: Patterns in the Distribution of Species*. Princeton: Princeton University Press.

MacArthur, R.H. and Wilson, E.O. 1967. *The Theory of Island Biogeography*. Princeton: Princeton University Press.

Macaulay, V. et al. 2005. "Single, rapid coastal settlement of Asia revealed by analysis of complete mitochondrial genomes." *Science*, 308: 1034-1036.

Mace, R. and Pagel, M. 1995. "A latitudinal gradient in the density of human languages in North America." *Proceedings of the Royal Society, London.B.*, 261: 117-121.

MacPhee, R.D.E. and Marx, P.A. 1997. The 40,000-year plague: humans, hyperdisease, and first-contact extinctions. In *Natural Change and Human Impact in Madagascar*, edited by S.M. Goodman and B.D. Patterson, 169-217. Washington, D.C.: Smithsonian Institution Press.

Maffi, L. 2005. "Linguistic, cultural and biological diversity." *Annual Review of Anthropology*, 34: 599-617.

Majumder, P.P. 2010. "The human genetic history of South Asia." *Current Biology*, 20: R184-R187.

Malthus, T.R. 1798. *An Essay on the Principle of Population*. London: Macmillan.

Malville, J.M. et al. 1998. "Megaliths and Neolithic astronomy in southern Egypt." *Nature*, 392: 488-491.

Mandryk, C.A.S. et al. 2001. "Late Quaternary paleoenvironments of Northwestern North America: implications for inland versus coastal migration routes." *Quaternary Science Reviews*, 20: 301-314.

Marciano, J.B. 2014. *Whatever Happened to the Metric System? How America Kept its Feet*. New York: Bloomsbury.

Marlar, R.A. et al. 2000. "Biochemical evidence of cannibalism at a prehistoric Puebloan site in southwestern Colorado." *Nature*, 407: 74-78.

Marlowe, F.W. 2005. "Hunter-gatherers and human evolution." *Evolutionary Anthropology*, 14: 54-67.

Marlowe, F.W. and Berbesque, J.C. 2009. "Tubers as fallback foods and their impact on Hadza hunter-gatherers." *American Journal of Physical Anthropology*, 140: 751-758.

Marsh, L.K. (Ed.). (2003). *Primates in Fragments. Ecology and Conservation*. New York: Kluwer Academic/Plenum Publishers.

Martin, P.S. 1973. "Discovery of America." *Science*, 179: 969-974.

Martin, R.D. et al. 2006. "Flores hominid: new species or microcephalic dwarf?" *The Anatomical Record, Part A*, 288A: 1123-1145.

Marwick, B. 2009. "Biogeography of Middle Pleistocene hominins in mainland Southeast Asia: a review of current evidence." *Quaternary International*, 202: 51-58.

Matisoo-Smith, E. and Robins, J.H. 2004. "Origins and dispersals of Pacific peoples: Evidence from mtDNA phylogenies of the Pacific rat." *Proceedings of the National Academy of Sciences, USA*, 101: 9167-9172.

Matthew, W.D. 1915. "Climate and evolution." *Annals of the New York Academy of Sciences*, 24: 171-318.

Matthew, W.D. 1939. "Climate and evolution." *Special Publications of the New York Academy of Sciences*, 1: 1-147.

Matthews, L.J. and Butler, P.M. 2011. "Novelty-seeking DRD4 polymorphisms are associated with human migration distance out-of-Africa after controlling for neutral population gene structure." *American Journal of Physical Anthropology*, 145: 382-389.

Maudlin, I. 2006. "African trypanosomiasis." *Annals of Tropical Medicine & Parasitology*, 100: 679-701.

McDougall, C. 2009. *Born to Run: A Hidden Tribe, Superathletes, and the Greatest Race the World Has Never Seen*. New York: Alfred A. Knopf.

McDougall, I. et al. 2005. "Stratigraphic placement and age of modern humans from Kibish, Ethiopia." *Nature*, 433: 733-736.

McGeachin, R.L. and Akin, J.R. 1982. "Amylase levels in the tissues and body fluids of several primate species." *Comparative Biochemistry and Physiology. A. Comparative Physiology*, 72: 267-269.

McHenry, H.M. and Coffing, K. 2000. "Australopithecus to Homo: transformations in body and mind." *Annual Review of Anthropology*, 29: 125-146.

McNab, B.K. 2002. *The Physiological Ecology of Vertebrates. A View from Energetics*. Ithaca, New York: Comstock Publishing Associates.

McNeill, J.R. 2010. *Mosquito Empires. Ecology and War in the Greater Caribbean, 1620-1914*. Cambridge: Cambridge University Press.

McNeill, W.H. 1976. *Plagues and Peoples*. New York: Anchor Press.

Meehan, B. 1977. "Hunters by the seashore." *Journal of Human Evolution*, 6: 363-370.

Meggers, B.J. 1977. "Vegetational fluctuation and prehistoric cultural adaptation in Amazonia: some tentative correlations." *World Archaeology*, 8: 287-303.

Meijer, H.J.M. et al. 2010. "The fellowship of the hobbit: the fauna surrounding *Homo floresiensis*." *Journal of Biogeography*, 37: 995-1006.

Meiri, M. et al. 2014. "Faunal record identifies Bering isthmus conditions as constraint to end-Pleistocene migration to the New World." *Proceedings of the Royal Society B*, 281: 20132167.

Meiri, S. and Dayan, T. 2003. "On the validity of Bergmann's rule." *Journal of Biogeography*, 30: 331-351.

Mellars, P. 1996. The emergence of biologically modern populations in Europe: a social and cognitive revolution? In *Evolution of Social Behaviour Patterns in Primates and Man*, edited by W.G. Runciman, et al., 179-201. Oxford: Oxford University Press.

Mellars, P. 2006. "Why did modern human populations disperse from Africa *ca.* 60,000 years ago? A new model." *Proceedings of the National Academy of Sciences, USA*, 103: 9381-9386.

Mellars, P. et al. 2013. "Genetic and archaeological perspectives on the initial modern human colonization of southern Asia." *Proceedings of the National Academy of Sciences, USA*, 110: 10699-10704.

Meltzer, D.J. 2004. Modeling the inital colonization of the Americas: issues of scale, demography, and landscape learning. In *The Settlement of the American Continents. A Multidisciplinary Approach to Human Biogeography*, edited by C.M. Barton, et al., 123-137. Tucson: University of Arizona Press.

Meyer, D. 1999. Medically relevant genetic variation of drug effects. In *Evolution in Health & Disease*, edited by S.C. Stearns, 41-49. Oxford: Oxford University Press.

Mielke, J.H. et al. 2006. *Human Biological Variation*. New York: Oxford University Press.

Migliano, A.B. et al. 2007. "Life history trade-offs explain the evolution of human pygmies." *Proceedings of the National Academy of Sciences, USA*, 104: 20216-20219.

Miller, G.H. et al. 2005. "Ecosystem collapse in pleistocene Australia and a human role in megafaunal extinction." *Science*, 309: 287-290.

Milton, K. 1991. "Comparative aspects of diet in Amazonian forest dwellers." *Proceedings of the Royal Society, London.B.*, 334: 253-263.

Molnar, R.E. 2004. *Dragons in the Dust. The Paleobiology of the Giant Monitor Lizard Megalania*. Bloomington: Indiana University Press.

Molnar, S. 2006. *Human Variation. Races, Types, and Ethnic Groups, 6th. ed.* Upper Saddle River: Pearson Prentice Hall.

Montgomery, S.H. 2013. "Primate brains, the 'island rule' and the evolution of *Homo floresiensis*." *Journal of Human Evolution*, 65: 750-760.

Moodley, Y. et al. 2009. "The peopling of the Pacific from a bacterial perspective." *Science*, 323: 527-530.

Moore, J.L. et al. 2002. "The distribution of cultural and biological diversity in Africa." *Proceedings of the Royal Society, London.B.*, 269: 1645-1653.

Moore, L.G. 1998. "Human adaptation to high altitude: regional and life-cycle perspectives." *Yearbook of Physical Anthropology*, 41: 25-64.

Moore, L.G. 2001. "Human genetic adaptation to high altitude." *High Altitude Medicine & Biology*, 2: 257-279.

Moore, L.G. et al. 2011. "Humans at high altitude: hypoxia and fetal growth." *Respiratory Physiology and Neurobiology*, 178: 181-190.

Moran, E.E. 2000. *Human Adaptability. An Introduction to Ecological Anthropology*, 2nd. ed. Boulder, CO, USA: Westview Press.

Moreau, C. et al. 2011. "Deep human genealogies reveal a selective advantage to be on an expanding wave front." *Science*, 334: 1148-1150.

Murray-McIntosh, R.P. et al. 1998. "Testing migration patterns and estimating founding population size in Polynesia by using human mtDNA sequences." *Proceedings of the National Academy of Sciences*, 95: 9047-9052.

Need, A.C. and Goldstein, D.B. 2009. "Next generation disparities in human genomics: concerns and remedies." *Trends in Genetics*, 25: 489-494.

Nettle, D. 1996. "Language diversity in West Africa: An ecological approach." *Journal of Anthropological Archaeology*, 15: 403-438.

Nettle, D. 1998. "Explaining global patterns of language diversity." *Journal of Anthropological Archaeology*, 17: 354-374.

Nettle, D. 1999. *Linguistic Diversity*. Oxford: Oxford University Press.

Nettle, D. and Romaine, S. 2000. *Vanishing Voices*. Oxford: Oxford University Press.

Nikolskiy, P.A. et al. 2011. "Last straw versus Blitzkrieg overkill: Climate-driven changes in the Arctic Siberian mammoth population and the Late Pleistocene extinction problem." *Quaternary Science Reviews*, 30: 2309-2328.

Nogués-Bravo, D. et al. 2008. "Climate change, humans, and the extinction of the woolly mammoth." *PLOS Biology*, 6: e79. 0685-0692.

Norton, H.L. et al. 2007. "Molecular Biology and Evolution." *Genetic evidence for the convergent evolution of light skin in Europeans and East Asians*, 24: 710–722.

Novembre, J. et al. 2008. "Genes mirror geography within Europe." *Nature*, 456: 98-101.

Nunn, C.L. et al. 2005. "Latitudinal gradients of parasite species richness in primates." *Diversity and Distributions*, 11: 249-256.

O'Connell, J.F. and Allen, J. 2004. "Dating the colonization of Sahul (Pleistocene Australia-New Guinea): a review of recent research." *Journal of Archaeological Science*, 31: 835-853.

O'Fallon, B.D. and Fehren-Schmitz, L. 2011. "Native Americans experienced a strong population bottleneck coincident with European contact." *Proceedings of the National Academy of Sciences, USA*, 108: 20444-20448.

O'Rourke, D.H. and Raff, J.A. 2010. "The human genetic history of the Americas: the final frontier." *Current Biology*, 20: R202-R207.

O'Loughlin, J. et al. 2012. "Climate variability and conflict risk in East Africa, 1990-2009." *Proceedings of the National Academy of Sciences USA*, 109: 18344-18349.

Okie, J.G. and Brown, J.H. 2009. "Niches, body sizes, and the disassembly of mammal communities on the Sunda Shelf islands." *Proceedings of the National Academy of Sciences, USA*, 106: 19679-19684.

Oppenheimer, C. 2003a. "Climatic, environmental and human consequences of the largest known historic eruption: Tambora volcano (Indonesia) 1815." *Progress in Physical Geography*, 27: 230-259.

Oppenheimer, S. 2003b. *Out of Eden. The Peopling of the World*. London: Constable.

Oppenheimer, S. 2006. *The Origins of the British*. London: Constable & Robinson ltd.

Oppenheimer, S. 2009. "The great arc of dispersal of modern humans: Africa to Australia." *Quaternary International*, 202: 2-13.

Osborne, A.H. et al. 2008. "A humid corridor across the Sahara for the migration of early modern humans out of Africa 120,000 years ago." *Proceedings of the National Academy of Sciences, USA*, 105: 16444-16447.

Ottaviano, G.I.P. and Peri, G. 2005. "Cities and cultures." *Journal of Urban Economics*, 58: 304-337.

Overy, R. 2010. *The Times Complete History of the World*. London: Times Books.

Padmaja, G. and Steinkraus, K.H. 1995. "Cyanide detoxification in cassava for food and feed uses." *Critical Reviews in Food Science and Nutrition*, 35: 299-339.

Pagel, M. et al. 2013. "Ultraconserved words point to deep language ancestry across Eurasia." *Proceedings of the National Academy of Sciences, USA*, 110: 8471-8476.

Palombo, M. 2007. "How can endemic proboscideans help us understand the 'island rule'? A case study of Mediterranean islands." *Quaternary International*, 169-170: 105-124.

Parmesan, C. 2006. "Ecological and evolutionary responses to recent climate change." *Annual Review of Ecology, Evolution, and Systematics*, 37: 637-669

Patin, E. et al. 2009. "Inferring the demographic history of African farmers and pygmy hunter-gatherers using a multilocus resequencing data set." *PLOS Genetics*, 5: e1000448.

Pearson, O.M. 2004. "Has the combination of genetic and fossil evidence solved the riddle of modern human origins?" *Evolutionary Anthropology*, 13: 145-159.

Peng, Y. et al. 2010. "The ADH1B Arg47His polymorphism in East Asian populations and expansion of rice domestication in history." *BioMed Central Evolutionary Biology*, 10: 8 pp. http://www.biomedcentral.com/1471-2148/1410/1415.

Penny, D. et al. 2002. "Estimating the number of females in the founding population of New Zealand: analysis of mtDNA variation." *Journal of Polynesian Society*, 111: 207-221.

Pepperell, C.S. et al. 2011. "Dispersal of *Mycobacterium tuberculosis* via the Canadian fur trade." *Proceedings of the National Academy of Sciences USA*, 108: 6526-6531.

Perry, G.H. and Dominy, N.J. 2009. "Evolution of the human pygmy phenotype." *Trends in Ecology and Evolution*, 24: 218-225.

Perry, G.H. et al. 2007. "Diet and the evolution of human amylase gene copy number variation." *Nature Genetics*, 39: 1256-1260.

Perry, P. 1969. "Marriage-distance relationships in North Otago 1875-1914." *New Zealand Geographer*, 25: 36-43.

Petraglia, M. et al. 2007. "Middle Paleolithic assemblages from the Indian sub-continent before and after the Toba super-eruption." *Science*, 317: 114-116.

Petraglia, M.D. et al. 2010. "Out of Africa: new hypotheses and evidence for the dispersal of *Homo sapiens* along the Indian Ocean rim." *Annals of Human Biology*, 37: 288-311.

Pickrell, J.K. et al. 2014. "Ancient west Eurasian ancestry in southern and eastern Africa." *Proceedings of the National Academy of Sciences, USA*, 111: 2632-2637.

Pinhasi, R. et al. 2012. "New chronology for the Middle Palaeolithic of the southern Caucasus suggests early demise of Neanderthals in this region." *Journal of Human Evolution*, 63: 770-780.

Pitblado, B.L. 2011. "A tale of two migrations: reconciling recent biological and archaeological evidence for the Pleistocene peopling of the Americas." *Journal of Archaeological Research*, 19: 327-375.

Pitulko, V.V. and Nikolskiy, P.A. 2012. "The extinction of the woolly mammoth and the archaeological record in Northeastern Asia." *World Archaeology*, 44: 21-42.

Pitulko, V.V. et al. 2004. "The Yana RHA site: humans in the Arctic before the last glacial maximum." *Science*, 303: 52-56.

Pollak, M.R. et al. 2010. "Association of trypanolytic ApoL1 variants with kidney disease in African-Americans." *Science*, 329: 841-845.

Pope, K.O. and Terrell, J.E. 2008. "Environmental setting of human migrations in the circum-Pacific region." *Journal of Biogeography*, 35: 1-21.

Poulin, R. and Morand, S. 2004. *Parasite Biodiversity*. Washington, D.C.: Smithsonian Institution Press.

Powell, A. et al. 2009. "Late Pleistocene demography and the appearance of modern human behavior." *Science*, 324: 1298-1301.

Prescott, G.W. et al. 2012. "Quantitative global analysis of the role of climate and people in explaining late Quaternary megafaunal extinctions." *Proceedings of the National Academy of Sciences, USA*, 109: 4527-4531.

Prideaux, G.J. et al. 2010. "Timing and dynamics of Late Pleistocene mammal extinctions in southwestern Australia." *Proceedings of the National Academy of Sciences, USA*, 107: 22157-22162.

Prüfer, K. et al. 2014. "The complete genome sequence of a Neanderthal from the Altai Mountains." *Nature*, 505: 43-49.

Pugach, I. et al. 2013. "Genome-wide data substantiate Holocene gene flow from India to Australia." *Proceedings of the National Academy of Sciences, USA*, 110: 1803-1808.

Qi, X. et al. 2013. "Genetic evidence of Paleolithic colonization and Neolithic expansion of modern humans on the Tibetan Plateau." *Molecular Biology and Evolution*, 30: 1761-1778.

Qiu, Q. et al. 2012. "The yak genome and adaptation to life at high altitude." *Nature Genetics*, 44: 946-949.

Rabbie, J.M. 1992. The effects of intragroup cooperation and intergroup competition on ingroup cohesion and out-group hostility. In *Coalitions and Alliances in Humans and other Animals*, edited by A.H. Harcourt and F.B.M. de Waal, 175-205. Oxford: Oxford University Press.

Rademaker, K. et al. 2013. "A Late Pleistocene/early Holocene archaeological 14C database for Central and South America: palaeoenvironmental contexts and demographic interpretations." *Quaternary International*, 301: 34-45.

Raghavan, M. et al. 2014. "Upper Palaeolithic Siberian genome reveals dual ancestry of Native Americans." *Nature*, 505: 87-91.

Ramachandran, S. et al. 2005. "Support from the relationship of genetic and geographic distance in human populations for a serial founder effect originating in Africa." *Proceedings of the National Academy of Sciences, USA*, 102: 15942-15947.

Ranciaro, A. et al. 2014. "Genetic origins of lactase persistence and the spread of pastoralism in Africa." *American Journal of Human Genetics*, 94: 496-510.

Ranson, M. 2014. "Crime, weather, and climate change." *Journal of Environmental Economics and Management*, 67: 274-302.

Rasmussen, M. et al. 2014. "The genome of a Late Pleistocene human from a Clovis burial site in western Montana." *Nature*, 506: 225-229.

Razafindrazaka, H. et al. 2010. "Complete mitochondrial DNA sequences provide new insights into the Polynesian motif and the peopling of Madagascar." *European Journal of Human Genetics*, 18: 575-581.

Read, D. 2008. "An interaction model for resource implement complexity based on risk and number of annual moves." *American Antiquity*, 73: 599-625.

Reich, D. et al. 2010. "Genetic history of an archaic hominin group from Denisova Cave in Siberia." *Nature*, 468: 1053-1060.

Reich, D. et al. 2012. "Reconstructing Native American population history." *Nature*, 488: 370-374.

Reséndez, A. 2007. *A Land So Strange. The Epic Journey of Cabeza de Vaca*. New York: Basic Books.

Reyes-Centeno, H. et al. 2014. "Genomic and cranial phenotype data support multiple modern human dispersals from Africa and a southern route into Asia." *Proceedings of the National Academy of Sciences USA*, 111: 7248-7253.

Richards, M. et al. 2000. "Tracing European founder lineages in the Near Eastern mtDNA pool." *American Journal of Human Genetics*, 67: 1251-1276.

Richards, M.P. et al. 2000. "Neanderthal diet at Vindija and Neanderthal predation: the evidence from stable isotopes." *Proceedings of the National Academy of Sciences USA*, 97: 7663-7666.

Rick, T.C. et al. 2012. "Flightless ducks, giant mice and pygmy mammoths: Late Quaternary extinctions on California's Channel Islands." *World Archaeology*, 44: 3-20.

Risch, N. et al. 2003. "Geographic distribution of disease mutations in the Ashkenazi Jewish population supports genetic drift over selection." *American Journal of Human Genetics*, 72 812-822.

Robb, G. 2007. *The Discovery of France*. New York: W.W. Norton & Co., Inc.

Roberts, D.F. 1978. *Climate and Human Variability, 2nd. ed*. Menlo Park: Cummings Publishing Co.

Roberts, P. et al. 2014. "Continuity of mammalian fauna over the last 200,000 y in the Indian subcontinent." *Proceedings of the National Academy of Sciences, USA*, 111: 5848-5853.

Roosevelt, A.C. et al. 2002. The migrations and adaptations of the first Americans. Clovis and pre-Clovis viewed from South America. In *The First Americans. The Pleistocene Colonization of the New World*, edited by N.G. Jablonski, 159-235. San Francisco: California Academy of Sciences.

Roullier, C. et al. 2013. "Historical collections reveal patterns of diffusion of sweet potato in Oceania obscured by modern plant movements and recombination." *Proceedings of the National Academy of Sciences USA*, 110: 2205-2210.

Rowe, N. and Myers, M. (2011). All the World's Primates. www.alltheworldsprimates.org

Ruff, C. 2002. "Variation in human body size and shape." *Annual Review of Anthropology*, 31: 211-232.

Ruggiero, A. 1994. "Latitudinal correlates of the sizes of mammalian geographical ranges in South America." *Journal of Biogeography*, 21: 545-559.

Rule, S. et al. 2012. "The aftermath of megafaunal extinction: ecosystem transformation in Pleistocene Australia." *Science*, 335: 1483-1486.

Salque, M. et al. 2013. "Earliest evidence for cheese making in the sixth millennium B.C. in northern Europe." *Nature*, 493: 522-525.

Sanchez, G. et al. 2014. "Human (Clovis)-gomphothere (*Cuvieronius* sp.) association ~13,390 calibrated yBP in Sonora, Mexico." *Proceedings of the National Academy of Sciences USA*, 111: 10972-10977.

Sandom, C. et al. 2014. "Global late Quaternary megafauna extinctions linked to humans, not climate change." *Proceedings of the Royal Society B*, 281: 20133254.

Savolainen, P. et al. 2004. "A detailed picture of the origin of the Australian dingo, obtained from the study of mitochondrial DNA." *Proceedings of the National Academy of Sciences, USA*, 101: 12387-12390.

Sax, D.F. and Gaines, S.D. 2008. "Species invasions and extinction: The future of native biodiversity on islands." *Proceedings of the National Academy of Sciences, USA*, 105: 11490-11497.

Scheinfeldt, L.B. et al. 2012. "Genetic adaptation to high altitude in the Ethiopian highlands." *Genome Biology*, 13: R1.

Schlebusch, C.M. et al. 2012. "Genomic variation in seven Khoe-San groups reveals adaptation and complex African history." *Science*, 338: 374-379.

Schloss, C.A. et al. 2012. "Dispersal will limit ability of mammals to track climate change in the Western Hemisphere." *Proceedings of the National Academy of Sciences, USA*, 109: 8606-8611.

Schmidt-Nielsen, K. 1997. *Animal Physiology. Adaptation and Environment. 4th. ed*. Cambridge: Cambridge University Press.

Schnorr, S.L. et al. 2014. "Gut microbiome of the Hadza hunter-gatherers." *Nature Communications*, 5: 3654.

Scholz, C.A. et al. 2007. "East African megadroughts between 135 and 75 thousand years ago and bearing on early-modern human origins." *Proceedings of the National Academy of Sciences, USA*, 104: 16416-16421.

Schroeder, K.B. et al. 2007. "A private allele ubiquitous in the Americas." *Biology Letters*, 3: 218-223.

Schurr, T.G. 2004. "The peopling of the New World: perspectives from molecular anthropology." *Annual Review of Anthropology*, 33: 551-583.

Searle, J.B. et al. 2009. "Of mice and (Viking?) men: phylogeography of British and Irish house mice." *Proceedings of the Royal Society, London.B.*, 276: 201-207.

Segal, B. and Duffy, L.K. 1992. "Ethanol elimination among different racial groups." *Alcohol*, 9: 213-217.

Serva, M. et al. 2012. "Malagasy dialects and the peopling of Madagascar." *Journal of the Royal Society Interface*, 9: 54-67.

Shaw, C.N. and Stock, J.T. 2013. "Extreme mobility in the Late Pleistocene? Comparing limb biomechanics among fossil *Homo*, varsity athletes and Holocene foragers." *Journal of Human Evolution*, 64: 242–249.

Shipman, P. 2012. "The eyes have it." *American Scientist*, 100: 198-201.

Simonson, T.S. et al. 2010. "Genetic evidence for high-altitude adaptation in Tibet." *Science*, 329: 72-75.

Simpson, G.G. 1961. *Principles of Animal Taxonomy*. New York: Columbia University Press.

Skoglund, P. et al. 2012. "Origins and genetic legacy of Neolithic farmers and hunter-gatherers in Europe." *Science*, 336: 466-469.

Soares, P. et al. 2010. "The archaeogenetics of Europe." *Current Biology*, 20: R174-R183.

Soares, P. et al. 2011. "Ancient voyaging and Polynesian origins." *American Journal of Human Genetics*, 88: 239-247.

Stanford, D.J. and Bradley, B.A. 2012. *Across Atlantic Ice: The Origins of America's Clovis Culture*. Berkeley: University of California Press.

Steadman, D.W. 1995. "Prehistoric extinctions of Pacific island birds: biodiversity meets zooarchaeology." *Science*, 267: 1123-1131.

Steadman, D.W. 2006. *Extinction and Biogeography of Tropical Pacific Birds*. Chicago: University of Chicago Press.

Steele, T.E. and Klein, R.G. 2008. "Intertidal shellfish use during the Middle and Later Stone Age of South Africa." *Archeofauna*, 17: 63-76.

Stephens, D.W. and Krebs, J.R. 1986. *Foraging Theory*. Princeton: Princeton University Press.

Stepp, J.R. et al. 2005. "Mountains and biocultural diversity." *Mountain Research and Development*, 25: 223-227.

Stewart, J.R. 2005. "The ecology and adaptation of Neanderthals during the non-analogue environment of Oxygen Isotope Stage 3." *Quaternary International*, 137: 35-46.

Stewart, J.R. and Stringer, C. 2012. "Human evolution out of Africa: the role of refugia and climate change." *Science*, 335: 1317-1325.

Stinson, S. 2000. Growth variation: biological and cultural factors. In *Human Biology. An Evolutionary and Biocultural Perspective.*, edited by S. Stinson, et al., 425-463. New York: Wiley-Liss.

Stinson, S. and Frisancho, A.R. 1978. "Body proportions of highland and lowland Peruvian Quechua children." *Human Biology*, 50: 57-68.

Stone, A.C. et al. 2002. "High levels of Y-chromosome nucleotide diversity in the genus Pan." *Proceedings of the National Academy of Sciences, USA*, 99: 43-48.

Stoneking, M. and Delfin, F. 2010. "The human genetic history of East Asia: weaving a complex tapestry." *Current Biology*, 20: R188-R193.

Stoneking, M. et al. 1990. "Geographic variation in human mitochondrial DNA from Papua New Guinea." *Genetics*, 124: 717-723.

Storey, A.A. et al. 2007. "Radiocarbon and DNA evidence for a pre-Columbian introduction of Polynesian chickens to Chile." *Proceedings of the National Academy of Sciences, USA*, 104: 10335-10339

Storey, M. et al. 2012. "Astronomically calibrated 40Ar/39Ar age for the Toba supereruption and global synchronization of late Quaternary records." *Proceedings of the National Academy of Sciences, USA*, 109: 18684-18688.

Strassman, B.I. and Dunbar, R.I.M. 1999. Human evolution and disease: putting the Stone Age in perspective. In *Evolution in Health & Disease*, edited by S.C. Stearns, 91-101. Oxford: Oxford University Press.

Straus, L.G. et al. 2005. "Ice Age Atlantis? Exploring the Solutrean-Clovis 'connection'." *World Archaeology*, 37: 507-532.

Stuart, A.J. 1991. "Mammalian extinctions in the Late Pleistocene of Northern Eurasia and North America." *Biological Reviews*, 66: 453-562.

Surovell, T. et al. 2005. "Global archaeological evidence for proboscidean overkill." *Proceedings of the National Academy of Sciences USA*, 102: 6231-6236.

Surovell, T.A. 2000. "Early Paleoindian women, children, mobility, and fertility." *American Antiquity*, 65: 493-508.

Surovell, T.A. 2003. "Simulating coastal migration in New World colonization." *Current Anthropology*, 44: 580-589.

Surovell, T.A. and Waguespack, N.M. 2009. Human prey choice in the late Pleistocene and its relation to megafaunal extinctions. In *American Megafaunal Extinctions at the End of the Pleistocene*, edited by G. Haynes, 77-105: Springer.

Sutherland, W.J. 2003. "Parallel extinction risk and global distribution of languages and species." *Nature*, 423: 276-279.

Swain, D.P. et al. 2007. "Evolutionary response to size-selective mortality in an exploited fish population." *Proceedings of the Royal Society B*, 274: 1015-1022.

Tamm, E. et al. 2007. "Beringian standstill and spread of Native American founders." *PLOS ONE*, 2: e829, 821-825.

Tanabe, K. et al. 2010. "*Plasmodium falciparum* accompanied the human expansion out of Africa." *Current Biology*, 20: 1283-1289.

Taylor, A.K. et al. 2011. "Big sites, small sites, and coastal settlement patterns in the San Juan Islands, Washington, USA." *Journal of Inland and Coastal Archaeology*, 6: 287-313.

Taylor, A.L. et al. 2004. "Combination of isosorbide dinitrate and hydralazine in blacks with heart failure." *New England Journal of Medicine*, 351: 2049-2057.

Taylor, N.A.S. 2006. "Ethnic differences in thermoregulation: genotypic versus phenotypic heat adaptation." *Journal of Thermal Biology*, 31: 90-104.

Terborgh, J. 1999. *Requiem for Nature*. Washington, D.C.: Island Press.

Terrell, J. 1986. *Prehistory in the Pacific Islands. A Study of Variation in Language, Customs, and Human Biology*. Cambridge: Cambridge University Press.

Thomson, J. 1887. *Through Masai Land: A Journey of Exploration Among the Snowclad Volcanic Mountains and Strange Tribes of Eastern Equatorial Africa, Being the Narrative of The Royal Geographical Society's Expedition to Mount Kenia and Lake Victoria Nyanza, 1883-1884*. London: Sampson Low, Marston, Searle, & Rivington.

Thomson, V.A. et al. 2014. "Using ancient DNA to study the origins and dispersal of ancestral Polynesian chickens across the Pacific." *Proceedings of the National Academy of Sciences USA*, 111: 4826-4831.

Tilkens, M.J. et al. 2007. "The effects of body proportions on thermoregulation: an experimental assessment of Allen's rule." *Journal of Human Evolution*, 53: 286-291.

Tishkoff, S.A. et al. 2009. "The genetic structure and history of Africans and African-Americans." *Science*, 324: 1035-1044.

Tishkoff, S.A. et al. 2007. "Convergent adaptation of human lactase persistence in Africa and Europe." *Nature Genetics*, 39: 31-40.

Tishkoff, S.A. and Verrelli, B.C. 2003. "Patterns of human genetic diversity: implications for human evolutionary history and disease." *Annual Review of Genomics and Human Genetics*, 4: 293-340.

Torrence, R. 1983. Time budgeting and hunter-gatherer technology. In *Hunter-Gatherer Economy in Prehistory*, edited by G. Bailey, 11-22. Cambridge: Cambridge University Press.

Townroe, S. and Callaghan, A. 2014. "British container breeding mosquitoes: the impact of urbanisation and climate change on community composition and phenology." *PLOS ONE*, 9: e95325.

Tumonggor, M.K. et al. 2013. "The Indonesian archipelago: an ancient genetic highway linking Asia and the Pacific." *Journal of Human Genetics*, 58: 165-173.

Turchin, P. et al. 2013. "War, space, and the evolution of Old World complex societies." *Proceedings of the National Academy of Sciences USA*, 110: 16384-16389.

Turner, A. 1992. "Large carnivores and earliest European hominids: changing determinants of resource availability during the Lower and Middle Pleistocene." *Journal of Human Evolution*, 22: 109-126.

Turney, C.S.M. et al. 2008. "Late-surviving megafauna in Tasmania, Australia, implicate human involvement in their extinction."*Proceedings of the National Academy of Sciences, USA*, 105: 12150-12153.

United States Environmental Protection Agency. (2014). Future climate change. http://www.epa.gov/climatechange/science/future.html#Ice

van Holst Pellekaan, S. 2013. "Genetic evidence for the colonization of Australia." *Quaternary International*, 285: 44-56.

van Riper, C. et al. 1986. "The epizootiology and ecological significance of malaria in Hawaiian land birds." *Ecological Monographs*, 56: 327-344.

Vigne, J.-D. 2008. Zooarchaeological aspects of the Neolithic transition in the Near East and Europe, and their putative relationships with the Neolithic demographic transition. In *The Neolithic Demographic Transition and its Consequences*, edited by J.-P. Bocquet-Appel and O. Bar-Yosef, 179-205: Springer.

Vigne, J.-D. et al. 2012. "First wave of cultivators spread to Cyprus at least 10,600 y ago." *Proceedings of the National Academy of Sciences, USA*, 109: 8445-8449.

Wadley, L. et al. 2011. "Middle Stone Age bedding construction and settlement patterns at Sibudu, South Africa." *Science*, 334: 1388-1391.

Walker, M.J. et al. 2011. "Morphology, body proportions, and postcranial hypertrophy of a female Neandertal from the Sima de las Palomas, southeastern Spain." *Proceedings of the National Academy of Sciences, USA*, 108: 10087–10091.

Walter, R.C. et al. 2000. "Early human occupation of the Red Sea coast of Eritrea during the last interglacial." *Nature*, 405: 65-69.

Warner, R.E. 1964. "The role of introduced diseases in the extinction of the endemic Hawaiian avifauna." *The Condor*, 70: 101-120.

Waters, M.R. et al. 2011. "The Buttermilk Creek Complex and the origins of Clovis at the Debra L. Friedkin Site, Texas." *Science*, 331: 1599-1603.

Waters, M.R. and Stafford, T.W. 2007. "Redefining the Age of Clovis: Implications for the Peopling of the Americas." *Science*, 315: 1122-1126

Weaver, T.D. and Roseman, C.C. 2008. "New developments in the genetic evidence for modern human origins." *Evolutionary Anthropology*, 17: 69-80.

Weaver, T.D. et al. 2007. "Were Neandertal and modern human cranial differences produced by natural selection or genetic drift?" *Journal of Human Evolution*, 53: 135-145.

Weaver, T.D. and Steudel-Numbers, K. 2005. "Does climate or mobility explain the differences in body proportions between Neandertals and their Upper Paleolithic successors?" *Evolutionary Anthropology*, 14: 218-223.

Weitz, C.A. et al. 2013. "Responses of Han migrants compared to Tibetans at high altitude." *American Journal of Human Biology*, 25: 169-178.

White, J.P. 2004. Where the wild things are: Prehistoric animal translocation in the circum New Guinea Archipelago. In *Voyages of Discovery. The Archaeology of Islands*, edited by S.M. FitzPatrick, 147-164: Praeger.

Wilde, S. et al. 2014. "Direct evidence for positive selection of skin, hair, and eye pigmentation in Europeans during the last 5,000 y." *Proceedings of the National Academy of Sciences USA*, 111: 4832-4837.

Wilmshurst, J.M. et al. 2008. "Dating the late prehistoric dispersal of Polynesians to New Zealand using the commensal Pacific rat." *Proceedings of the National Academy of Sciences, USA*, 105: 7676-7680.

Witze, A. and Kanipe, J. 2015. *Island on Fire. The Extraordinary Story of a Forgotten Volcano that Changed the World*. New York: Pegasus Books.

Wollstein, A. et al. 2010. "Demographic history of Oceania inferred from genome-wide data." *Current Biology*, 20: 1983-1992.

Wood, B. 2010. "Reconstructing human evolution: Achievements, challenges, and opportunities." *Proceedings of the National Academy of Sciences, USA*, 107: 8902-8909.

Wood, R.E. et al. 2013. "Radiocarbon dating casts doubt on the late chronology of the Middle to Upper Palaeolithic transition in southern Iberia." *Proceedings of the National Academy of Sciences USA*, 110: 2781-2786.

Wrangham, R. and Carmody, R. 2010. "Human adaptation to the control of fire." *Evolutionary Anthropology*, 19: 187-199.

Wroe, S. 2002. "A review of terrestrial mammalian and reptilian carnivore ecology in Australian fossil faunas, and factors influencing their diversity: the myth of reptilian domination and its broader ramifications." *Australian Journal of Zoology*, 50: 1-24.

Yi, X. et al. 2010. "Sequencing of 50 exomes reveals adaptation to high altitude." *Science*, 329: 75-78.

Young, J.H. 2007. "Evolution of blood pressure regulation in humans." *Current Hypertension Reports*, 9: 13-18.

Young, J.H. et al. 2005. "Differential susceptibility to hypertension is due to selection during the out-of-Africa expansion." *PLOS Genetics*, 1: 0730-0738.

Yule, J.V. et al. 2014. "A review and synthesis of Late Pleistocene extinction modeling: progress delayed by mismatches between ecological realism, interpretation, and methodological transparency." *Quarterly Review of Biology*, 89: 91-106.

Zazula, G.D. and Froese, D.G. 2003. "Ice-age steppe vegetation in east Beringia." *Nature*, 423: 603.

Zerjal, T. et al. 2003. "The genetic legacy of the Mongols." *American Journal of Human Genetics*, 72: 717-721.

Zhang, D.D. et al. 2011. "The causality analysis of climate change and large-scale human crisis." *Proceedings of the National Academy of Sciences, USA*, 108: 17296-17301.

Zhang, Z. et al. 2010. "Periodic climate cooling enhanced natural disasters and wars in China during A.D. 10-1900." *Proceedings of the Royal Society B*, 277: 3745-3753.

Zhao, Y. et al. 2010. "Out of China: distribution history of *Ginkgo biloba* L." *Taxon*, 59: 495-504.

SOME SUGGESTED READING

A selection of readings—books and the occasional article—that might be of interest. Some I have cited for specific topics, but I repeat them here as general reading.

GENERAL

Janet Browne. 1983. The Secular Ark. Studies in the History of Biogeography. Yale University Press. *A very readable, erudite history of biogeography as a field of study, starting with attempts to find the animals transported by Noah.*

Charles Darwin. 1871. The Descent of Man, and Selection in Relation to Sex. John Murray, London. *The great man was a biogeographer as well as a naturalist. It is the first half of this book that touches on many of the themes that I address, human origins in Africa, geographic barriers to our spread, variation in form across regions (although he argued that environment did not explain the variation), and competition between human populations.*

Jared Diamond. 1997. Guns, Germs, and Steel: The Fates of Human Societies. W.W. Norton. *A fascinating account of how biology (in the form of disease, for example) and geography (the shape of continents) can determine the outcome of competition between human societies.*

Exploratorium. http://www.exploratorium.edu/evidence/. *An interesting and enlightening discussion about how science works—with pictures, videos, interviews, and demonstrations.*

Alexander Harcourt. 2012. Human Biogeography. University of California Press. *Go here for all the quantitative science, the graphs, the statistics, and more sources behind this book's account of the topic of the biogeography of the human species.*

Mark Lomolino and co-authors. 2010. Biogeography. Sinauer Associates. *If anyone is stimulated by either this book or my scientific book to go further into the general subject of biogeography, this is the place to go to. This fourth edition has so many colorful graphs and pictures, the reader would not even have to read the text to get the gist of what the field is all about.*

John B. Marciano. 2014. Whatever Happened to the Metric System? How America Kept its Feet. Bloomsbury Press. *An amusing, erudite account of how the rest of the world eventually changed (mostly) to the metric system of weights and measures, but the USA (mostly) did not.*

Patrick O'Brien. 2002. Atlas of World History. Oxford University Press. *Any atlas of world history is an account of the movement of humans around the world, of origins, expansions, and disappearances of human populations, much of it affected by competition between peoples, in other words of aspects of the biogeography of humans.*

Pat Shipman. 1994. The Evolution of Racism. Simon & Schuster. *From chapter 6 on, an account of racism in the USA, and how easy it is for scientists to be insensitive and misunderstood about race, as well as wrong about it. The first five chapters are about the development of the theory of evolution by natural selection—Darwin, Wallace, Huxley, etc.*

Alfred Russel Wallace. 1869. The Malay Archipelago. Macmillan. *Wallace's two-volume* The Geographical Distribution of Animals *is the cornerstone of the field of biogeography.* The Malay Archipelago *is a highly readable account of his eight years collecting natural history specimens in Indonesia and New Guinea. Subtitled* The Land of the Orangutan, and the Bird of Paradise. A Narrative of Travel, with Sketches of Man and Nature, *it is full of sympathetic descriptions of the region's local people, as well as of its animals, plants, and geography, along with accounts of Wallace's sometimes hair-raising adventures.*

CHAPTER 1. PROLOGUE

Kenneth Miller. 2008. Only A Theory: Evolution and the Battle for America's Soul. Penguin. *A thorough dissection of creationism, or, as the proponents liked to call it, "intelligent design," and of creationists' misrepresentation of not only the evidence for evolution, but of their very own views. I am not quite sure what the "soul" of the title is, but I like to think of it as America's scientific soul, in other words Americans' ability to think for themselves and question authority. Science is fundamentally a democratic process—anyone can and should question received wisdom.*

CHAPTERS 2, 3, 4. FROM AFRICA TO THE REST OF THE WORLD

Luigi Cavalli-Sforza. 2000. Genes, Peoples, and Languages. University of California Press. *A brief popular account of the geography of human population genetics. If there is a founding father of the field of genetical biogeography of humans, Cavalli-Sforza is the man. His work has been crucial to our understanding of humans' movements around the globe.*

Brian Fagan. (2003). The Great Journey. The Peopling of Ancient America. Gainesville: University of Florida Press. *Despite the title, the book begins with our evolutionary origins in Africa. It ends with the arrival of Europeans. Fagan considers only archeological evidence.*

Richard Klein. 2009. The Human Career. University of Chicago Press. *This is far from a popular book, but absolutely everything anyone could possibly want to know about human origins from a mostly paleontological and archeological perspective is here. If any human ancestors that I mention might interest you, you will find all about them in* The Human Career.

I. Ness. (2013). The Encyclopedia of Global Human Migration. Wiley-Blackwell. *Everything you always wanted to know about the movement of humans around the world—in five volumes, 3444 pages.*

John Parkington. (2006). Shorelines, Strandlopers and Shell Middens. Krakadouw Trust. *A popular account of shellfish use by Stone Age peoples of southern Africa.*

Stephen Oppenheimer. 2003. Out of Eden. The Peopling of the World. Constable. *A detailed account of the spread of humans out of Africa around the world, based mostly on genetical information.*

Stephen Oppenheimer. 2006. The Origins of the British. Constable & Robinson Ltd. *A detailed but readable account of the entry of humans into Britain, based mostly on genetical information.*

Chris Stringer. 2011. The Origin of Our Species. Allen Lane, published in 2012 in the USA as Lone Survivors: How We Came To Be the Only Humans on Earth, Times Books. *A popular account of our origins in Africa and our spread around the world by a scientist who has had a major role in producing the information and ideas on the topic. Particularly useful accounts of the methods by which we obtain and interpret the information from bones, stones, and DNA.*

A. Witze and J. Kanipe. 2015. Island on Fire. The Extraordinary Story of a Forgotten Volcano that Changed the World. New York: Pegasus Books. *An account of the Icelandic volcano Laki and its global effects.*

And for anyone who wants a detailed summary on our movements around the world, the 23 February 2010 issue of the journal *Current Biology* has an article per continent, each by an expert on the continent.

CHAPTER 5. VARIETY IS THE SPICE OF LIFE

Robert Boyd and Joan Silk. 2011. How Humans Evolved, 6th edition. Prentice Hall. *My chapter 4 covers a fundamental topic in almost any textbook of physical anthropology. I have picked one of the most readable and recent of these by authors whom I know personally. Conversely, the topic barely appears in any book on biogeography, even though the topic is fundamentally biogeographical. Biogeography can benefit from anthropology's immense amount of data and understanding on the issue.*
Nina Jablonski and George Chaplin. October 2002. Skin deep. Scientific American. *A popular account of one of the most thorough studies yet on why people from different regions have different skin color.*

CHAPTER 6. GENE MAPS AND ROADS LESS TRAVELED

Graham Robb. 2007. The Discovery of France. A Historical Geography. W.W. Norton & Co. *A fascinating account of the cultural diversity of France, maintained into the nineteenth century because people traveled so little.*

CHAPTER 7. IS MAN MERELY A MONKEY?

Paul Lewis. 2009. Ethnologue: Languages of the World. http://www.ethnologue.com/. SIL International. *As the website describes it, 'an encyclopedic reference work cataloging all of the world's 6,909 known living languages.' The map on the top page of "Country index with maps" shows with no words needed the amazing tropical diversity of the world's languages.*
Mark Lomolino and co-authors. 2010. Biogeography, Ch. 15. Sinauer Associates. *I repeat this book here (it is also a general suggestion) because the topic is a centuries-old one in biogeography, and understood in detail, even if still argued. Yet the topic barely registers in any general anthropology text, or text on human evolution. Anthropology can benefit from biogeography's immense amount of data and understanding on the issue, and biogeography from anthropology's immense amount of data on variation within one species.*

CHAPTER 8. ISLANDS ARE SPECIAL

Dean Falk. 2011. The Fossil Chronicles. University of California Press. *An account of the sometimes unpleasantly expressed arguments over the Hobbit, and also over the "Taung baby," the interpretation of which as a hominin was for long rejected by establishment scientists. Scientists should be objective, and welcome criticism, but we're only human.*
Mark Lomolino and co-authors. 2010. Biogeography, Ch. 14. Sinauer Associates. *I repeat this book here for the same reasons as I repeated it for Ch. 5.*
D. Quammen. 1996. The Song of the Dodo: Island Biogeography in an Age of Extinctions. Scribner. *Highly readable, like all of Quammen's books, a thorough account of both biogeography and hence of why island species are particularly prone to extinction.*

CHAPTER 10. WHAT DOESN'T KILL US HALTS US OR MOVES US

J.R. McNeill. 2010. Mosquito Empires. Ecology and War in the Greater Caribbean, 1620-1914. Cambridge University Press. *A highly readable account of how disease, particularly yellow fever and malaria, affected foreign domination in the Americas and the Caribbean, largely by killing more invaders than the defenders killed.*
W.H. McNeill. 1976 (reprinted 1998). Plagues and People. Anchor Books. *Terrifying accounts of the nonstop plagues that have affected humanity since the beginnings of history, with an added section on AIDS.*

CHAPTER 11. MAD, BAD, AND DANGEROUS TO KNOW

P. Crane. (2013). *Ginkgo. The Tree that Time Forgot.* New Haven, USA: Yale University Press. *All you ever wanted to know about the ginkgo, including its worldwide spread from remnant forests in China.*

D. Hart & R.W. Sussman. (2005). Man the Hunted. Primates, Predators, and Human Evolution. New York: Westview Press. *A detailed but popular account of the effect of predators on monkeys, apes, and humans.*

D. Quammen. 1996. The Song of the Dodo: Island Biogeography in an Age of Extinctions. Scribner. *Humans are particularly good at killing off island animals.*

CHAPTER 12. CONQUEST AND COOPERATION

Brian M. Fagan. 2000. The Little Ice Age: How Climate Made History, 1300-1850. Basic Books. *Poor climate reduces harvests, increases competition for food, causes war, and hence affects the movement and survival of people. This book is a general account of, as its title says, the effect of climate on human history.*

Luisa Maffi. 2001. On Biocultural Diversity. Linking Language, Knowledge, and the Environment. Smithsonian Institute Press. *A large (578 pp.) edited book giving a wide variety of perspectives on linguistic and biological diversity and their actual and potential connections.*

Daniel Nettle & Suzanne Romaine. 2000. Vanishing Voices. The Extinction of the World's Languages. Oxford University Press. *A must-read for anyone interested in the tragedy of disappearing cultures. The extinction of species at the hand of humans seems never to be out of the news. The loss of languages and cultures seems never to make it to the news. Perhaps that's because so often the language disappears with a whisper, as the last speaker dies of old age. Mostly a depressing book, of course. But the last chapter discusses what can be done and is being done throughout the world to save disappearing cultures.*

CHAPTER 13. EPILOGUE

Robert Hinde. 2011. Changing How We Live. Society from the Bottom Up. Spokesman. *Group-living animals, such as humans, have to be as cooperative as they are competitive. We must build on our cooperative natures to ensure that moral behavior means working for the good of society and the planet, not working for ourselves.*

INDEX

GEOGRAPHIC INDEX

Maps: *Frontispiece, 24, 29, 39, 47, 51, 164, 219 (and italicized pg. numbers, below)*

Africa, 9-10, 14-23, *47*, 64, 87-88, 92-93, 97, 99-103, 107-110, 132, 140-141, 153, 184, 199-202, 211-212, 214-215, 217-221, 225, 238, 251

Amazon, 57, 137, 143-144, 163

Americas, 28-*29*-31, 53-58, 65, 70, 225, 234, 237-239, 244, 258

Andes, 102, 113-120, 138, 153

Antarctica, 21, 111

Arabia, 15, 23, 45-*47*-48, 199, 201

Arctic, 27-28, 101-102, 110-111, 137-138, 153, 158, 159, 160, 208-211, 253-254

Argentina, 57, 150

Asia, 24-26, 34, 37, 51-53, 153, 168, 174-182, 184-186, 196, 199, 211-212, 238, 255, 258

Australia, 23-*24*-25, 45, 48-50, 65, 70, 88, 94, 135-136, 153, 166, 181, 187, 230, 233, 236, 237, 241-243, 252, 264

Bering Strait, 28, 53, 136

Bhutan, 114, 137, 147, 203-204, 239

Bolivia, 116-119

Britain (UK), 26-27, 32-33, 37, 112, 138, 150-151, 167, 209, 216-217, 222-223, 266-267, 276

California, 161-162, *164*-167, 171, 238-239, 246, 264

Cameroon, 14, 22, 23, 162

Canada, 31

Caribbean, 222, 224

Caucasus, 199, 229, 255-256

Chile, *29*, 57-58, 65

China, 52, 199, 222, 225-226, 246, 253, 270

Colombia, 160

Congo Basin, 162-163

Costa Rica, 247-248

Cuba, 269

East Africa, 33

Ecuador, 150-151, 154, 198

Egypt, 222

Eritrea, 45, 201

Ethiopia, 14, 22, 23, 44, 115-116, 120, 121, 245

Eurasia, 15, 25, 27, 33, 38, 50, 225, 234, 236-239, 255

Europe, 17, 31, 32-34, 37, 50-*51*-52, 64-66, 89, 92-94, 107-111, 116-118, 136-137, 140-142, 145, 196-197, 199-202, 215, 224, 227, 237-238, 253, 255, 257-258, 269

Everest, Mt., 113-114

Falklands, 194

Finland, 122-123, 250-151

Flores (*see* Islands)

France, 56, 76, 80-81, 89, 139, 145, 194, 209, 269

Galapagos, 240

Georgia, 15

Greenland, 19, 21, 31, 37

Hawaii, *39*, 40, 63, 192, 243

Himalayas, 113-121, 163

Holland, 122-123, 254

Iceland, 21, 36-37, 194, 196

India, 24-25, 48, 50, 143

Indonesia, 41, 48, 58-60, 63, 73, 136, 174-177, 179-181

Ireland, 145, 239-240

Israel, 15

Italy, 143, 196

Japan, 26, 52-53, 196, 207

Java, 182, 184-186, 192, 255

Karakorams, 163

Kenya, 100-101

Korea, 26
Krakatau/Krakatoa, 19, 182
Lake Chad, 161
Laki, 21
Madagascar, 41-42, 63, 136, 162-163, 190-191, 231, 235, 238, 242, 243
Malawi, 17, 20
Mauritius, 231
Mediterranean, 33, 37, 217
Melanesia, *39*-40, 58-61
Micronesia, *39*-40, 61
Middle East, 9, 15-16, 23, 45-46-*47*, 204
Mongolia, 25, 52, 59, 93-94, 203
New Guinea, 23-*24*, 48-49, 58, 87, 135, 142, 152, 163
New Zealand, *39*, 40, 62, 88, 136, 190, 231, 232-233, 235, 244, 252
Nigeria, 23, 161
North America (*see also* USA), 7, 20, 28-31, 36, 38, 45, 53-57, 61, 63, 65, 66, 67, 93, 94, 122, 141-142, 151, 153-154, 163-*164*-167, 171, 176, 209, 211, 223-225, 227, 231, 233, 234-235, 238-239, 244, 249-250, 252, 264-265, 271-272, 276
Pacific, 34, *39*-40, 58-61, 88, 112-113, 182, 183, 185, 187-189, 192-193, 224-225, 258
Pakistan, 37, 199
Panama, 223
Peru (*see also* Andes), 30, 114
Philippines, 58-59, 168
Pitcairn Island, 67
Poland, 204
Polynesia, *39*-40, 58-63, 88, 193
Red (Reed) Sea, *47*, 134-135
Russia, 26-27, 52-53, 93-94
Sahara, 18, 132
Sahul, *24*
Samoa, 112-113
Scandinavia, 31, 33, 35
Siberia, 26-28, 31, 52-53, 228, 247
South Africa, 45, 130
South America, 17, 28-*29*-30, 56-58, 62, 65, 86, 137, 143-144, 158, 220-221, 224, 225, 234, 244, 253
Spain, 104-105, 222-223, 255
St. Helena, 194
Sumatra, 19, 20, 186, 192
Taiwan, 58, 88, 266
Tambora volcano, Indonesia, 20-21, 274
Tanzania, 201-202, 225
Tasmania, 23-*24*, 49, 94, 112, 187, 242, 246-247, 263-264
Tibet, 25, 114-121, 138, 246
Toba volcano, Sumatra, 19-20
Tropics, 107-112
Turkey, 31
USA (*see also* North America), 7, 36, 38, 61, 66, 112, 141-142, 151, 163-*164*-167, 176, 209, 223, 225, 227, 245, 249-250, 253, 264-265, 271-272, 276

SUBJECT INDEX
Accident (genetic), 127-129
Adaptation, altitude, 102, 110, 113-121; diet, 195-212; disease, 214, 215, 218, 222, 259; general, 90-125; sunlight, 91-96; temperature, 97-112
African-Americans, 211-212, 218
Agriculture (*incl.* Livestock), 20, 31, 33-35, 91, 94-95, 168, 197, 199, 200-204, 245, 253, 269-270, 272
Alcohol, acetaldehyde and alcoholism, 196-198
Allen effect, 98-107
Altitude, 30-31, 102, 110, 113-121
Amylase, Amy1 (*see* Starch)
Anatomy (*see* Body shape, size)
Ancestors, 10-15, 73, 279
Anglo-Saxons, 32, 35
Animals, 25, 62-63, 70, 86, 88-89, 91-92, 96, 97-98, 105-107, 113, 120-121, 134, 145, 149, 153-154, 174, 176-185, 189-191, 200-206, 208-210, 214, 216, 218-*219*, 226-228, 230-255, 269, 280
Archeology, 75, 80
Arctic peoples (Eskimo, Inuit), 93-94, 110, 208-210
Area (*see* Diversity, Geographical range size, Islands)
Ashkenazi Jews, 129
Assyrians, 220
Athletes, 97, 99-101
Aurignacian, 80-81
Australopithecus, 10-13, 73
Bacteria, 87, 203, 207, 254-255
Bantu language area, 162
Barriers, cultural, 141-147; geographic, 45-54, 57, 132-141, 155, 163, 168-171
Bergmann effect, 98-107
Bilharzia (Schistosomiasis), 219-220, 222
Biodiversity (*see* Diversity)
Blood groups, 86, 127-128
Boats (sea-crossing), 53
Body shape, size (*see also* Islands), 97-107, 112-113, 116-118, 121-123; extinctions, 230-231, 234-236, 238
Brain size, 10, 206
Cane toad, 70
Carbon, etc. dating, 76-77
Cattle, 91
Caucasians, 199, 229
Celts, 32, 128
Chagas disease, 221
Chromosome, 36-37
Climate (*see also* Agriculture, Global warming, Temperature), agriculture, 20, 94-95, 245, 269-270, 272; diasporas, 16-22, 26-28, 31-32; extinction, 234-240, 245, 255-256; skin (fur, feather) color, 91-96; violence, war, 269-272
Clovis, 29-30, 55, 82-83
Coastal routes around the world, 43-64
Colonialism, 169, 171-172
Commonness (see also Rarity), 273-275
Competition, 191-192, 256-258, 261-267, 269-272
Cooperation, 276, 280

Corridor, ice-free, 134
Crops (*see* Agriculture)
Cultures (*see* Diversity)
Dairying (*see* Milk)
Darwin, Charles, 3, 83, 90-91, 127
Dating, 75-82
Denisovans, 15, 119, 258-259
Desert (*see* Barriers)
Diet, 94-95, 101-103, 143-144, 195-212
Dingo, 25, 242-243
Disease (parasites, pathogens), 3-5, 87, 103, 108-109, 129-130, 141, 145-146, 169-172, 175-176, 181, 192-194, 211, 213-226, 228-229, 243, 245-246, 249-252, 259
Dispersal (*see also* Barriers), 14-42, 43-*51*-71, 128-130, 134-136, 169-172, 265, 276-277
Distance and diversity, 185-186
Distances moved, 65-71, 139, 186, 210
Diversity (*see also* Endemism, Forster effect, Islands), area, 182-189; cultures, 131, 139, 148-*164*-172, 183-190; disease, 250-151; species, 148-160, 162-163, 168-170, 182-186, 190-192
Dodo, 177, 231
Donner Party, 122-123
Drugs, 4, 130, 211
Dutch Hunger, 122
Ear wax, 128
Ebola, 251
El Niño-La Niña, 270-271
Elk, Irish, 239-240
Endemism, 173-174, 190
Eskimo (*see* Arctic peoples)
Evolution, 2, 73, 90-125, 195-207, 212, 222
Extinction, 182, 230-251, 255-258, 260; causes, 231-250, 255-258; knock-on effects, 247-251; languages, 263-265, 268-269; rarity, 274-275
Famine, 122-123, 269-270
Farming (*see* Agriculture)
Fat, body mass index, 112-113; food, 208-210
Flores, 174-182, 184-186
Flu, 251
Forster effect, 155-172
Founder effect, 127-129
Future for humans, 279-280
Gelada baboon, 245
Genes (DNA), 35-37, 40, 52, 54, 58-60, 64, 78-79, 84-86, 89, 96, 103, 108, 116-121, 127-130, 136-137, 201-203, 205, 206-208, 214-218, 224-225, 255, 258-259
Genocide, 262-263
Geographical range size (*see also* Forster effect, Islands), 152-154, 156-162, 164-170; expansion, 251-254, 272; extinction, 230-250, 255-258; reduction to edge, 246-247, 265-266
Gigantism (*see* Islands)
Ginkgo, 253
Glaciers (*see* Ice cap/sheet)
Global dispersal, 9-42, 43-71, 85-89

Global warming (*see also* Climate), 33-34, 232, 238-239, 244-245, 253-254, 271-272
Gourds, 62-63
Hadza, 201-202, 205-206, 225, 255
Han Chinese peoples, 115-118
Helicobacter, 87
History, 19, 20-21, 23, 32-34, 40-41, 64, 122-124, 136, 166, 169, 171, 220-226, 262-267, 269-272, 276-277
HIV, 151
Hobbit, 174-178, 181-182
Hominin, 11, 182, 255-260
Homo, 10-15, 73, 174
Hotspots of diversity, 152, 162-168
Hunter-gatherers, 14, 31, 34-36, 64, 66, 67, 69, 71, 161, 168, 201-202, 208-210
Ice cap/sheet (Glaciers), 28, 54-55, 133-134, 157
Illness (not disease) (*see* Diet)
Imperialism, 169
"Indian" reservations, 265
Inuit (*see* Arctic peoples)
Invasions, 220-221
Islands, 173-194; biodiversity, 182-185, 189-192; cultural diversity, 183-189; disease, 192-194; extinctions, 238, 240-241, 246-248; Flores (Flo, Hobbit), 174-182; gigantism, 177-182; miniaturization, 176-182
Java Man, 73, 174
Kalahari, 110
Komodo dragon, 177, 180-181
Kon-Tiki, 62
Kudzu, 252
Language/linguistics, 34-37, 50, 58, 63-64, 74, 131, 139, 141-143, 150-153, 263-269, 275
Lapita culture, 40
Last speakers, 264-268
Latitude (*see* Forster effect)
Limb (leg) length, 97-107
Little Ice Age, 19, 269-270
Livestock (*see* Agriculture)
Lusitania and *Titanic*, 123
Maasai, 23, 91
Mal'ta boy, 28, 52, 54
Malaria, 87, 214-217, 221, 222, 223, 226
Mammoths, 176, 228, 238-239
Maori, 40, 62, 235
Marathons, 97
Marriage distances, 139
Mayflower, 123-124
Measles, 224
Measuring systems, 7
Megalania, 181
Metabolism, altitude, 113-121; diversity, 162; islands, 112-113, 180, 182; Samoans, 182; sexes, 121-125; starvation, 112-113; temperature, 109-113
Milk (lactose, lactase, cheese, yogurt), 198-204
Miniaturization (*see* Islands)
Moa, 131, 231, 235

Models, 17-18, 45-48, 66-68, 70, 189, 233, 236-238
Monte Verde site, *29*-30, 57, 82-83
Mormons, 61
Mosquito (disease), 87, 192, 216, 218, 221, 222
Musk ox, 239
Myxomatosis, 252
Native Americans, 53-54, 161-162, *164*-167, 224, 265, 268
Natural selection (*see also* Sexual selection), 83, 90-125, 128, 212, 215, 218
Neanderthal, 15-16, 25, 51, 103-106, 255-258-259
New York, Central Park, 133
Nutrition, anatomy, physiology, sexes, 101-103, 122
Oldupai, Olduvai, 75-76
Olympics, 97, 100-101
Origins, 9-14, 84-85, 128-129
Out of Africa (diaspora), 15-23, 44-*47*-48, 85-88
Oysters (*see* shellfish)
Painting, 81-82
Paleolithic, 80-81
Panama Canal, 223
Parasites, pathogens (*see* Disease)
Peninsulas and diversity, 186
Phlorizin, 201
Physiology, altitude, 102, 110, 113-121; diet, 196-212; sexes, 121-125; sunlight, 91-96; temperature, 97-112
Plague, 226
Plants (*see also* Agriculture, Forster effect, Productivity), 62, 82, 150, 162-163, 204-207, 249, 252-254, 257, 269-270
Plasmodium, 87
Population density, size, 159-160, 190-194, 279
Predators, 226-227, 248
Prey, 227
Primates (monkeys, apes), 97, 152-153, 163
Productivity, cities, 167; dispersal, 17, 44-46; diversity, 158-162, 164-167, 248-249
Protein, 208-210
Pygmy peoples, 23, 99, 102-103
Rabbit, 252
"Race," 2-3, 38
Racism, 100
Radioactive decay, 76-77
Rarity, 273-275
Religion, 145, 197-198, 245-246
Rinderpest, 246
Routes around the world, 43-*47*-*51*-64, 87-88
Saami, 94
Salt loss, 107-109
San peoples, 14, 38, 64, 79, 225
Schistosomiasis (Bilharzia), 219-220
Science, disagreement, 12-13, 22, 36, 44, 82-84, 95-96, 104, 106, 145-146, 149, 259
Science, methods, 72-89, 95-96, 124, 132-133, 149-150, 186, 192, 193, 202, 227-228, 231-233, 237-238, 254-255, 256, 259-260
Scientists, too old, 83
Sea crossing, 33, 49, 50, 53, 58-63

Sea level, 24, 28, 48, 52, 135, 173
Seaweed, 206-207
Sexes, 36, 40, 59, 106, 121-125
Sexual selection, 127
Sharon Stone, 181
Shellfish (middens), 18, 21, 45-46, 54
Sherpas, 113-114
Sickle-cell anemia, 215
Silk Road, 52
Skin (fur, feather) color, 91-96
Slaughter, 262-264
Slavery, 38
Sleeping sickness, 217-*219*-221
Smallpox, 223, 224, 245
Solutrean (peopling of America), 56-57
Species (*see also* Animals, Endemism, Islands), diaspora, 86-89; diversity, 148-160, 162-163, 168-170, 182-186, 190
Speed of global spread, 65-70
Starch, 204-206
Stegodon, 176-177
Stone Age, 80-81
Summer Institute of Linguistics, 277
Survival International, 8, 277
Sweat (salt, water loss), 107-109
Sweet potato, 62, 88
Tasmania effect, 187-189
Temperate regions (*see* Tropics)
Temperature, anatomy, 97-107; physiology, 107-113; sexes, 121-125
Territoriality, 154, 166
Thalassemia, 217
Thrifty genes, 113
Thylacine, 242-243
Tibetan mastiff, 121
Titanic and *Lusitania*, 123
Toad, speed of spread, 70
Tools, 30, 31-32, 189
Toxins, 206
Transport and dispersal, 140
Tropics, 57, 91-93, 97-99, 101, 103, 107-108, 110, 138, 140-141, 150-164, 166-172
Trypanosome (sleeping sickness), 217-*219*, 221
Tuskegee experiment, 211
Vikings, 37
Vitamins, 92-94, 202, 206, 210
Volcanoes, 19-21, 182
Wallace's Line, 59
War, 222-223, 225, 261-262, 269-272
Weather (*see* Climate)
Weight (*see* Body)
West Nile virus, 171, 249-250
Worldwide spread (diaspora), 9-71
Xenophobia as barrier, 141-147
Yak, 239
Yana Rhinoceros Horn site, 27
Yellow fever, 221, 222, 223
Zebrafish, 96